Building and Surveying Series

Sub-contracting under the JCT Standard Forms of Building Contract
  Jennie Price
Urban Land Economics and Public Policy, fifth edition
  P. N. Balchin, G. H. Bull and J. L. Kieve

Macmillan Building and Surveying Series
Series Standing Order ISBN 0–333–69333–7

You can receive future titles in this series as they are published by placing a
standing order. Please contact your bookseller or, in the case of difficulty, write to us at
the address below with your name and address, the title of the series and the ISBN
quoted above.

Customer Services Department, Macmillan Distribution Ltd
Houndmills, Basingstoke, Hampshire, RG21 6XS, England.

Building Quantities Explained

IVOR H. SEELEY

Formerly Emeritus Professor of Nottingham Trent University

In collaboration with

Roger Winfield

Director of Studies
Department of Surveying
The Nottingham Trent University

Fifth Edition

palgrave
macmillan

First published by
PALGRAVE MACMILLAN
Houndmills, Basingstoke, Hampshire RG21 6XS and
175 Fifth Avenue, New York, N.Y. 10010
Companies and representatives throughout the world

PALGRAVE MACMILLAN is the global academic imprint of the Palgrave
Macmillan division of St. Martin's Press, LLC and of Palgrave Macmillan Ltd.
Macmillan® is a registered trademark in the United States, United Kingdom
and other countries. Palgrave is a registered trademark in the European
Union and other countries.

ISBN 0–333–71972–7

This book is printed on paper suitable for recycling and
made from fully managed and sustained forest sources.

A catalogue record for this book is available from the British Library.

14 13 12 11 10   9   8
10 09 08 07 06 05 04

Printed in China

This book is dedicated to those numerous graduates, diplomats and students in quantity surveying of the former Trent Polytechnic and its predecessor colleges, with whom I had the privilege and pleasure to be associated, and from whom I learnt a great deal.

Contents

1 General Introduction   **1**

Historical background of quantity surveying; functions of bill of
quantities; contract documentation; processes used in quantity
surveying work; other measurement approaches; Standard
Method of Measurement of Building Works; coordinated project
information; other functions of the quantity surveyor; changes
in quantity surveying techniques.

2 Measurement Procedures   **16**

General rules: basic principles; tabulated rules; dimensions paper;
measurement procedures; entering dimensions; spacing of
items; waste; order of dimensions; timesing; abbreviations and
symbols; grouping of dimensions; deductions; measurement of
irregular figures; alterations to dimensions; figured dimensions;
numbering and titles of dimension sheets; order of taking off;
adjustment of openings and voids; descriptions; use of metric
symbols; extra over items; deemed to be included items; accuracy in dimensions; falls, crossfalls and slopes; drawn information;
use of schedules; take off lists; query sheets; preambles; prime
cost items; provisional sums; work of special types; fixing, base
and background; composite items; general definitions; services
and facilities; plant items; standard components.

3 Mensuration Applications   **37**

Introduction; girth of buildings: rectangular buildings; buildings of
irregular outline; measurement of areas: irregular areas; trape-
zoids; segments; bellmouths; measurement of earthworks: sloping site excavation; cuttings and embankments; measurement
of pitched roofs: lengths of rafters; lengths of hips and valleys;
roof coverings.

List of Figures

List of Examples

List of Tables

Preface to the Fifth Edition

The primary aim of this book continues to be to meet the needs of students studying the subject of building measurement in the earlier years of degree courses in quantity surveying and building, and those preparing for the appropriate examinations of the relevant professional bodies, the Business and Technician Education Council and National Vocational Qualifications. Although the needs of some of these students differ, it is believed that they all will benefit from the fundamental and yet practical approach that has been adopted.

This book describes and illustrates the measurement of relatively simple building work in accordance with the principles laid down in the *Standard Method of Measurement of Building Works: Seventh Edition (SMM7)* supported by the associated *Code of Procedure*. The book contains a careful selection of worked examples of taking off, accompanied by extensive explanatory notes, encompassing all the basic work sections of the *Standard Method*. Its main aim is to explain simply and to illustrate the measurement of building work and to amplify and clarify the basic principles contained in the *Standard Method*, for the benefit of quantity surveying and building students.

In preparing this edition a prime objective has been to increase its usefulness to students. This has been achieved by introducing new examples which incorporate modern constructional techniques and updating the remaining drawings. The format of the descriptions has been changed from traditional prose to a structured and standardised approach, encompassing classification tables, which follows closely the arrangements contained in the *SMM7 Library of Standard Descriptions*. This approach is more akin to current quantity surveying office practice and is complementary to the majority of standard libraries of descriptions used in conjunction with the production of computerised bills of quantities, and hence will better meet the needs of present day students.

The explanatory notes to the worked examples have been printed for ease of reading, but retaining the handwritten dimensions and descriptions to illustrate good practice. They have proved popular in earlier editions, as giving students a benchmark against which they can gauge their own performance. The texts of many chapters have been extended to give greater guidance and contain more diagrams, where they were felt to be both relevant and useful.

Many of the worked examples contain alternative constructional methods to illustrate a range of approaches and descriptions. For example, the pitched roof worked examples in chapter 8 cover both traditional cut and trussed rafter types and both tiled and slated roofs, encompassing roof void ventilation systems, while the flat roofs embrace both timber and concrete construction and coverings of built up felt, asphalt and lead. In the external works worked example, the road is taken in bituminous macadam and in concrete, the footpaths in precast concrete flags and clay cobbles, and the grassed area as seeded and turfed.

The worked examples covering wood casement windows and internal doors have been extended to embrace a number of windows and doors, to resemble more fully measurement work in practice. The internal finishes example relates to an eleven roomed bungalow, with a variety of ceiling, wall and floor finishes, while the drainage example has been extended to include separate foul and surface water drainage systems, incorporating drains in two different materials and three types of inspection chambers, to give the widest possible coverage. There are also new worked examples covering gas flue blocks and stone rubble walling, and joinery examples encompass both purpose made and stock pattern items.

A new chapter has been added to cover the measurement of simple electrical services and the former plumbing chapter has been more approp- riately retitled 'water, heating and waste services' to indicate its wider scope. As the measurement of building services is more widely practised, it was considered desirable to extend the basic examples in this area to give improved coverage. Although students requiring a more detailed knowledge of this class of work are referred to *Measurement of Building Services*, also in the Macmillan Building and Surveying series.

The format and use of schedules are given much greater attention than hitherto, and worked examples including them cover the measurement of internal finishes, windows, internal doors, cold water services, drains and inspection chambers. There are also more examples of dimensioned diagrams, typical take off lists and a query sheet. The information on the principles, aims and application of coordinated project information has been retained as it is still of great relevance, despite its somewhat haphazard use in practice. In the final chapter the emphasis on abstracting and billing by the traditional method has been reduced and replaced by computerised taking off and billing, in keeping with current quantity surveying practice.

Following previous practice, the symbols 'mm' and 'm' are omitted from the drawings and measurement item descriptions, except where fractions of millimetres are involved. The decimal marker provides the distinguishing feature between metres and millimetres and hence there can be no misunderstanding, despite their use in the *SMM7 Measurement Code*. The procedure adopted follows the recommendation of the British Standards Institution, when advising on the change to metric in the construction industry in 1968, and is the approach that is frequently used in practice. It avoids the unnecessary repetition of metric symbols and reduces the length of descriptions in bills of quantities and on drawings.

Nottingham

IVOR H. SEELEY

A revised edition of the *Standard Method of Measurement of Building Works* (SMM7) was published in 1998. The need for the revision was mainly due to the introduction of a new document entitled UNICLASS which has had the effect of changing a significant number of the Common Arrangement work sections into which SMM7 is subdivided. Consequently a second edition of the Common Arrangement of Work Sections for Building Works (CAWS) has also been published. In publishing the revised edition of SMM7, the opportunity has been taken to amend existing rules and introduce new rules of measurement on a minimal basis.

This book has been prepared in accordance with the revised editions of SMM7 and CAWS. However further amendments to SMM7 will be issued in 1999. In particular these will affect the existing rules relating to active, toxic and hazardous materials which appear in a number of different work sections.

Nottingham,
1998

ROGER WINFIELD

Acknowledgements

Publisher's note: The late Ivor Seeley prepared the following paragraphs before his untimely death in 1997. The Publishers would like to express appreciation to Roger Winfield for joining this project at a critical stage in order to ensure its successful publication.

The author expresses his thanks to the Standing Joint Committee for the Standard Method of Measurement of Building Works for kind permission to quote from the *Standard Method of Measurement of Building Works: Seventh Edition, Revised 1998 (SMM7), A Code of Procedure for the Measurement of Building Works*, and the *SMM7 Library of Standard Descriptions*.

Norman Wheatley FRICS, Honorary Secretary of the Standing Joint Committee, gave helpful advice in his customary friendly way.

Roger Winfield ARICS and Peter Holden ARICS Dip Ed (FE), both principal lecturers in quantity surveying in the Department of Surveying at Nottingham Trent University, gave me much sound advice based on their long experience of teaching this subject. I cannot praise too highly the invaluable help and support which I received from Roger Winfield throughout the preparation of the fifth edition, as a result of which the format and scope of the book have changed dramatically for the benefit of readers.

Holly Nixon undertook the artwork with great care and skill, and I am very much indebted for her outstanding contribution.

I am grateful to Nottingham City Council, Department of Design and Property Services, for providing me with computerised contract documents, to David Windsor FRICS of Masterbill Micro Computers Systems Ltd for supplying extensive information on the Masterbill '97 system of taking off and billing and to Steve Pittard FRICS of Elstree Computing Ltd for providing details of the CATOPro taking off and billing system.

Brian Williams Dip QS (Nottm) FRICS and Clive West BSc FRICS of the Building Design and Management Unit of North Yorkshire County Council kindly supplied me with operational notes relating to the use of the Masterbill system in practice, for which I am much indebted.

Kiernan Larkin MCIOB of Willmot Dixon, London assisted in the preparation of the cover illustration.

Thanks are also due to the numerous manufacturers of building materials and components who provided a wealth of technical information so readily and which has helped in formulating many measured item descriptions.

1 General Introduction

The quantity surveying profession has largely developed over the last century, but has now grown to such an extent that it forms the second largest sector or specialism in the membership of the Royal Institution of Chartered Surveyors, aided by the amalgamation with the former Institute of Quantity Surveyors (IQS) in 1983. Quantity surveyors are employed in private practices, public offices and by contractors, and they undertake a great diversity of work, as described by the author in *Quantity Surveying Practice*.

In more recent times, quantity surveyors are engaged increasingly in the financial management of contracts and ensuring that clients secure value for money and that the completed projects provide substantial added value to the client's property asset. In addition to being construction cost consultants as described in *Building Economics*, quantity surveyors are playing an increasingly important role in project management, value management and facilities management, as described in *Quantity Surveying Practice*. Furthermore, they are sometimes engaged as lead consultants for large projects, where they are responsible for the delivery of all professional services from inception to completion.

The earliest quantity surveying firm of which records are available is a Reading firm which was operating in 1785. There is little doubt that other firms were in existence at this time and a number of Scottish quantity surveyors met in 1802 and produced the first method of measurement. Up to the middle of the nineteenth century it was the practice to measure and value the building work after it had been completed and bills of quantities were not prepared.

The need for quantity surveyors became evident as building work increased in volume and building clients became dissatisfied with the method adopted for settling the cost of the work.

In the seventeenth century the architect was responsible for the erection of buildings, as well as their design, and he employed a number of master craftsmen who performed the work in each trade. Drawings were of a very sketchy nature and much of the work was ordered during the course of the job. On completion each master craftsman submitted an account for the materials used and labour employed on the work.

It later became the practice for many of the master craftsmen to engage 'surveyors' or 'measurers' to prepare these accounts. One of the major problems was to reconcile the amount of material listed on invoices with the quantity measured on completion of the work. Some of the craftsmen's surveyors made extravagant claims for waste of material in executing the work on the site and the architects also engaged surveyors to contest these claims.

General contractors became established during the period of the industrial revolution and they submitted inclusive estimates covering the work of all trades. Furthermore they engaged surveyors to prepare bills of quantities on which their estimates were based. As competitive tendering became more common the general contractors began to combine to appoint a single surveyor to prepare a bill of quantities, which all the contractors priced. In addition, the architect on behalf of the building owner usually appointed a second surveyor, who collaborated with the surveyor for the contractors in preparing the bill of quantities, which was used for tendering purposes.

In later years it became the practice to employ one surveyor only who prepared an accurate bill of quantities and measured any variations that arose during the progress of the project. This was the origin of the independent and impartial quantity surveyor as he operates today.

An excellent account of the development of quantity surveying in the UK between 1936 and 1986 is provided by Nisbet (1989).

FUNCTIONS OF BILL OF QUANTITIES

Frequently, one of the principal activities of the client's quantity surveyor is the preparation of bills of quantities, although he does also perform a number of other functions which will be described later in this chapter. Consideration will now be given to the main purposes of a bill of quantities:

(1) First and foremost it enables all contractors tendering for a contract to price on exactly the same information.
(2) It limits the risk element borne by the contractor to the rates he enters in the bill and thereby results in more realistic and competitive tenders.
(3) It prompts the client and design team to finalise most project particulars before the bill is prepared, and ideally based on full production drawing and project specification.
(4) After being priced it provides a satisfactory basis for the valuation of variations and adjustments to the final account.
(5) Priced bills also provide a useful basis for the valuation of certified stage payments throughout a contract.
(6) It gives an itemised list of the component parts of the building, with a full description and the quantity of each part, and could form an approximate

checklist for the successful contractor in ordering materials and components and assessing his requirements of labour and other resources, and in programming the work.

(7) After being priced, it provides a good basis for the preparation of cost analyses for use in the cost planning of future projects.

(8) If prepared in annotated form, it will help in the locational identification of the work.

It will be apparent that with the increasing size and complexity of building operations, it would be impossible for a contractor to price a medium to large sized project without a bill of quantities. For this reason it has been the practice for contractors to refrain from tendering in competition for all but the smallest contracts without bills of quantities being supplied. This approach does not apply to contracts for repairs or painting and decorating, where schedules of rates are usually more appropriate.

Furthermore, building projects even when they are concerned with the same type of building, usually vary considerably in detailed design, size, shape, materials used, site conditions and other aspects. For this reason a contractor could not readily give a price for building work, such as an office block, hospital or shop, based on the cost of a previous contract of similar type.

In the absence of a bill of quantities being prepared on behalf of the client, each contractor would have to prepare his own bill of quantities in the limited amount of time allowed for tendering. This places a heavy burden on each contractor and also involves him in additional cost which must be spread over the contracts in which he is successful. It could also result in higher cost to the client as contractors may feel compelled to increase their prices to cover the increased risks emanating from this approach.

CONTRACT DOCUMENTATION

It will probably help the student at this stage to describe briefly the form of the contract documents on a traditional building contract incorporating a bill of quantities. The principal documents are as follows:

(a) *Conditions of Contract*: The most common are the standard sets of conditions published by the Joint Contracts Tribunal (JCT). They define the terms under which the work is to be undertaken, the relationship between the client, architect and contractor, the duties of the architect and contractor, and terms of payment. These forms include the main form of JCT 80, the intermediate form of IFC 84 and minor building works, form MW 80.

(b) *Specification*: This amplifies the information given in the contract drawings and bill of quantities, and describes in detail the work to be executed under the contract and the nature and quality of the materials, components and workmanship. Where there is a bill of quantities, the specification will not be a contract document unless so prescribed, and it may be incorporated in the bill of quantities in the form of preambles, as described in chapters 2 and 17.

(c) *Bill of Quantities*: This consists of a schedule of items of work to be carried out under the contract with quantities entered against each item, prepared in accordance with the Standard Method of Measurement of Building Works (currently SMM7). Part of a bill is illustrated in chapter 17.

(d) *Contract Drawings*: These depict the details and scope of the works to be executed under the contract. They must be prepared in sufficient detail to enable the contractor to satisfactorily price the bill of quantities.

(e) *Form of Tender*: This constitutes a formal offer to construct and complete the contract works in accordance with the various contract documents for the tender sum.

Several alternative building procurement systems evolved in the 1980s and 1990s, giving greater choice and flexibility, and often resulting in faster completion and the transfer of greater risk to the contractor. These included design and build, management contracting, construction management, and design and manage. Readers requiring information on these alternative methods are referred to *Quantity Surveying Practice* by the present author.

PROCESSES USED IN QUANTITY SURVEYING WORK

The traditional method of preparation of a bill of quantities can conveniently be broken down into two main processes:

(1) 'Taking off', in which the dimensions are scaled or read from drawings and entered in a recognised form on specially ruled paper, called 'dimensions paper' (illustrated on page 18); and
(2) 'Working up' which comprises squaring the dimensions, as described in chapter 17, transferring the resultant lengths, areas and volumes to the abstract, where they are arranged in a convenient order for billing and reduced to the recognised units of measurement; and finally the billing operation, where the various items of work making up the complete project are listed in full, with the quantities involved in a suitable order under work section or elemental headings. Later developments which eliminated much of the traditional 'working up' process are described in chapter 17.

The most common approach is the group system (London method) whereby the work is measured in groups, each representing a particular section of the building without regard to the order in which the items will appear in the bill, as illustrated in the worked examples in this book. The alternative method is known as the trade by trade system (Northern method) when the taking off is carried out in trade order ready for direct billing, and thus eliminating the need for an abstract.

The term 'quantities' refers to the amounts of the different types of work fixed in position which collectively give the total requirements of the building contract.

These quantities are set down in a standard form on 'billing paper', as illustrated on page 351, which has been suitably ruled in columns, so that each item of work may be conveniently detailed with a description of the work, the quantity involved and a suitable reference. The billing paper also contains columns in which a contractor tendering for a particular project enters the rates and prices for each item of work. These prices added together give the 'Contract Sum'.

The recognised units of measurement are detailed in the *Standard Method of Measurement of Building Works*, as listed in the Bibliography at the end of this book. This document is extremely comprehensive and covers the majority of items of building work that are normally encountered. Many items are measured in metres and may be cubic, square or linear. Some items are enumerated and others, such as structural steelwork and steel reinforcing bars are measured by the tonne. The abbreviation SMM is used extensively throughout this book and refers to the *Standard Method of Measurement of Building Works: Seventh Edition, Revised 1998 (SMM7)*.

The bill of quantities thus sets down the various items of work in a logical sequence and recognised manner, so that they may be readily priced by contractors. A contractor will build up in detail a price for each item contained in the bill of quantities, allowing for the cost of the necessary labour, materials and plant, together with the probable wastage on materials and generally a percentage to cover establishment charges and profit. It is most important that each billed item should be so worded that there is no doubt at all in the mind of a contractor as to the nature and extent of the item which he is pricing. Contractors often tender in keen competition with one another and this calls for very skilful pricing to secure contracts.

The subject of estimating for building contracts is outside the scope of this book, but detailed information on this subject can be found in the books listed in the Bibliography.

Civil engineering work is measured in accordance with the *Civil Engineering Standard Method of Measurement* (CESMM) and a useful textbook on this subject has been written by the present author, entitled *Engineering Quantities*.

Where a bill of quantities is prepared in connection with a building contract, it will almost invariably form a contract document to the exclusion of the specification, although much or all of the contents of a specification may be found in the preambles in a bill of quantities. The successful contractor is fully bound by the contents of all the contract documents when he signs the contract. The other contract documents on a normal building contract are the JCT Articles of Agreement and Conditions of Contract, Contract Drawings and Form of Tender.

Worked examples of more complicated types of building work including reinforced concrete, structural steelwork, mechanical and electrical services, and alteration work are contained in *Advanced Building Measurement* by the present author.

OTHER MEASUREMENT APPROACHES

It would be misleading to imply that all measurement is based on the application of formalised rules of measurement. The use of SMM7 rules in the late 1990s was mainly associated with the traditional procurement system of bills of quantities contracts. This occurs in the initial preparation of the bills of quantities by the client's quantity surveyor, and also in remeasurement for final account purposes.

Other procurement methods, including specification and drawings, design and build, management contracting, etc., usually place the responsibility for preparing documentation, to facilitate pricing, upon the contractor or subcontractor. In these cases, measurement is still necessary as it forms the most frequently used mechanism for preparing a price. To achieve speed and reduce the cost of presentation in such circumstances, it is normal to concentrate the measurement on the cost significant items, which are also made inclusive of the associated peripheral aspects of the work. This practice departs from the notion of a standard method of measurement, as the cost significant items vary from project to project. However, the risk of mis-understandings arising are minimised by the measurer and estimator either being one and the same person or at least in close contact with one another.

With the progressive adoption of various alternative forms of procurement, there has been a major shift in the point of measurement from the client's quantity surveyor to contractors and subcontractors.

Measurement can also be applied to materials purchasing, where the emphasis is on quantification relevant to the units of purchase. A further simplified approach to measurement is used in connection with bonus pay-ments, where an assessment is required of the amount of work done by an individual or gang over a given period of time. Yet another measurement technique is the compiling and pricing of approximate, rough or abridged quantities by the client's quantity surveyor, when preparing preliminary estimates of the cost of a project, early in the design stage, as described in *Building Economics* by the present author.

Although it has been established that a number of measurement approaches are employed within the construction industry, this does not alter the need to learn initially how to measure using the SMM7 rules. Once this technique has been mastered, it is a relatively straightforward matter to adapt to other forms of measurement.

STANDARD METHOD OF MEASUREMENT OF BUILDING WORKS

The *Standard Method of Measurement of Building Works*, issued by the Royal Institution of Chartered Surveyors and the Building Employers Confederation, forms the basis for the measurement of the bulk of building work. The first edition was issued in 1922 with the expressed object of providing a uniform method of measurement based on the practice of the leading London quantity surveyors. Prior to the introduction of the first edition of the *Standard Method*, a large diversity of practice existed, varying with local custom and even with the idiosyncrasies of individual surveyors. This lack of uniformity afforded a just ground for complaint on the part of the contractors – that the estimator was often left in doubt as to the true meaning and intent of items in the bill of quantities which he was called upon to price, a condition which militated against scientific and accurate tendering.

It is interesting to note that this first nationally recognised *Standard Method of Measurement of Building Works* was prepared by representatives of the quantity surveyors and the building industry and that this Joint Committee also had consultations with representatives of certain trades. Building contractors have to price the bills of quantities and it is very desirable that they should be represented on the body which formulates the rules for measurement.

Further editions were issued in 1927, 1935, 1948, 1963, 1968, 1978 and 1988. A revised edition of SMM7 was issued in 1998. All the references to the *Standard Method* in this book relate to the revised 1998 edition (SMM7).

The Co-ordinating Committee for Project Information, sponsored by the Association of Consulting Engineers, Building Employers Confederation, Royal Institute of British Architects and Royal Institution of Chartered Surveyors produced a *Common Arrangement of Work Sections for Building Works* (CAWS) to be used in drafting specifications and bills of quantities. The 1988 edition of SMM7 was structured on the basis of work sections derived from the Common Arrangement as opposed to traditional trade sections. More recently the authors of CAWS have been succeeded by the Construction Project Information Committee who have produced a new document entitled UNICLASS. This contains numerous amendments to the work sections contained in CAWS. These changes have in turn required alterations to the same work section subdivision in SMM7. The opportunity was also taken to introduce a small number of changes to the measurement rules in SMM7.

The presentation of the measurement rules in the form of tabulated classification tables, which was first introduced in the original version of SMM7 instead of the previous prose, is maintained in the 1998 edition. This approach enables a quicker and more systematic use to be made of the measurement rules and readily lends itself to the use of standard phraseology and computerisation. The change does not, however, inhibit the use of traditional prose in the writing of bills of quantities where so desired. However, by the mid-1990s the majority of bills of quantities were produced in a structured format, encompassing the SMM7 classification tables, and the examples in the present edition of this book incorporate the same approach. In addition the rules in SMM7 have generally been simplified to produce shorter bills and the contents updated to conform to modern practice.

The first section of the SMM incorporates *General Rules* which are of general applicability to all works sections and these are considered in detail in chapter 2.

Section A of the SMM is devoted to *Preliminaries and General Conditions* which incorporate general particulars of the project and the contract, including contractor's obligations, general arrangements relating to work by nominated subcontractors, goods and materials from nominated suppliers and works by public bodies, and a list of general facilities and services which are included for convenience of pricing. The Preliminaries Bill is considered in greater detail in chapter 17.

COORDINATED PROJECT INFORMATION

Research by the Building Research Establishment has shown that the biggest single cause of quality problems on building sites is unclear or missing project information. Another significant cause is uncoordinated design, and on occasions much of the time of site management can be devoted to searching for missing information or reconciling inconsistencies in the data supplied.

The crux of the problem is that for most building projects the total package of information provided to the contractor for tendering and construction is produced in a variety of offices of different disciplines.

To overcome these weaknesses, the Co-ordinating Committee for Project Information (CCPI) was formed on the recommendation of the Project Information Group, sponsored by the four bodies listed previously (ACE, BEC, RIBA, and RICS). Its brief was to clarify, simplify and coordinate the national conventions used in the preparation of project documentation. In 1987, the Co-ordinating Committee published a useful guide on coordinated project information for building works, showing the interrelationship of the different documents with illustrated examples, and this proved very useful at the introductory stage.

The following five documents were published either by CCPI or by the separate sponsoring bodies, during 1987 and 1988:

(1) *Common Arrangement of Work Sections for Building Works.*
(2) *Project Specification* – a code of procedure for building works.
(3) *Production Drawings* – a code of procedure for building works.
(4) *Code of Procedure for Measurement of Building Works.*
(5) *SMM7.*

It is, however, unlikely that any single discipline office will require all these documents. For example, SMM7 conforms to the *Common Arrangement* and so quantity surveyors using the *Standard Method* will not of necessity require the latter document. Similarly, users of the National Building Specification (NBS) and the National Engineering Specification (NES) will not require the *Common Arrangement*. However, the construction industry has been slow to adapt to the use of coordinated project information (CPI) in its entirety.

Common Arrangement of Work Sections for Building Works (CAWS)
This document plays a major role in coordinating the arrangement of drawings, specifications and bills of quantities. It reflects the current pattern of sub-contracting and work organisation in building. To avoid problems of overlap between similar or related work sections, each section contains a comprehensive list of what is included in the section and what is excluded, stating the appropriate section of the excluded items.

SMM7 uses the same work sections and this will eliminate any inconsistencies between specifications and bills of quantities, where the quantity surveyor structures the bill on SMM7.

The *Common Arrangement* has a hierarchical arrangement in three levels, for example:

Level 1: E *In-situ* concrete/Large precast concrete
Level 2: E1 *In-situ* concrete
Level 3: E10 Mixing/Casting/Curing *in-situ* concrete (work section).

It lists 24 level 1 group headings and about 300 work sections, roughly equally divided between building fabric and services. However, no single project will encompass more than a relatively small number of them. Only levels 1 and 3 will normally be used in specifications and bills of quantities, while level 2 allows for the insertion of new works sections if required later, without recourse to extensive renumbering.

Common Arrangement describes how a work section is a dual concept, involving the resources being used and also the parts of the work being constructed, including their essential functions. The category is usually influenced and characterised by both input and output of walling, while an input of mastic asphalt could have an output of tanking.

Section numbers are kept short for ease of reference. The widespread use of cross references to the specification should encourage designers to be more consistent in the amount of description which they provide on drawings. The SMM7 Measurement Code describes how bills of quantities prepared in accordance with CAWS derive the greatest benefit and ease of use.

Project Specification – a code of procedure for building works
This code draws a distinction between specification information and the project specification. Information may be provided on drawings, in bills of quantities or schedules, but the project specifications should be the first point of reference when details of the type and quality of materials and work are required. Hence drawings and bills of quantities should identify kinds of work but not specify them. Instead, simple cross references should ideally be made to the specification as, for example, Ledkore damp-proof course F30:2.

The project specification should be prepared by the designer and the use of a standard library of specification clauses will make the task easier. Specifications should be arranged on the basis of the *Common Arrangement*. Both the National Building Specification (NBS) library of clauses and the National Engineering Specification (NES) conform to the *Common Arrangement*.

The code provides extensive checklists for the specification of each work section to ensure that project specifications are complete, and it also gives advice on specification preparation by reference to British Standards or other published documents or by description.

Production Drawings – a code of procedure for building works
This code deals with the management of the preparation, coordination and issue of sets of drawings, and with the programming of the design and communication operation, and thus complements BS 1192 (Construction Drawings Practice).

The following criteria should preferably be adopted in the preparation of drawings:

(1) Use of common terminology. If the content of a drawing coincides with a *Common Arrangement* work section, then the *Common Arrangement* title should be used on that drawing.
(2) Annotate drawings by cross reference to specification clause numbers, for example, concrete mix A, E10:4 and lead flashing, H71:320.

SMM7 and Code of Procedure for Measurement of Building Works
The arrangement of SMM7 is based on the *Common Arrangement* and the rules of measurement for each work section are in the same sequence.

If descriptions in the bills of quantities are cross referenced to clause numbers in the specification, for example concrete mix A, E10:4, as for the drawings, then the coordination of drawings, specifications and bills of quantities will be improved, and the risk of inconsistent information will be reduced. If required, the specification can be incorporated into the bill of quantities as preambles.

The code of procedure explains and enlarges upon the SMM as necessary and gives guidance on the arrangement of bills of quantities. It should be emphasised that the code is for guidance only and does not have the mandatory status of SMM7. The code has not yet been revised to accompany the revised edition of SMM7 and whilst the guidance which it contains continues to be largely relevant, there are now inevitably some inconsistencies in cross referencing.

Coordinated Project Information in Use
Since the *Common Arrangement* is based on natural groupings of work within the building industry, it is likely to provide benefits in the management of the construction stages. Not only will it be much easier to find the required information, but it will also be structured by the *Common Arrangement* in a manner which conforms to normal subcontracting and specialist contracting practice. Thus in obtaining estimates from subcontractors, it will be a straightforward task to assemble the correct set of drawings, specification clauses and bill items. In management contracting, the *Common Arrangement* is likely to provide a convenient means of identifying separate work packages. Similarly, construction programmes based on the *Common Arrangement* will provide direct links to other project information, thus bringing together quantity, cost and time data into an integrated information package.

Further standardisation has been introduced through the publication of *SMM7 Library of Standard Descriptions*, jointly sponsored by the former Property Services Agency (DoE), Royal Institution of Chartered Surveyors and Building Employers Confederation. However, quantity surveyors are not obliged to refer to any of the Coordinated Project Information (CPI) documents apart from SMM7 when producing bills of quantities. Hence, in practice, a variety of approaches can be adopted when framing billed descriptions including cross references to project specification clauses, use of the *SMM7 Library of Standard Descriptions* as used in the worked examples contained in this book, individual descriptions built up from SMM7 by quantity surveyors, in traditional prose, given as another acceptable option in the preface to SMM7, or the use of *Shorter Bills of Quantities: The Concise Standard Phraseology and Library of Descriptions*.

SMM7 Library of Standard Descriptions
In the fourth edition of this book, published in 1988, it was decided to prepare the measurement descriptions in traditional prose as an acceptable option contained in the preface to SMM7. It was felt, at that time, that students would prefer this approach to the more stereotyped version emanating from

the application of the *Common Arrangement of Work Sections for Building Works* and the *SMM7 Library of Standard Descriptions*, and that it would be more user friendly.

When preparing the fifth edition in 1996/97, consideration was given to the common practice of using computerised methods of bill production, largely based on the *Common Arrangement* (CAWS) and the *SMM7 Library of Standard Descriptions*, as described in chapter 17. It seemed highly desirable to adopt a format which more closely resembled current quantity surveying office practice and which is used in the compilation of the major building price books. Hence this standardised and structured approach, encompassing classification tables instead of traditional prose, has been adopted throughout the book. It is believed that this approach will better meet the needs of present day students.

The *SMM7 Library of Standard Descriptions* follows the *Common Arrangement* and is essentially a tiered formation of phrases based on SMM7 rules of measurement from which selections are made and organised to describe items of work in a structured priceable format. The Library is designed to make full use of computers with three options:

(1) self coding using the former Property Services Agency (PSA) mnemonic coding printed down the left hand side of each page in the Library;
(2) self coding with the users' own codes;
(3) purchasing a commercial software package.

Each page of the Standard Library is headed with the work group, such as F: Masonry and the work section, i.e. F10: Brick/block walling. Each page is divided into five columns: code, level, description, unit and notes. The backbone to the Library is the level and description columns, in that:

level 1 is the work section in bold, large type;
level 2 includes specification information and is in bold type but smaller than
     level 1;
level 3 is the basic item in normal type;
level 4 is variable information such as size in lower case and indented.

The worked examples in the book follow the Library format and incorporate levels 2, 3 and 4, with level 2 information underlined and level 4 information suitably indented.

OTHER FUNCTIONS OF THE QUANTITY SURVEYOR

The *client's quantity surveyor* performs a variety of functions as now listed, and the underlying theme of a quantity surveyor's work is one of cost management rather than the preparation of bills of quantities and settlement of final accounts, whether he be engaged in private practice or in the public sector.

(1) Preparing approximate estimates of cost in the very early stages of the formulation of a building project, and giving advice on alternative materials and components and types of construction and the financial aspects of contracts, and assisting with feasibility studies.

(2) Cost planning and value analysis during the design stage of a project to ensure that the client obtains the best possible value for his money, including added value to his property asset, preferably having regard to total costs using life cycle costing techniques, that the costs are distributed in the most realistic way throughout the various sections or elements of the building and that the tender figure is kept within the client's budget, as described in *Building Economics*.

(3) Advising on the most appropriate form of building procurement, having regard to the type of project, quality, speed of construction, apportionment of risk and price certainty.

(4) Preparation of bills of quantities and other contract documents relating to the project.

(5) Examining tenders and priced bills of quantities and reporting his findings.

(6) Negotiating rates with contractors on negotiated contracts and dealing with cost reimbursement contracts, design and build, management and other forms of contract.

(7) Valuing work in progress and making recommendations as to payments to be made to the contractor, including advising on the financial effect of variations.

(8) Preparing the final account on completion of the contract works.

(9) Advising on the financial and contractual aspects of contractors' claims.

(10) Giving cost advice and information at all stages of the contract and preparing cost analyses and cost reports to clients.

(11) Technical auditing, valuations for fire insurance, giving advice on funding, grants, capital allowances and taxation, risk analysis and management, and other related matters including health and safety and quality control. Many of these activities are described and illustrated in *Quantity Surveying Practice*.

The *contractor's quantity surveyor* performs a rather different range of functions, and these are now described, since there can be few of the large or medium size contracting firms who do not employ quantity surveyors. Usually the contractor's organisation will include a quantity surveying department controlled by a qualified quantity surveyor who is normally a senior executive and may have director status.

The duties of the contractor's quantity surveyor will vary according to the size of the company employing him; tending to be very wide in scope with the smaller companies, but rather more specialised with the larger firms.

In the smaller company, his activities will be of a general nature and include preparing bills of quantities for small contracts; agreeing measurements with the client's quantity surveyor; collecting information about the cost of various operations from which the contractor can prepare future estimates; preparing precise details of the materials required for the projects in hand; compiling target figures so that the operatives can be awarded production bonuses; preparing interim costings so that the financial position of the project can be ascertained as the work proceeds and appropriate action taken where necessary; planning contracts and preparing progress charts in conjunction with the general foremen/site manager; making application to the architect for variation orders if drawings or site instructions vary the work; agreeing subcontractors' accounts; placing subcontract orders and comparing the costs of alternative methods of carrying out various operations so that the most economical procedure can be adopted; and advising on the implementation of contract conditions and different contractual methods.

In larger companies, the contractor's quantity surveyor will not usually cover such a wide range of activities since different departments handle specific activities. During his training period, the trainee quantity surveyor will probably spend some time in each department.

Readers requiring more detailed information about quantity surveying functions are referred to *Quantity Surveying Practice*.

CHANGES IN QUANTITY SURVEYING TECHNIQUES

A number of developments have taken place in the method of preparation and form of bills of quantities in recent years.

The traditional method of taking off, abstracting and billing was both lengthy and tedious in the extreme. Alternative systems have accordingly been introduced with a view to speeding up the process, lowering cost and reducing the requirements for working up staff, who are in short supply, as the large workforce of quantity surveying technicians envisaged in the early 1970s has not materialised, mainly because of the lack of adequate status. However, developments in the mid-1990s with increased integration within the RICS, the introduction of the term 'technical surveyor' and the use of specific designatory letters (Tech. Surv. RICS) have done much to enhance their status and appeal.

The two principal improved methods of bill preparation are cut and shuffle and the use of computers; both these methods will be considered in some detail in chapter 17. As long ago as 1962, a working party of the Royal Institution of Chartered Surveyors was of the opinion that the quantity surveying profession could and ought to take advantage of mechanical and other aids which were available or could be developed, to economise in the use of labour and accelerate the production of bills of quantities, and there have been substantial developments since that time.

The introduction of modern systems for preparing bills of quantities has led to the transfer of staff in quantity surveyors' offices from working up to other types of work. Different methods of training quantity surveyors have been developed and increasing numbers of students are attending full-time and thick sandwich quantity surveying degree courses at universities and other colleges. There has also been a progressive increase in the number of part-time degree, advanced diploma and post-graduate courses and significant improvements in the scope and form of distance learning, and continuing professional development facilities.

Greater standardisation in the presentation of bills of quantities, both as regards order of billing and method of presenting items, is now considered vitally important and this, coupled with more uniformity in contractors' methods of estimating, costing and programming, enables the fullest use to be made of computerised systems.

In the post-war years a number of different formats of bills of quantities were used, in an endeavour to produce a bill of quantities which would be of greater value to the contractor than the normal work section order bill of quantities. Some of these newer bill formats, such as operational bills and elemental bills proved unacceptable, as despite their inherent advantages, they also had serious limitations. The different bill formats are examined in some detail in chapter 17.

2 Measurement Procedures

Basic Principles

Some of the general principles to be followed in taking off building quantities are detailed in the General Rules in the first section of the *Standard Method of Measurement of Building Works,* of which the following statements are particularly important.

> This Standard Method of Measurement provides a uniform basis for measuring building works and embodies the essentials of good practice. Bills of quantities shall fully describe and accurately represent the quantity and quality of the works to be carried out. More detailed information than is required by these rules shall be given where necessary in order to define the precise nature and extent of the required work (General Rules 1.1 in SMM7). The format and coding of the bill of quantities should desirably follow the *Common Arrangement of Work Sections* (CPI), as illustrated in chapter 17.
>
> Rules of measurement adopted for work not covered by these rules shall be stated in a bill of quantities. Such rules shall, as far as possible, conform with those given in this document for similar work (General Rules 11.1).

The billed descriptions are to be reasonably comprehensive and sufficient to enable the estimator fully to understand what is required and to give a realistic price. All quantities must be as accurate as the information available permits, as inaccurate bills cause major problems.

It is most important that all work whose extent cannot be determined with a reasonable degree of accuracy should be described as approximate quantities, and items of this kind should be kept separate from those which contain accurate quantities (General Rules 10.1). In this way the contractor is made aware of the uncertain nature of the quantity entered and that it will be subject to remeasurement on completion and valuation at billed rates. This can apply to any work where the architect is unable to give full details at the time of measurement.

In General Rules 3:1–3, it is emphasised that measurements are to relate to work net as fixed in position, except where otherwise described in a measurement rule applicable to the work. Measurements are to be taken to the nearest 10 mm (5 mm and over shall be regarded as 10 mm and less than 5 mm shall be disregarded). Lengths are entered in the dimension column in metres to two places of decimals. When billing in metres the quantity is billed to the nearest whole unit, but where the unit of billing is the tonne, quantities shall be billed to two places of decimals.

Where a measurement rule provides that the area or volume comprising a void is not deducted from the area or volume of the surrounding material, for example $\leq 1.00\,m^2$ for roof coverings, this shall refer only to openings or wants which are within the boundaries of the measured areas. Openings or wants which are at the boundaries of measured areas shall always be the subject of deduction irrespective of size (General Rules 3.4).

Billed items are generally deemed to include, that is, without the need for specific mention: labour, materials, goods and plant, including all associated costs such as assembling, fitting and fixing, waste of materials, square cutting; establishment and overhead charges and profit (General Rules 4.6).

Each work section of a bill of quantities shall begin with a description stating the nature and location of the work, unless evident from the drawn or other information required to be provided by the SMM rules (General Rules 4.5).

Four categories of drawing are listed in General Rules 5:1–4 (location drawings, component drawings, dimensioned diagrams and schedules). The student will be particularly concerned with the third category – dimensioned diagrams – to show the shape and dimensions of the work covered by an item, and they may be used in a bill of quantities in place of a dimensioned description, but not to replace an item otherwise required to be measured. These drawings will be considered in more detail later in the chapter and subsequently in the relevant worked examples that follow.

Tabulated Rules

The rules prescribed in SMM7 are set out in the form of tables and these comprise classification tables and supplementary rules. Horizontal lines divide the classification table and supplementary rules into zones to which different rules apply. Where broken horizontal lines appear within a classification table, the rules entered above and below these lines may be used as alternatives (General Rules 2:1–3). As, for example, metal sheet flashings, aprons, cappings and the like which may be measured either with a dimensioned description or with a dimensioned diagram (H70:10–18.1–2). Within the supplementary rules everything above the horizontal line, which is immediately below the classification table heading, is applicable throughout that table (General Rules 2.8).

The left hand column of a classification table lists descriptive features commonly encountered in building works. The next or second column lists subgroups into which each main group shall be divided and the third column provides for further subdivision, although these lists are not intended to be exhaustive. The relevant unit of measurement is also indicated. Each item description shall identify the work relating to one descriptive feature drawn from each of the first three columns in the classification table, and as many of the features in the fourth or last column as are appropriate. Where the abbreviation (nr) is given in the classification table, that quantity shall be stated in the item description (General Rules 2:5–7).

The supplementary rules form an extension of the classification tables and are subdivided into the following four columns:

(1) measurement rules prescribe when and how work shall be measured;
(2) definition rules define the extent and limits of the work contained in the rules and subsequently used in the preparation of bills of quantities;
(3) coverage rules draw attention to incidental work which is deemed to be included in appropriate items in the bill of quantities to the extent that such work is included in the project, and where coverage rules include materials they shall be mentioned in item descriptions;
(4) supplementary information contains rules covering any additional information that is required (General Rules 2:9–12).

Cross references within the classification tables encompass the numbers from the four columns, such as D20:2.6.2.0: excavating trenches; width >0.30 m; maximum depth ≤1.00 m. The digit 0 indicates that there are no entries in the column in which it appears, while an asterisk represents all entries to the column in which it occurs (General Rules 12:2–4).

DIMENSIONS PAPER

The normal ruling of dimensions paper on which the dimensions, as scaled or taken direct from drawings, are entered, is now indicated. This ruling conformed to the requirements of BS 3327: 1970 *Stationery for Quantity Surveying,* which has since been withdrawn, showing the face side of the sheet with a binding margin on the left hand side.

	1	2	3	4	1	2	3	4

Each dimension sheet is split into two identically ruled parts, each consisting of four columns. The purpose of each column will now be indicated for the benefit of those readers who are unfamiliar with the use of this type of paper.

Column 1 is called the 'timesing column' in which multiplying figures are entered when there is more than one of the particular item being measured.

Column 2 is called the 'dimension column' in which the actual dimensions, as scaled or taken direct from the drawings, are entered. There may be one, two or three lines of dimensions in an item depending on whether it is linear, square or cubic.

Column 3 is called the 'squaring column' in which the length, area or volume obtained by multiplying together the figures in columns 1 and 2 is recorded, ready for transfer to the abstract or bill.

Column 4 is called the 'description column' in which the written description of each item is entered. The right hand side of this wider column is frequently used to accommodate preliminary calculations, sometimes termed 'sidecasts', and other basic information needed in building up the dimensions, which is referred to as 'waste'. Locational notes are more often inserted on the left hand side of the description column inside a bracket.

In the worked examples that follow in succeeding chapters the reader will notice that one set of columns only is used on each dimension sheet, with the remainder used for explanatory notes, but in practice both sets of columns will be used for taking off.

Dimensions paper is almost universally of international paper size A4 (210 mm × 297 mm).

An alternative approach is to use some form of cut and shuffle sheets as described and illustrated in chapter 17.

MEASUREMENT PROCEDURES

Entering Dimensions

Spacing of Items
It is essential that ample space is left between all items on the dimension sheets so that it is possible to follow the dimensions easily and to enable any items, which may have been omitted when the dimensions were first taken off, to be subsequently inserted without cramping the dimensions unduly. The cramping of dimensions is a common failing among examination candidates and does cause loss of marks. The items contained in the worked examples in this book are often closer than is ideal, solely to conserve space and keep down the price of this student textbook.

Waste

The use of the right hand side of the description column for preliminary calculations, build up of lengths, explanatory notes and the like should not be overlooked. All steps that have been taken in arriving at dimensions, no matter how elementary or trivial they may appear, should be entered in the waste section of the description column. Following this procedure will do much to prevent doubts and misunderstandings concerning dimensions arising at some future date. It also enables all calculations for dimensions to be checked.

Order of Dimensions

A constant order of entering dimensions should be maintained throughout in accordance with General Rules 4.1, that is, (i) length, (ii) width or breadth and (iii) height or depth, even although the SMM requirement strictly relates only to dimensions in descriptions. In this way there can be no doubt as to the shape of the item being measured. When measuring a cubic item of concrete 3.500 long, 2.500 wide and 0.500 deep, the entry in the dimension column could be

3.50	*In-situ* conc., class A	
2.50	Isoltd. fdns.	
0.50	poured on or	
	against earth	

It will be noted that dimensions are usually recorded in metres to two places of decimals with a dot between the metres and fractions and a line drawn across the dimension column under each set of figures.

Timesing

If there were twelve such items, then this dimension would be multiplied by twelve in the timesing column, as in the following example:

12/	3.50	*In-situ* conc., class A	
	2.50	Isoltd. fdns. a,b.	
	0.50		

If it was subsequently found that four more foundation bases of the same dimensions were to be provided, then a further four could be added in the timesing column by the process known as 'dotting on', as indicated in the next example.

12/	3.50	*In-situ* conc., class A
4̇	2.50	Isoltd. fdns. a,b.
	0.50	

Abbreviations and Symbols

Many of the words entered in the description column are abbreviated in order to save space and time spent in entering the items by highly skilled professional technical staff. Many abbreviations have become almost standard and are of general application; for this reason a list of the more common abbreviations is given in appendix 1 at the end of this book. A considerable number of abbreviations are obtained by merely shortening the particular words, such as the use of 'conc.' in place of concrete and 'rad.' for radius. With some measurement techniques, such as cut and shuffle, it may be considered desirable to avoid the use of abbreviations where bill descriptions are to be typed direct from the initial measurement or dimension sheets or slips. Abbreviations save time in examinations and are incorporated in the worked examples contained in this book.

An extensive list of symbols is given in SMM General Rules 12.1 and includes m (metre), m^2 (sq.m), m^3 (cu.m), mm (millimetre), nr (number), kg (kilogram), t (tonne), h (hour), > (exceeding), $\geq$ (equal to or exceeding), $\leq$ (not exceeding), < (less than), % (percentage) and − (hyphen; often used to denote range of dimensions).

Grouping of Dimensions

Where more than one set of dimensions relate to the same description, the dimensions should be suitably bracketed so that this shall be made clear. The following example illustrates this point:

		Clay pipewk.
		S & S mech. jts.
		to BS EN 295−1
		Pipes
		nom. size 100;
		in trs.
		Mhs.
	18.60	(1−2
	25.00	(2−3
	42.60	(3−4
	36.00	(4−5

Note also the location particulars entered in the description column which readily identify the location of each length of drain.

Where the same dimensions apply to more than one item, the best procedure is to separate each of the dimensions by an '&' sign and to bracket the descriptions, as illustrated in the following example. This process is sometimes described as 'anding on'.

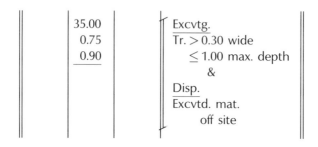

Deductions

After measuring an item it is sometimes necessary to deduct for voids or openings in the main area or volume. This is normally performed by following the main item by a deduction item as shown in the following example:

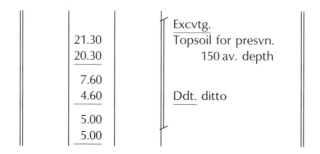

(Note the underlining of the word <u>Ddt.</u> [Deduct])

Measurement of Irregular Figures

It is sometimes necessary to measure the areas of triangles and circles, the circumferences of circles, and the volumes of cylinders and the like, and the usual method of entering the dimensions is illustrated in the following examples.

$\frac{1}{2}$/	4.00			Area of triangle
	3.00			with a base of 4 m
				and a height of
				3 m.
				(area = base ×
				$\frac{1}{2}$ height)
$\frac{22}{7}$/	2.00			Area of circle, 2 m
	2.00			radius.
				(area = πr^2)
$\frac{1}{2}$/2/$\frac{22}{7}$/	2.00			Circumference of
				semi-circle, 2 m
				radius.
				(circumference of
				whole circle = $2\pi r$)
$\frac{22}{7}$/	0.50			Volume of cylinder,
	0.50			1 m diameter and
	3.00			3 m high.
				(area of circle ×
				height of cylinder)

Alterations to Dimensions

It is sometimes necessary to substitute amended dimensions in place of those which have already been entered on the dimensions paper. The student is advised never to alter the original figures, because, apart from looking most untidy, it is often extremely difficult to decipher the correct figures. If it is necessary to amend figures one procedure is to cross out the original figures and neatly write the new figures above them, but probably a better approach is to NIL the item as next described, unless there is insufficient space.

Where it is required to omit dimensions which have previously been recorded the easiest method is to write the word 'NIL' in the squaring column as shown in the following example:

	23.50			Excvtg,
	0.75			Tr. > 0.30 wide
	0.80			≤1.00 max. depth
				&
	8.20	↑		Excvtd. mat.
	0.90	NIL		Fillg. to excvns.
	0.85	↓		> 250 av. th.;
				arisg. from excvns.

Figured Dimensions

When taking off the best procedure is to use figured dimensions on the drawings in preference to scaling, since the drawings are almost invariably in the form of prints, which are not always true to scale. It is sometimes necessary to build up overall dimensions from a series of figured dimensions and this work must be set down in waste, on the right hand side of the description column.

Numbering and Titles of Dimension Sheets

Each dimension sheet should be suitably headed with the title and taking off section of the project at the top of each sheet and with each sheet numbered consecutively at the bottom. Some prefer to number the columns on each dimension sheet rather than the pages. The practice of consecutive numbering ensures the early discovery of a missing sheet. A typical heading for a dimension sheet follows:

CHARLESWORTH HOSPITAL EXTENSION   Fdns. 8

At the top of the first dimension sheet for a project it is good practice to enter a list of the drawings from which the measurements have been taken, with the precise drawing number of each contract drawing carefully recorded.

The importance of listing the contract drawings from which the dimensions have been obtained in this way, is that in the event of changes being made to the work as originally planned, resulting in the issue of amended drawings, it will be clearly seen that these changes occurred after the bill of quantities was prepared and that variations to the quantities can be expected. It is in fact a *Standard Method* requirement to include in the Preliminaries Bill a list of the drawings from which the bill of quantities was prepared (A11:1.1).

It is good practice to hole all dimension sheets at their top left hand corner and to fasten together with treasury tags.

Order of Taking Off

The order of taking off largely follows the order of construction to simplify the work and to reduce the risk of items being missed, but it is not necessarily that adopted in SMM7. The measured items will subsequently be sorted into bill order which can embrace the work sections in SMM7, to secure uniformity and assist with computerisation. For instance, foundation work will be spread over a number of SMM work sections, such as D20 (Excavating and Filling), E10 (Mixing/Casting/Curing *in situ* concrete), E20 (Formwork for *in-situ* concrete), E30 (Reinforcement for *in-situ* concrete), E41 (Worked finishes/cutting to *in-situ* concrete), F10 (Brick/block walling) and F30 (Accessories/sundry items for brick/block/stone walling). In a simple building the order of taking off could take the form shown in the following schedule, although it will be appreciated that this may be varied to suit individual preferences and specific locations.

Sections of Work	*Broad Classifications*
1. Groundwork and substructure or foundation work up to and including damp-proof course; 2. Brickwork, including facework; 3. Blockwork; 4. Fireplaces, chimney breasts and stacks (where appropriate); 5. Floors (solid and suspended); 6. Roofs (pitched and flat, including coverings and rainwater installations);	Carcass (structure)
7. Windows, including adjustment of openings; 8. Doors, including adjustment of openings; 9. Fittings and fixtures; 10. Stairs; 11. Finishes (walls, ceilings and floors); 12. External works, including roads, paths, fences and grassed areas; 13. Drainage work; 14. Water, heating and waste services; 15. Other services.	Finishes and Services

Adjustment of Openings and Voids

When measuring areas of excavation, concrete oversite, brickwork and blockwork, the most convenient practice is usually to measure the full area in the first instance, and to subsequently adjust for any voids or openings. The adjustments for the brickwork and finishes to the window and door openings are usually taken at the same time as taking off the windows or doors. This is a more logical and satisfactory method of measuring, while all the relevant dimensions are to hand, and results in a smaller overall error occurring if the very worst happens and a window or door is inadvertently omitted from the dimensions.

Descriptions

General Requirements
Considerable care and skill are required to frame adequate, and yet at the same time, concise descriptions. This is probably the most difficult aspect of taking off work and one which the student should take great pains to master. The vetting of descriptions forms an important part of editing the bill.

In addition to covering all the matters detailed in the *Standard Method of Measurement of Building Works*, the descriptions must include all the information which the estimator will require to build up a realistic price for the item in question. Where there is doubt in the mind of the estimator as to the full nature and/or extent of the item being priced, then the description is lacking in some essential feature. Descriptions can often be shortened significantly by references to clauses in the project specification as described in chapter 1, and use may be made of the SMM7 *Library of Standard Descriptions*, as illustrated in the worked examples in this book.

Order and Form of Wording
The first few words of a description should clearly indicate the nature of the item being described. The description is badly worded if the reader has to wait almost to the end of the description to determine the subject of the item. The following example serves to illustrate this point and the first type of description may sometimes be produced by students when commencing their studies in this subject.

'Bit. felt, lapped 100 mm at jts. b. & p. in ct. and laid on 102 mm bk. walls, with a width not exceeding 225 mm as dpc.'

This description would be far better worded as follows. 'Dpc, width ≤ 225, hor., single layer of hessian base bit. felt, to BS 743 type A & bedded in c.m. (1:3).'

The second description indicates at the outset the nature of the item under consideration, including the width, range and plane in which the damp-proof course is to be laid in accordance with F30:2.1.3.0, followed by a full description of the materials used as listed in F30:S4–6. It will further be noted from F30:C2 that pointing the exposed edges of damp-proof courses is deemed to be included and does not require specific mention, and that no allowance is made for laps (F30:M2).

An alternative approach to the description, as adopted in the worked examples in this book, is to use a structured approach encompassing classification tables instead of traditional prose following the arrangement depicted in the SMM7 *Library of Standard Descriptions*. In this case, the description will appear as follows:

Dpc of single lyr. hess. base bit. felt to BS 743, type A; bedded in c.m. (1:3)
On surfs.
   ≤ 225 wide; hor.

The third description with its structured approach contains particulars of the materials in an underlined heading, which is akin to the specification provisions. This is followed by the measurement description as F30:2.1.3.0, set out in the form illustrated in the *SMM7 Library of Standard Descriptions*, with the level 4 variable information relating to size and plane indented.

The use of a hyphen between two dimensions in a description, such as 150–300, shall mean a range of dimensions exceeding the first dimension stated but not exceeding the second (General Rules 4.4). A dimensioned description for an item shall define and state all the dimensions necessary to identify the shape and size of the work (General Rules 4.7).

Practical Implementation of Description Preparation
The wording of billed descriptions can vary considerably and it is possible to interpret and implement the provisions of SMM7 in differing ways. The main advantages to be gained by adopting the structured, classification approach used in this book are that it conforms more closely to the wording of SMM7 and the *SMM7 Library of Standard Descriptions*, permits greater rationalisation, facilitates computerisation, and is more akin to normal quantity surveying office practice. Similarly, some surveyors may prefer to use the traditional terms 'not exceeding' and 'exceeding' instead of the symbols $\leq$ and $>$. However, these symbols are used throughout SMM7, have the merit of brevity and clarity and have gained general recognition and usage.

It will be apparent that there will, in practice, be a variety of different methods adopted for framing billed descriptions, despite the extensive work undertaken by the Building Project Information Committee and the sponsoring bodies, and the wealth of published integrated documentation described in chapter 1. In the late 1990s many architects' drawings and specifications continued to be prepared without reference to the codes of procedure for production drawings and project specifications and the national specifications, and some quantity surveyors followed their own personal preferences with regard to bill preparation, so that one universal procedure is unlikely to emerge. Furthermore, the preface to SMM7 permits some flexibility in writing bills of quantities and does not prohibit the use of standard prose.

The student may find all this rather bewildering but must not lose sight of the prime objective: namely, to produce bills of quantities which fully and accurately represent the quantity and quality of the works to be carried out, founded on a uniform basis for measuring building works emanating from SMM7 and embodying the essentials of good practice as defined in General Rules 1.1.

Number of Units
In some cases it is necessary to give the number of units involved in a superficial or linear item, in order that the estimator can determine the average area or length of unit being priced. For instance L40:1.1.1.0 requires

the number of panes of glass, not exceeding 0.15 m², to be included in the description of the item as indicated in the following example:

3/6/	0.20	Std. pl. glass;
	0.32	4 clear flt.
		to BS 952
		To wd. w.l.o. putty & sprigs
		in panes areas: ≤ 0.15 m²
		(In 18 nr panes)

Measurement of Similar Items
Where an entry on the dimensions paper is to be followed immediately by a similar item, the use of the words 'ditto' or 'do.', meaning 'that which has been said before', will permit the description of the next item to be reduced considerably, as shown in the following example. The number of panes is not stated as their area is >0.15 m².

2/4/	0.40	Ditto.
	0.65	in panes areas:
		0.15–4.00 m²

Another practice is to use the expressions 'a.b.' (as before) or 'a.b.d.' (as before described), to refer to a description which has occurred at some earlier point in the taking off. Care must be taken in the use of both 'ditto' and 'a.b.' to ensure that there can be no misunderstanding as to meaning and content.

Use of Metric Symbols

The use of metric symbols in measurement descriptions and drawings in this book has been largely omitted to avoid extensive repetition and to shorten the descriptions, without any possibility of misunderstandings arising. The decimal marker forms the division or demarcation line between metres and millimetres, hence 3.250 is 3 metres and 250 millimetres, while 300 on its own represents millimetres.

The omission of metric symbols on drawings and in specifications and bills of quantities was recommended by the British Standards Institution (BSI) in their publication PD 6031 in 1968 (Use of the metric system in the construction industry). This publication has since been withdrawn, but the principle is still as relevant today as ever. The main exceptions are kilometres (km) and items involving fractions of millimetres, as with thin sheet metal. The author remembers it well as he was a member of the BSI lecturing panel on the change to metric in the construction industry.

It is recognised that the billed example in Appendix 4 of the SMM7 *Measurement Code*, with its inclusion of numerous mm's, could raise doubts in the minds of students. However, the symbols are omitted from the descriptions of many bills of quantities in practice, in the principal computerised billing systems and in the SMM7 *Library of Standard Descriptions*; all dimensions are deemed to be in millimetres unless otherwise stated, thereby eliminating the continual use of the metric symbol 'mm'.

Extra Over Items

When measuring certain types of work they are described as being extra over another item of work which has been previously measured. The estimator will price for the extra or additional cost involved in the second item as compared with the first. A typical example is the measurement of rainwater pipe and gutter fittings as extra over the cost of the pipe or gutter in which they occur, and which has been measured over the fittings, as illustrated in examples 12 and 15.

Deemed to be Included Items

In SMM7 the expression 'deemed to be included' is used extensively and indicates that this particular work is covered in the billed item without the need for specific mention. It is essential that the estimator is fully aware of all these items since he must include for them when building up the unit rates.

Typical examples are all rough and fair cutting which is deemed to be included in brickwork and blockwork (F10:C1b), roof coverings in slates or tiles are deemed to include underlay and battens and work in forming voids $\leq 1.00 \, m^2$ other than holes (H60:C1&M1), while excavating drain trenches is deemed to include earthwork support, consolidation of trench bottoms, trimming excavations, filling with and compaction of general filling materials and disposal of surplus excavated materials (R12:C1).

Accuracy in Dimensions

It is essential that all dimensions shall be as accurate as possible since inaccurate dimensions are worthless. A generally accepted limit of permissible error is around 1 per cent based on full working drawings, and so the student must exercise the greatest care in arriving at dimensions. Work in waste calculations should be to the nearest millimetre.

Falls, Crossfalls and Slopes

In SMM7, sections M10 relating to screeds and M40 and M50 covering floor finishes, the normal description reads as 'level or to falls only ≤158 from horizontal'. This indicates falls in one direction only and conflicts with the terminology used in sections Q22–25, encompassing roads and pavings taken to falls and crossfalls and to slopes ≤158 from horizontal. While in sections H71, J21 and J41, embracing flat roof coverings, such as lead, mastic asphalt and built up felt, and in sections H60–66, encompassing pitched roof coverings, such as tiling and slating, the term 'pitch' is used, which is the angular measurement of the slope of the roof.

It is surprising to see these variations in terminology which must confuse the student as much as it does the author. The various terms are now examined in an attempt to clarify the situation.

Fall: the amount of drop in the length of a surface from one end to the other and is sometimes used rather loosely described as a slope or inclination of a drain, gutter or any flat surface to throw off water. Alternatively, it may be expressed as the rate of fall/grade or gradient. For example, a drain may require a fall of 1 metre in 60 metres, i.e. a gradient of 1 in 60.

Crossfall: a fall across a surface as opposed to its length, as for example on a road from crown to channel, and often cambered in the case of 'black top' roads, and to paths as illustrated in example 27.

Slope: this is the same as gradient or grade and represents the normal measure of inclination from the horizontal, usually expressed as one vertical unit divided by the number of horizontal units needed to give that vertical rise. Examples are a drain laid at 1 in 60 (one metre rise or fall in a length of 60 metres) or earth slopes of 1 in 2 as illustrated in figure 10.

Hence roads and pavings can have falls in two directions (along the length and across the paved surface) and they normally have a slope or gradient ≤158 from the horizontal along the length of the paved surface.

The description of floor finishes and screeds as being 'level or to falls only ≤158 from horizontal' is therefore badly expressed, as it is really referring to a gradient or slope ≤158 from horizontal.

The term 'pitch' is used to express the angle of inclination of the roof to the horizontal, as for example, 458 with clay plain tiles and 308 with slates. However, in the case of flat roof coverings it is usual to give the gradient, such as 1 in 80.

Drawn Information

As outlined earlier in the chapter, the use of drawn information when measuring building work can take the following forms:

(1) *Location drawings* comprising block plans, site plans, and plans, sections and elevations, which together make up the contract drawings (General Rules 5.1).

(2) *Component drawings* showing the information necessary for the manufacture and assembly of a component (General Rules 5.2).

(3) *Dimensioned diagrams* showing the shape and dimensions of the work covered by a measured item, which may be used in place of a dimensioned description (General Rules 5.3). In some cases SMM rules specifically require the use of dimensioned diagrams, as in the instance of formwork to irregularly shaped beams or columns (E20:13–16). These diagrams simplify the descriptions and help to give the estimator a clearer picture of what is required. This book contains the following examples of dimensioned diagrams, which are all kept quite simple and consist of single line dimensioned sketches:

> stone jambs: example 5
> trussed rafter roof: example 15
> wood casement windows: example 19
> external door: example 22
> kitchen fitments: figure 36

(4) *Schedules* providing the required information are deemed to be drawings (General Rules 5.4), and these are described in the next section of this chapter, by reference to specification schedules.

Use of Schedules

When measuring a number of items with similar general characteristics but of varying components, It is advisable to use schedules as a means of setting down all the relevant information in tabulated form. This materially assists the taking off process and reduces the possibility of error. It would be a very lengthy process indeed to take off each item in detail separately and would involve the repetition of many similar items.

The use of schedules is particularly appropriate for the measurement of a considerable number of doors, windows or manholes, a number of lengths of drain and internal finishes to a series of rooms with different finishes. In some instances schedules are used to collect together specification information to assist in speedy taking off, while in other instances schedules are used for recording measurements and are in effect the taking off. An example of the latter would be a drain schedule. The following examples of schedules are provided in this book: internal finishes – table 9.1; windows – table 10.1; internal doors – table 11.1; kitchen fitments – figure 36; cold water services – table 13.1; electrical distribution sheet – table 14.1; drain schedule – table 15.1; and inspection chamber schedule – table 15.2.

Take Off Lists

Take off lists provide the quantity surveyor with the opportunity to look at a particular category of work in its entirety prior to measurement; the components are listed in a logical sequence, and they provide a checklist as the detailed measurement proceeds. This approach is particularly useful where work is fragmented, and the preparation of a take off list reduces the risk of omissions and ensures the entry of measured items in a logical sequence. Examples of take off lists are included in example 1 (foundations) and example 21 (internal doors).

Query Sheets

When taking off in practice the quantity surveyor will enter any queries for the architect on query sheets, normally divided down the centre to accommodate the queries on the left hand side of the sheet and the answers on the right hand side. In the examination the candidate will often have to decide the queries as they arise, when it will be desirable for him to indicate briefly in waste why he has adopted a certain course of action, and where appropriate to prepare a query sheet adopting a similar approach to that used in the office.

A typical query sheet follows with both questions by the quantity surveyor and answers by the architect inserted:

ARCHITECT'S QUERY SHEET

Questions (J T Smithson, 14 July 1998)	*Answers* (P M Arthurs, 18 July 1998)
1. Is the spacing of the cavity wall ties to be 900 horizontally and 450 vertically and staggered?	Yes
2. What is the construction at the head of the cavity wall?	One course of concrete blocks 215 wide × 140 thick
3. How far are facing bricks to be taken below finished ground level?	One course
4. What mix of concrete is to be used in filling the base of the cavity?	1 : 6

Preambles

Preambles are clauses usually inserted at the head of each work section bill and principally contain descriptions of materials and workmanship, as found in specifications, together with any other relevant information of which the contractor should be aware in pricing. In practice the full requirements of SMM7 are frequently not given in descriptions and the remaining information is contained in preambles, thus reducing the length of billed item descriptions. Indeed much of the information frequently found in preambles is of a specification type, and there is a distinct advantage in inserting it in the bill of quantities which is always a contract document, whereas the specification is probably not. Many government contracts use a separate specification document which largely replaces the preambles, and this does have some advantages for site management. Project specifications can either be separate documents or written into bills of quantities as preamble clauses. Some typical preamble clauses are given in chapter 17.

Another procedure which has been used on occasions is to combine all preamble clauses in a separate bill, following preliminaries. The contents of the preambles bill are often extracted from sets of standard clauses, such as those prepared by the former Greater London Council in *Preambles to Bills of Quantities*.

Prime Cost Items

The term 'prime cost sum' (often abbreviated to pc sum) is a sum provided for works or services to be executed by a nominated subcontractor, or for materials or goods to be obtained from a nominated supplier. Such sums are exclusive of any profit required by the main contractor and provision is made for its addition following the pc sum.

Thus the term includes specialist work carried out by persons other than the main contractor and for materials or components to be supplied to him by persons nominated by the architect. A typical example of a prime cost item follows:

Provide the pc sum of £3500 for the supply of 12 nr sanitary appliances, as specification clauses Y38–43		3500	00
Add main contractor's profit	%		

General attendances by the main contractor on nominated subcontractors are provided as items in the Preliminaries Bill (A42:1.16–17.1–2.0). Special attendance items required by the subcontractor, such as scaffolding, hardstandings, storage and power, are also inserted in the Preliminaries Bill. Opportunity must be given to the contractor to price such items on the basis of fixed and time related charges, depending on whether costs are incurred at a specific time or whether they are spread over a period of time (A51:1.3.1–8.1–2).

Provisional Sums

Where the work cannot be described and given in accordance with SMM7 rules it shall be given as a provisional sum and identified as for either defined or undefined work. In defined work items, a description and indication of the amount of work can be given, and the contractor will be deemed to have made due allowance in programming, planning and pricing the preliminaries. Where these details cannot be supplied, the work is classified as undefined and the contractor will be deemed not to have made any allowance in programming, planning and pricing preliminaries (General Rules 10:2–6).

Work by local authorities and statutory undertakings are the subject of provisional sums (A53:1.1–2). An example of an undefined provisional sum, which would be inserted in the preliminaries Bill, follows:

Provide the general provisional sum of £20 000 to cover the cost of any unforeseen works. This sum to be expended at the discretion of the architect.	20 000	00

Work of Special Types

Work of each of the following special types shall be separately identified (General Rules 7.1):

(a) Work on or in an existing building. Work to existing buildings is work on, in or immediately under work existing before the current project. A description of the additional preliminaries/general conditions appertaining to the work to the existing building shall be given, drawing attention to any specific requirements (General Rules 13).
(b) Work to be carried out and subsequently removed (other than temporary works).
(c) Work outside the curtilage of the site.
(d) Work carried out in or under water shall be so described stating whether canal, river or sea water and (where applicable) the mean Spring levels of high and low water.
(e) Work carried out in compressed air shall be so described stating the pressure and the method of entry and exit.

Fixing, Base and Background

Method of fixing shall only be measured where required by the rules in each work section. Where fixing through vulnerable materials is required to be identified, vulnerable materials are deemed to include the materials defined in General Rules 8.3e. Where the nature of the background is required to be identified, they shall be identified in the following classifications:

(a) Timber which shall be deemed to include manufactured building boards.
(b) Masonry which shall be deemed to include concrete, brick, block and stone.
(c) Metal.
(d) Metal faced material.
(e) Vulnerable materials which shall be deemed to include glass, marble, mosaic, tiled finishes and the like (General Rules 8:1–3).

Where the nature of the base is required, each type of base must be given separately (General Rules 8.2).

Composite Items

Where work which would otherwise be measured separately may be combined with other work in the course of off-site manufacture, it may be measured as one combined composite off-site item. The item description shall identify the resulting composite item and it shall be deemed to include breaking down for transport and installation and subsequent reassembly (General Rules 9.1).

General Definitions

Where the SMM rules require 'curved, radii stated', details shall be given of the curved work, including if concave or convex, if conical or spherical, if to more than one radius, and shall state the radius or radii. The radius is the mean radius measured to the centre line unless otherwise stated (General Rules 14:1–2).

Services and Facilities

Preliminaries Bill items include such services and facilities as power, lighting, safety, health and welfare, storage of materials, rubbish disposal, cleaning, drying out, protection and security (A42:1.1–15.1–2.0). There are no longer protection items in each work section as was the case with SMM6.

Plant Items

The contractor is given the opportunity in the Preliminaries Bill to price the various items of mechanical plant that generate costs which are not proportional to the quantities of permanent work. The principal items of mechanical plant are listed in A43:1.1–9.1–2.0, and range from cranes and hoists to earthmoving and concrete plant.

Standard Components

Many items of measured work, including timber roof trusses, joinery items and sanitary appliances, are standard, stock, or catalogued items. In these cases a cross reference to a catalogue or standard specification will form an adequate item description (General Rules 6.1). A number of items in this category are included in the worked examples in this book.

Worked Examples

The student is advised to proceed carefully through the worked examples in this book, comparing the measured items with the drawings, taking particular note of the comments accompanying the taking off and referring to the appropriate clauses of SMM7.

3   Mensuration Applications

INTRODUCTION

Mensuration is concerned with the measurement of areas and volumes of triangles, rectangles, circles and other figures, and some basic knowledge of this subject is required by all quantity surveying students. This chapter sets out to explain how the principles of mensuration are used in the measurement of building quantities.

A list of mensuration formulae is included in appendix 2 for reference purposes. On the figures that follow, dimensions containing a decimal marker are in metres and all others are in millimetres. Readers who are not familiar with metric dimensions may find the conversion table in appendix 3 helpful.

GIRTH OF BUILDINGS

Rectangular Buildings

One of the most common mensuration problems with which the quantity surveying student is concerned is the measurement of the girth or perimeter of a building. This length is required for foundations, external walls and associated items.

The length may be calculated on a straightforward rectangular building by determining the total external length of walling and making a deduction for each external angle equivalent to the thickness of the wall. Alternatively, the internal length might be taken and an addition made for each of the external angles. The example shown in figure 1 will serve to illustrate this point:

<u>Taking external dimensions</u>

	15.000
	6.000
Sum of one long and one short side	2/21.000
Sum of all four sides (measured externally)	42.000
<u>Less</u> corners 4/255	1.020
Mean girth of external wall	40.980
(centre line of cavity)	

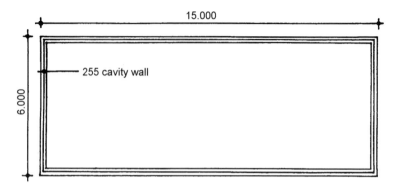

Figure 1   *Girth of rectangular building.*

<u>Taking internal dimensions</u>

15.000 <u>less</u> 2/255	14.490
6.000 <u>less</u> 2/255	5.490
	2/19.980
Sum of all four sides (measured internally)	39.960
<u>Add</u> corners 4/255	1.020
Mean girth of external wall	40.980

Figure 2 illustrates why it is necessary to take the full thickness of the wall when adjusting for corners.

Assume that in this particular case the external dimensions have been supplied and these have to be adjusted to give the girth on the centre line through the intersection point O. The procedure is made clearer if the centre lines are extended past O to meet the outer wall faces at Y and Z. It is then apparent that the lengths to be deducted are XY and XZ, which are equal to OZ and OY, respectively. These are equivalent to half the thickness of the wall in each case and so together are equal to the full thickness of the wall.

This centre line measurement is extremely important, since it provides the length to be used in the dimensions for trench excavation, concrete in

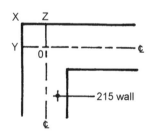

Figure 2   *Adjustment for corner.*

foundations, brickwork and damp-proof course. In the case of hollow walls with a faced outer skin, three different lengths will be required when taking off the dimensions of the brickwork, since the two skins and the forming of the cavity have each to be measured separately.

The centre line of wall measurement will be required for the formation of the cavity and associated work, and an addition or deduction will be needed for each corner, on a plain rectangular building, to give the centre line measurement of each skin. In a 255 hollow wall, to obtain the centre line length of each skin and the centre line length of the cavity, adjustments of 102.5 and 255 respectively are required for each corner relative to either the external or internal lengths of the wall.

Buildings of Irregular Outline

The position is a little more confusing if the building has an irregular outline as shown in figure 3.

In this case the internal and external angles, E and D, at the set-back cancel each other out, and the total length is the same as if there had been no recess and the building was of plain rectangular outline (ABCG), as shown by the broken lines.

The length on centre line of the enclosing walls can be found as follows:

	20.000
	9.000
	2/29.000
Length of all enclosing walls (measured externally)	58.000
Less corners 4/255	1.020
Length on centre line (₵)	56.980

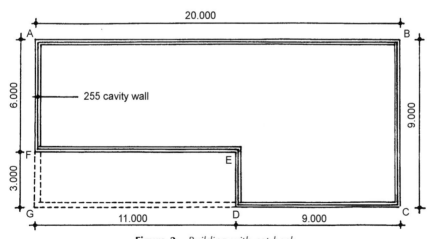

Figure 3 *Building with set-back.*

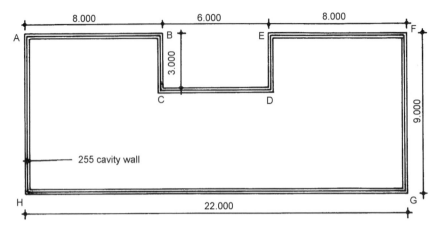

Figure 4   *Building with recess.*

On occasions buildings are planned with recesses and these involve further additions when arriving at the girth of the enclosing walls, as illustrated in the example in figure 4.

In this case twice the depth of the recess, that is, BC + ED, will have to be added to the lengths of the sides of the enclosing rectangle, AFGH. The internal and external angles at C, D, B and E, cancel each other out and the length of CD is equal to BE, plus twice the wall thickness, thus requiring no adjustment to the length of the external wall. The length of enclosing walls measured on centre line is found as follows:

	22.000
	9.000
	2/31.000
	62.000
<u>Add</u> twice depth of recess (2/3.000)	6.000
	68.000
<u>Less</u> corners 4/255	1.020
Total length of walls on ₵	66.980

MEASUREMENT OF AREAS

The quantity surveyor is often called on to calculate the areas of buildings, sites, roads and other features, and some of the basic rules of mensuration need to be applied to the problems that arise. Some of the more common cases encountered in practice will now be illustrated.

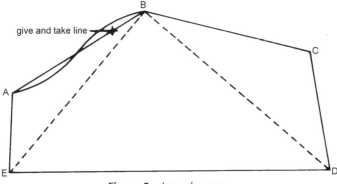

Figure 5 *Irregular area.*

Irregular Areas

In the measurement of irregular areas the best procedure is generally to break down the area into a number of triangles, as indicated in figure 5 by the broken lines EB and BD, giving three triangles, EAB, EBD and BCD. The area of each triangle is found by multiplying the base by half the height. Where irregular boundary lines are encountered as between A and B, the easiest method is to draw in a straight give-and-take line with a set square.

Trapezoids

Another type of irregular area which has to be measured on occasions, particularly with cuttings and embankments, is the trapezoid, which comprises a quadrilateral (four-sided figure) with unequal sides but with two opposite sides parallel. The term trapezium is often used in place of trapezoid.

In this case the area is found by multiplying the average width ($\frac{1}{2}$/top + bottom) by the height (vertical distance between the top and bottom of the trapezoid).

The area of the trapezoid illustrated in figure 6 is:

$$\tfrac{1}{2}(6.000 + 18.000) \times 3.000 = 12 \times 3 = \underline{\underline{36\,\text{m}^2}}$$

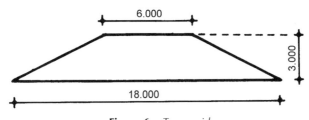

Figure 6 *Trapezoid.*

Segments

The area of a segment (part of circle bounded by arc and chord) is sometimes required in the measurement of an arch. The normal rule for the measurement of the area of a segment is to take the area of the sector and deduct the area of the triangle, the area of the sector being found from the following formula

$$\tfrac{1}{2}(\text{length of arc} \times \text{radius})$$

The area of the segment shown hatched below the arch in figure 7 can be determined in this way, and will be required for deduction purposes as part of the window or door opening. The length of arc BD can be found by its proportion of the circumference of the circle as a whole, related to the angle which it subtends at the centre of the circle ($\angle\alpha$). Thus

$$\frac{\text{arc BD}}{\text{circumference of whole circle}} = \frac{\angle\alpha}{360°}$$

$$\text{arc BD} = \angle\alpha \times 2\pi r/360°$$

Alternatively the length BD may be scaled off the drawing. The length of the arch itself will be taken on its centre line XY, since the mean length will be required. The same alternative methods of measurement are available.

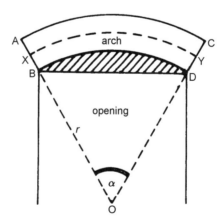

Figure 7   *Segmental arch.*

Another formula often used by quantity surveyors for obtaining the area of a segment follows:

$$\frac{H^3}{2C} + \frac{2}{3}CH$$

where C is length of chord and H is height of segment.

Bellmouths, as at Road Junctions

Difficulty may be experienced in measuring the irregular areas which arise at road junctions. The following example should clarify this problem.

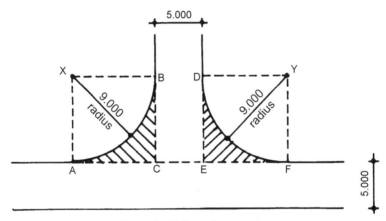

Figure 8 *Bellmouth to road.*

The carriageways are first measured through to their intersections for their full widths, 5 m in the example covered in figure 8. This leaves the hatched irregular-shaped areas ABC and DEF still to take. The area in each case is $\frac{3}{14}$ radius^2, and is equivalent to the area of a square, whose side is equal to the radius, less the area of a quadrant or quarter circle of the same radius.

In this case the area on both sides of the bellmouth would be entered on the dimension sheet as

$$2/\frac{3}{14}/ \quad \frac{9.00}{9.00}$$

MEASUREMENT OF EARTHWORKS

The student frequently experiences difficulty in measuring the volume of earthworks, particularly on sloping sites. The following examples are designed to indicate the main principles involved and generally clarify the method of approach.

Sloping Site Excavation

The quantity surveyor is often called on to calculate the volume of excavation and/or fill required on a sloping site and the following example indicates a comparatively simple method of approach.

Assume that in the example illustrated in figure 9 it is required to excavate down to a level of 2.000, including excavating topsoil to a depth of 150. In this case the whole of the site is to be excavated, whereas if fill had been required on part of the site, it would have been necessary to have plotted the reduced level contour line on the drawing, and this line would have formed the demarcation line between the areas of excavation and fill respectively.

The average depth of excavation over the site is most conveniently found by suitably weighting the depth at each point on the grid of levels, according to the area that each level affects. This involves taking the depths at the

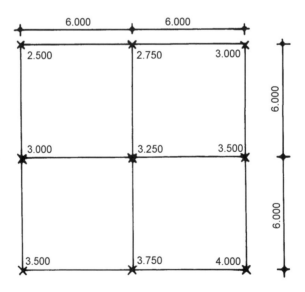

Figure 9   *Sloping site excavation.*

extreme corners of the area once, intermediate points on the boundary twice and all other intermediate points four times. The sum of the weighted depths is divided by the total number of weightings (number of squares × 4) to give the average weighted depth for the whole area. This method can only be used when the levels are spaced the same distance apart in both directions.

The volume in this example is now calculated:

Corner depths		2.500
		3.000
		4.000
		3.500
Depths at intermediate	2/2.750 =	5.500
points on boundary	2/3.000 =	6.000
	2/3.500 =	7.000
	2/3.750 =	7.500
Depth at centre point	4/3.250 =	13.000
Sum of weighted depths		16)52.000
Average total depth		3.250
Less reduced level excavation		
and topsoil (2.000 + 150)		2.150
Average adjusted depth		1.100

The dimensions would then appear as follows:

12.00		Excvtg.
12.00		Topsoil for preservn.
		150 av. depth
12.00		Disp.
12.00		Excvtd mat
0.15		on site; 20.00 av.
		dist. in sp.hps.
12.00		Excvtg.
12.00		To reduce levs.
1.10		max. depth ≤ 2.00
		&
		Disp.
		Excvtd. mat.
		off site

Cuttings and Embankments

The volumes of cuttings and embankments are generally calculated from the cross sectional areas taken from plotted cross sections, often prepared at 30.000 intervals along the line of the cutting or embankment. Certain intermediate cross sectional areas are often weighted by using Simpson's rule or the prismoidal formula.

Furthermore, allowance must be made for the sloping banks on either side as illustrated in the example shown in figure 10. The bank slopes may be described as say 1 in 2 or 2 to 1, indicating that the bank rises 1.000 vertically for every 2.000 in the horizontal plane.

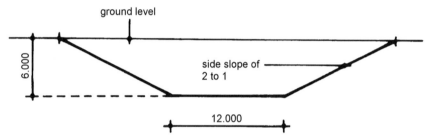

Figure 10 *Excavation to cutting. Note that this figure is drawn to a natural scale, whereas in practice cross sections are often drawn to an exaggerated vertical scale.*

When calculating the volume of excavation and fill for cuttings and embankments, Simpson's rule can often be used to advantage and a simple example follows to illustrate this point.

Using Simpson's rule the area at intermediate even cross sections (nrs. 2, 4, 6, etc.) are each multiplied by 4, the areas at intermediate uneven cross sections (nrs. 3, 5, 7, etc.) are each multiplied by 2 and the end cross sections taken once only. The sum of these areas is multiplied by 1/3 of the distance between the cross sections to give the total volume. To use this formula it is essential that the cross sections are taken at the same fixed distance apart and that there are an odd number of cross sections (even number of spaces between cross sections).

For instance, take a cutting to be excavated for a road, 180.000 in length and 12.000 in width, to an even gradient, with mean depths calculated at 30.000 intervals as indicated and side slopes of 2 to 1:

Cross section	1	2	3	4	5	6	7
Mean depth (m)	1	3	5	7	6	4	2

The width at the top of the cutting can be found by taking the width at the base, that is, 12.000 and adding 2/2/the depth to give the horizontal spread of the banks (the width of each bank being twice the depth with a side slope of 2 to 1).

Cross section	Depth (m)	Width at top of cutting (m)	Mean width (m)	Weighting
1	1	$12 + 4/1 = 16$	$(16 + 12)/2 = 14$	1
2	3	$12 + 4/3 = 24$	$(24 + 12)/2 = 18$	4
3	5	$12 + 4/5 = 32$	$(32 + 12)/2 = 22$	2
4	7	$12 + 4/7 = 40$	$(40 + 12)/2 = 26$	4
5	6	$12 + 4/6 = 36$	$(36 + 12)/2 = 24$	2
6	4	$12 + 4/4 = 28$	$(28 + 12)/2 = 20$	4
7	2	$12 + 4/2 = 20$	$(20 + 12)/2 = 16$	1

The dimensions can now be entered on dimensions paper in the following way:

	14.00		Excvtg.	
	1.00		(CS1 To reduce levs.	*Note:* A great deal of
			max. depth ≤ 8.00	laborious and un-
4/	18.00			necessary labour in
	3.00		(CS2	squaring has been
				avoiding by entering
				all the dimensions
2/	22.00			as superficial items,
	5.00		(CS3	to be subsequently
				cubed by multiplying
				the sum of the
4/	26.00			areas by 1/3 of the
	7.00		(CS4 &	length between the
				cross sections. (Total
2/	24.00		Excvtd. mat.	weighting is 18 and
	6.00		(CS5	number of 30.000 long
			Fillg. to make up levs.	sections of excavation
4/	20.00		>250 av. th.	is 6, so that 6/18
	4.00		(CS6 arisg from excvns.	or 1/3 of the distance
				of 30.000 must be the
	16.00		Cube × $\frac{1}{3}$/30.00 = ___ m³	timesing factor re-
	2.00		(CS7	quired.) It is assumed
				that the excavated
				material will be used
				as filling elsewhere on
				the road project.

In simpler cases involving three cross sections only, the prismoidal formula may be used, whereby

$$\text{volume} = \frac{1}{6}\left\{\begin{array}{c}\text{total}\\\text{length}\end{array}\right\} \times \left\{\begin{array}{c}\text{area of}\\\text{first section}\end{array} + \begin{array}{c}4 \text{ times area of}\\\text{middle section}\end{array} + \begin{array}{c}\text{area of last}\\\text{section}\end{array}\right\}$$

These formulae can also be used to calculate the volume of banks which vary in cross sectional area throughout their lengths.

MEASUREMENT OF PITCHED ROOFS

Lengths of Rafters

Where roof sections are drawn to a sufficiently large scale the easiest method is to scale the length of the rafter off the drawing, taking the length from one extremity to the other of the rafter.

Another alternative is to calculate the length by multiplying the natural secant of the angle of pitch by half the total span of the roof. The natural secants of the more usual pitches of roof are as follows:

Pitch of roof	158	308	408	458	508
Natural secant	1.036	1.155	1.305	1.414	1.555

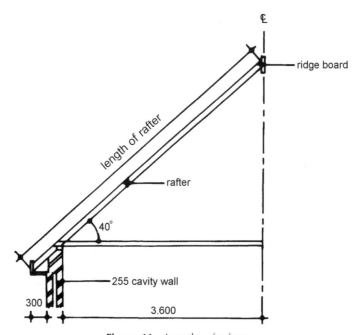

Figure 11   *Lengths of rafters.*

The student is referred to four-figure mathematical tables for values of natural secants relating to other angles of pitch. The example shown in figure 11 illustrates the method of calculation of the lengths of rafters.

Half total span of roof =   3.600   +   255 + 300   = 4.155

(half effect-   (wall   (over-
ive span)   thick-   hang at
ness)   eaves)

Length of rafter = 4.155 × 1.305 (secant 40°)

= 5.422

(to which a small addition of 75 should be made for a tapered end to be precise).

Lengths of Hips and Valleys

The length of a hip or valley is most conveniently found by plotting and scaling from the roof plan as shown in figure 12.

The length AB represents the length of the hip on plan, while the length on slope is actually required. To obtain the length on slope, the height of the roof is set out at right angles to AB on the line AC. BC then represents the length of the hip to the slope of the roof.

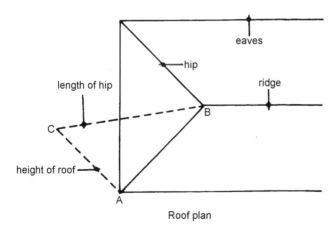

Roof plan

Figure 12 *Lengths of hips and valleys.*

Roof Coverings

The area of roof covering is measured in the same way whether the roof be hipped or gabled at the ends, provided that the angle of pitch is constant throughout, that is, twice the length of the roof multiplied by the length on slope. The length of the roof will probably vary a little with each of these two forms of construction, owing to the different amounts of overhang at the ends of the roof.

4   Groundwork and Foundations

Preliminary Investigations

Before taking off any dimensions the quantity surveyor normally makes a careful study of all the drawings relating to the project to obtain the overall picture and become familiar with the main details, at the same time checking to ensure that all the drawings are complete and, at there are adequate specification particulars, to enable the measurement to proceed unhindered by inadequate information. The next step is usually a visit to the site to obtain details and measurements of any work required on the site, often termed siteworks. These works include breaking up paving, taking down boundary walls and fences, and possibly demolishing existing buildings, felling trees, grubbing up hedges, and similar work.

Some contracts involve alterations to existing buildings and these are sometimes termed 'spot items' and are normally kept together in a separate section of the bill of quantities possibly headed 'Demolitions and Alterations', with some of the items covered by Provisional Sums, where the full extent of the work involved cannot be accurately determined. Many of the details relating to spot items will also need to be obtained on the site. A worked example of the measurement of demolition and alteration work is provided in *Advanced Building Measurement*.

When visiting the site the quantity surveyor should also be on the look-out for any unusual items which affect cost, and which should accordingly be included in the billed description. A check on the type of soil and groundwater level comes into this category, unless the information is supplied by the architect, probably in the form of particulars from trial pits or boreholes excavated on the site. With landfill or brownfield sites, it is also necessary to check on possible land contamination, with the likely consequential high cost of removing contaminated soil and remedial measures, including the extraction of landfill gases such as methane.

General Items

It is necessary to give details and locations of any trial pits or boreholes, the groundwater level on the site (at a prescribed precontract date), and details and locations of existing services. Alternatively, in the absence of any trial pits or boreholes, a description of the ground and strata to be assumed should be given. Information should also be provided of any features on the site which are to be retained, such as mature trees (D20:P1e).

Site Preparation

Removal of trees and tree stumps is measured as an enumerated item, including grubbing up roots, disposal of materials and filling voids, classified in the girth ranges listed in D20:1.1–2.1–3.0. Tree girths are measured at a height of 1.000 above ground and stump girths at the top of the stumps (D20:M1–2). Clearing site vegetation is measured in m^2 with a description sufficient for identification purposes (D20:1.3.4.0). Site vegetation embraces bushes, scrub, undergrowth, hedges, trees and tree stumps ≤600 girth (D20:D1).

The first building operation is normally the excavation of the topsoil for preservation over the whole area of the building and this usually forms the first excavation item in the Excavating and Filling section of the bill of quantities. The area is measured to the outer extremities of the foundations in m^2 and the average depth, often 150, is included in the description (D20:2.1.1.0). Disposal of topsoil on the site in temporary spoil heaps for reuse is covered by a separate cubic item stating the location of any spoil heaps (D20:8.3.2.1). Spreading soil on the site to make up levels is measured in m^3, distinguishing between average thicknesses ≤ and >250. Disposal of excavated material off the site is measured in m^3, giving details where appropriate of specified locations or handling and where active, toxic or hazardous materials are involved (D20:8.3.1.1–4). If the existing turf over the site of the building is to be preserved, then this forms a separate billed item measured in m^2, stating the method of preserving the turf, such as stacking in rolls in a specific location (D20:1.4.1.0).

Excavation to Reduce Levels

Where the site is sloping then further excavation is required to reduce the level of the ground to the specified formation level; this excavation is measured in m^3 as excavation to reduce levels in accordance with D20:2.2.1–4.0, giving the appropriate maximum depth range. The excavation rules in SMM7 are based on all excavation being carried out by mechanical plant.

Excavation of Foundation Trenches

Foundation trench excavation is measured in m^3, stating the commencing level where >250 below existing ground level and the maximum depth range in accordance with D20:2.5.1–4.1, namely ≤250, 1.000 (1 m), 2.000 and thereafter in 2.000 stages. It is necessary to distinguish between trenches ≤300 wide and those >300 wide.

All excavation is measured net with no allowance for increasing in bulk after excavation or for the extra space required for working space or to accommodate earthwork support (D20:M3). Breaking out rock; concrete; reinforced concrete; brickwork, blockwork or stonework; shall each be described and measured separately in m^3 as extra over any types of excavating (D20:4.0.1–4.1), while breaking out existing hard pavings is measured in m^2, stating the thickness, as extra over excavating (D20:5.0.5.1). Rock is defined as any material which is of such size or position that it can only be removed by wedges, special plant or explosives (D20:D5). Examples of 'special plant' are given in the *SMM7 Measurement Code*. Excavating below groundwater level is given in m^3 as extra over any types of excavating (D20:3.1.0.1–2).

Working space allowance to excavations, categorised in four types of excavation as D20:6.1–4, is measured in m^2, where the face of the excavation is <600 from the face of formwork, rendering, tanking or protective walls (D20:M7).

Excavating next to existing services is measured in metres as extra over any types of excavating, stating the type of service, such as gas or water mains, electricity or BT cables or sewers (D20:3.2.1.1–2). While that around existing services crossing excavation is an enumerated extra over item (D20:3.3.1.1–2), since it is likely to entail hand digging, most other excavation will be carried out by machine.

Three sets of levels will be required before foundation work can be measured: (1) bottom of foundations, (2) ground levels and (3) finished floor levels.

When measuring foundation trenches it is advisable to separate the trenches into external and internal walls. Where the external wall foundation is of constant width and the site is level, its measurement presents no real difficulty with the length being obtained by the normal girthing method, as outlined in chapter 3. Where the site is sloping and stepped foundations are introduced the process of measurement of the foundation trench excavation is more complex, since each length of trench will have to be dealt with separately. It is good policy to use a schedule for this purpose giving the lengths, levels, average depths within maximum depth ranges, and widths of trench for each section between steps. The sum of the lengths of all the individual sections will need to be checked against the total calculated length to avoid any possibility of error. Furthermore, the individual brickwork lengths will not always coincide with the lengths of excavation and concrete, as illustrated in *Advanced Building Measurement*.

After measuring the excavation for external wall foundations, the internal wall foundations will be taken, and this will often involve a number of varying lengths and widths of foundations, which are best collected together in waste and the drawing suitably marked as each length is extracted. Care must be taken to adjust for the overlap of trenches at the intersection of the external and internal walls, as shown in figure 13. The external wall foundation trench

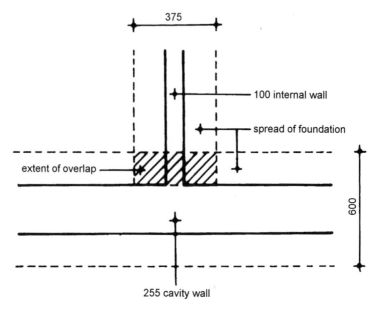

Figure 13   *Intersection of internal and external walls.*

will have been measured around the whole building in the first instance, and the hatched section will have to be deducted at each intersection when arriving at the length of internal wall foundation trench.

Disposal of Excavated Material

The subsequent disposal of excavated material forms a separate billed item in m³, either of soil to be stored on site, used as filling to make up levels, filling to excavations, or to be removed off the site. In the first instance, when measuring the trench excavation, it may be simplest and most convenient to take the full volume as filling to excavations and subsequently to adjust as disposal of excavated materials off site with the measurement of the concrete and brickwork, as illustrated in figure 14 for a traditional strip foundation. However, in other circumstances different approaches are more appropriate, such as for the trench fill foundation in figure 15 and the alternative strip foundation in figure 16. In both cases the best approach is to take the volume of trench excavation as disposal of excavated material off site, followed by any necessary adjustments. In all cases the excavation of topsoil for preservation, and its disposal on site will form separate items as described earlier.

Handling of excavated material is normally at the discretion of the contractor.

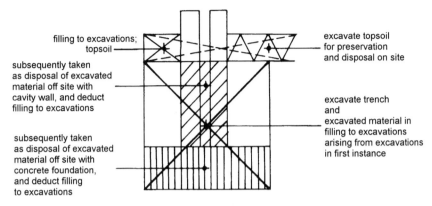

filling to excavations; topsoil

subsequently taken as disposal of excavated material off site with cavity wall, and deduct filling to excavations

subsequently taken as disposal of excavated material off site with concrete foundation, and deduct filling to excavations

excavate topsoil for preservation and disposal on site

excavate trench and excavated material in filling to excavations arising from excavations in first instance

Figure 14 *Excavating and filling: traditional strip foundation.*

Surface Treatments

The measurement of the excavation is generally followed by a superficial item for compacting the bottom of the excavation (D20:13.2.3.0). Surface treatments may alternatively be given in the description of any superficial item (D20:M17). Compacting is deemed to include levelling and grading to falls and slopes ≤158 from horizontal (D20:C5).

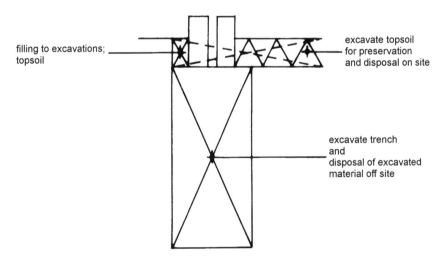

filling to excavations; topsoil

excavate topsoil for preservation and disposal on site

excavate trench and disposal of excavated material off site

Figure 15 *Excavating and filling: trench fill foundation.*

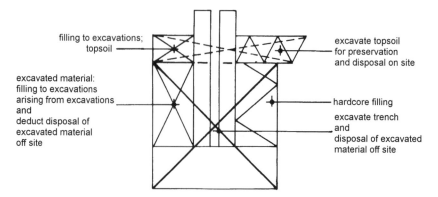

filling to excavations;
topsoil

excavate topsoil
for preservation
and disposal on site

excavated material:
filling to excavations
arising from excavations
and
deduct disposal of
excavated material
off site

hardcore filling

excavate trench
and
disposal of excavated
material off site

Figure 16 *Excavating and filling: strip foundation with hardcore backfill internally.*

Basement Excavation

Basement excavation is measured to the outside of foundations in m^3 stating the maximum depth range. Working space allowance is measured in m^2 on the external face of the basement walls, where the face of the excavation is >600 from the face of the tanking or protective walls (D20:M7). The area measured is calculated by multiplying the girth of the tanking or protective walls by the depth of excavation below the commencing level of the excavation (D20:M8). Additional earthwork support, disposal, backfilling, work below groundwater level and breaking out are deemed to be included (D20:C2).

It is usual to take the basement excavation as disposal of excavated material off site in the first instance and to later adjust the filling to excavation where appropriate. It was decided to omit the basement example of earlier editions, as the structure is likely to be of reinforced concrete construction, and this was considered to be outside the scope of the book.

Earthwork Support

Support to the sides of excavation is measured in m^2 to trenches (excluding pipe trenches where it is deemed to be included in the linear trench excavation description), pits, and the like, where >250 in depth, whether the support will actually be required on the project or not. The maximum depth is given in stages in accordance with D20:7.1–3. Earthwork support is also classified by the distance between opposing faces in stages ≤2.000, 2.000 to 4.000, and >4.000. Earthwork support left in, curved, next to roadways or existing buildings (illustrated in the *SMM7 Measurement Code*), below groundwater level or to unstable ground shall be described and separately measured in accordance with the rules contained in D20:7.1–3.1–3.1–6.

Concrete Foundations

Concrete particulars are to include the kind and quality of materials, mix details, tests of materials and finished work, methods of compaction and curing and other requirements (E05/10:S1–5), but much of this information may be included in preamble clauses or cross references to project specification clauses. Concrete poured on or against earth or unblinded hardcore shall be so described (E10:1–8.1–3.0.5). Concrete foundations include attached column bases and attached pile caps, while isolated foundations include isolated column bases, isolated pile caps and machine bases. Beds include blinding beds, plinths and thickening of beds.

In-situ concrete is measured in m^3 and the degree of difficulty in placing the concrete is reflected by giving the thickness ranges ≤150, 150–450 and >450 in the case of beds, slabs and walls.

On a sloping site the concrete foundations will probably be stepped and it will be necessary to measure the additional concrete at the step and a linear item of formwork to the face of the step (E20:1.1.2–4) classified in three stages of depth: ≤250, 250–500 and 500–1.000.

In the example shown in figure 17 it will be necessary to add a 450 length of foundation, 225 thick, and to make the necessary excavated material filling and disposal adjustments as shown in the following entry on the dimensions paper:

	Add *In-situ* conc. 21 N/mm^2 (20 agg.) Fdns. a.b.	A full description of the concrete foundation is not required as it has been given earlier.
0.75 0.45 0.23	(step & Ddt. Excvtd. mat. Fillg. to excvns. a.b.	Note the adjustment of excavated material filling and disposal that is also required. It will also be necessary to take a 750 length of
	& Add Disp. Excvtd. mat. off site	formwork to sides of foundations, plain vertical, height ≤250.

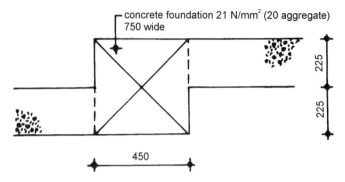

Figure 17 *Stepped foundation.*

Note: Concrete has been classified by strength, i.e. 21 N/mm² (20 aggregate) as an alternative to a designed mix to BS 5328 or a specified mix of concrete, such as 1 : 3 : 6/20 mm aggregate.

If the concrete foundations are reinforced with fabric reinforcement, the reinforcement is measured net in m² stating the mesh reference, weight per m², minimum laps and strip width, where placed in one width (E30:4.1.0.2). Bar reinforcement is billed in tonnes, keeping each diameter (nominal size) separate, although it will be entered by length on the dimensions sheet, distinguishing between straight, bent and curved bars (E30:1.1.1–3). Hooks and tying wire, and spacers and chairs which are at the discretion of the contractor are deemed to be included (E30:C1).

Figure 18 shows the thickening of an *in-situ* concrete bed to support an internal wall, assuming a length of 10 metres. The thicker section of concrete bed falls into a different thickness classification as E10:4.2.0. The volume occupied by the 150 thick bed has to be deducted as shown in the following dimensions. Some surveyors may however consider that no adjustment is necessary, taking the view that this is covered by E10:M2, whereby the thickness range stated in descriptions excludes projections.

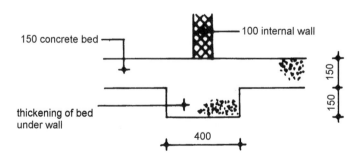

Figure 18 *Thickening of concrete bed under internal wall.*

	10.00 0.40 0.30		Conc. (21 N/mm^2) Beds 150–450 th.
	10.00 0.40 0.15		Ddt. conc. a.b. Beds $\leq$150 th.

Other adjustments will probably be needed to the excavation and hardcore quantities.

Other Substructure Work

It has been customary to include all substructure work in a separate section of the bill, including brickwork up to and including damp-proof course, which may be subject to remeasurement. However, SMM7 in line with *Common Arrangement of Work Sections for Building Works*, subdivides this work into several work sections: D20 (Excavating and Filling), E10 (Mixing/Casting/Curing *in-situ* concrete), E20 (Formwork for *in-situ* concrete), E30 (Reinforcement for *in-situ* concrete), F10 (Brick/block walling) and F30 (Accessories/sundry items for brick/block/stone walling). Hence the substructure work will probably be subdivided into these work sections in the bill of quantities, albeit rather fragmented. The measurement of brickwork is covered in detail in chapter 5, but it is considered desirable to include some brickwork in the worked example in the present chapter to follow normal taking off practice.

Brick Walling
Brick walling is measured in m^2, stating the nominal thickness, such as one brick thick, and whether there is facework (fair finish) on one or both sides (F10:1.1–3.1.0). It should be noted that all brickwork is deemed to be vertical unless otherwise described (F10:D3). The skins of hollow walls and the formation of the cavity, including wall ties, are each separately measured in m^2 (F30:1.1.1.0).

The projecting brickwork in footings, which are now little used except in very thick walls, is separately measured as horizontal projections in metres, stating the width and depth of projection (F10: 5.1.3.0). The example shown in figure 19 illustrates the measurement of brick footings.

The 1$\frac{1}{2}$B wall is measured down to the base of the wall and the projections are measured as an additional linear item, taking the combined average projection on each face of the wall, and assuming a 10.000 length of wall.

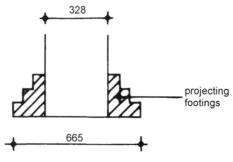

Figure 19   *Footings.*

| 2/ | 10.00 | | ftgs.
width of projs. on ea.
face.
top. cos. $\frac{1}{4}$B
bottom cos. $\frac{3}{4}$B
                    2)1B
             av. $\frac{1}{2}$B

Comm. bwk. to
BS 3921 in c.m. (1:3)
Projs.
   225 × (av.) 103
   proj.; hor. | Average combined
projection measured
over three courses
on each wall face. |

Facework

Brick facework is included in the measurement of the brickwork on which it occurs, with a description of the kind, quality and size of bricks, type of bond, composition and mix of mortar and type of pointing (F10:S1–4). In practice these particulars could alternatively be included in preamble clauses or be cross referenced to a projct specification. Specifying the precise type of brick is a better approach as it simplifies the task of the estimator in determining its relative hardness and ease of laying, and should lead to more realistic pricing.

Damp-proof Courses

Damp-proof courses are measured in m², distinguishing between those ≤ and >225 in width, for example half brick, one brick and block partitions are all ≤225 wide. Vertical, raking, horizontal and stepped work are so described. There is a further classification of cavity trays (F30:2.1–2.1–4.1). Curved work is so described (F30:M1), although the extra materials for curved work are deemed to be included (F30:C1c), and it is not necessary to give the radius of the work.

The description of the damp-proof course contains particulars of the materials used, including the gauge, thickness or substance of sheet materials, number of layers and composition and mix of bedding materials (F30:S4–6). Pointing of exposed edges is deemed to be included and does not require specific mention (F30: C2), and no allowance is made for laps (F30:M2).

WORKED EXAMPLE

A worked example follows covering the foundations to a small building, to illustrate the method of approach in taking off this class of work and the application of the principles laid down in the *Standard Method of Measurement of Building Works* (SMM7).

The importance of a logical sequence in taking off cannot be over-emphasised. It simplifies the taking off process, reduces the risk of omission of items and gains the student additional marks in the examination. For example, in taking off foundations to a small building a satisfactory order of items could be as follows:

(1) excavating and disposal of topsoil;
(2) excavating foundation trenches and filling/disposal of excavated material;
(3) compacting bottoms of trenches;
(4) earthwork support to faces of excavation;
(5) disposal of surface water;
(6) *in-situ* concrete in foundations;
(7) brickwork/blockwork and associated items;
(8) damp-proof courses;
(9) adjustment of faced brickwork for common brickwork;
(10) filling to excavations.

On the drawings decimal markers in dimensions indicate measurements in metres. Where there is no decimal marker, the dimensions are normally in millimetres.

Take Off List
It is good practice to prepare a take off list preparatory to measuring where the work is complex or fragmented. This ensures that careful thought is given to the character and scope of the work to be measured and its sequence, thereby reducing the liability to errors and omissions and providing a useful checklist as an aid to measurement.

A take off list for example 1 could read as follows:

Excavating topsoil
Disposal of excavated material
Excavating trenches (external and internal walls)
Disposal of excavated material
Surface treatments (trenches)
Earthwork support
Disposal of surface water
Concrete in foundations
Brickwork to dpc
Blockwork to dpc
Forming cavities and providing insulation
Concrete fill to base of cavity
Damp-proof courses
Adjustment of brickwork and facework
Filling to excavations
Surface treatments (floor area)
Filling to make up levels
Damp-proofing
Concrete beds

This page has been left intentionally blank for make-up reasons.

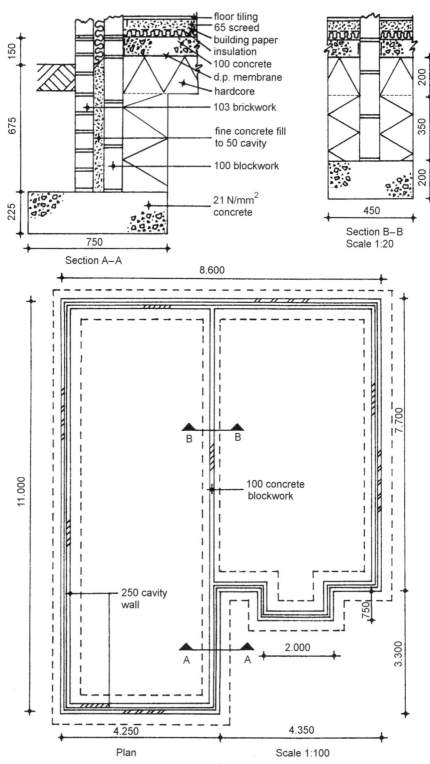

floor tiling
65 screed
building paper
insulation
100 concrete
d.p. membrane
hardcore
103 brickwork
fine concrete fill
to 50 cavity
100 blockwork
21 N/mm^2
concrete

150
675
225
750

Section A–A

200
350
200
450

Section B–B
Scale 1:20

8.600
7.700
11.000

100 concrete
blockwork

250 cavity
wall

2.000
3.300
750

4.250
4.350

Plan
Scale 1:100

Drg.1	Foundations 1 Example 1	

Work up to dpc

width of ext. fdn.	750	
less wall	253	
	2)497	
fdn. sprd.	248.5	

Note the extensive use of 'waste' calculations in the build up of dimensions with full descriptive notes.

width of int. fdn.	450	
less wall	100	
	2)350	
fdn. sprd.	175	

The dimensions in waste are expressed in metres and millimetres. The symbol 'mm' has been omitted from descriptions, as it appears superfluous and should not cause any confusion in practice.

	11.000	4.250
add fdn. sprd.		
2/248.5	497	497
	11.497	x 4.747

	7.700	4.350
add fdn. sprd.		
2/248.5	497	—
	8.197	x 4.350

The references that follow relate to the relevant clauses in SMM7.

Projn.	2.000	750
add fdn. sprd.		
2/248.5	497	—
	2.497	x 0.750

Excavating topsoil to be preserved is measured in m², stating the average depth (D20:2.1.1.0). In this case the best approach is to measure each of the three rectangular areas separately.

Excvtg.

11.50	Topsoil for presvn.
4.75	
	150 av. depth
8.20	
4.35	projn.) &
2.50	
0.75	

It will be noted that the dimensions in the dimension column are expressed in metres to two places of decimals (nearest 10 mm).

The disposal of the excavated soil is measured in accordance with D20:8.3.2.1 as a cubic item and can be linked with the previous superficial item by the method shown.

Disp.

Excvtd. matl.

on site; 30.00 av. dist. in sp. hps.

Cub x 0.15 = _____ m³

Redistribution of the topsoil from spoil heaps will be covered subsequently by a separate item, probably as part of external works, as D20:10.1.1.3.

In the case of a sloping site, reduce level excavation is taken in m³, giving the maximum depth ranges as D20:2.2.1.0.

Foundations 2

Ext. Wall Trenches

		length
2/	11.000	22.000
2/	8.600	17.200
2/	750	1.500
	ext. gth.	40.700

less cnrs.

4/253	1.012
	39.688

	depth
	675
	225
	900
less topsoil	150
	750

Note the method of building up the length of the trench excavation on its centre line and also the relevant depth.

The internal and external angles at the set back and projection cancel themselves out and no adjustments are necessary.

Note the order of length, width and height following the procedure prescribed for dimensions in descriptions in General Rules: 4.1.

Int. Wall Trenches

		length
		7.700

less ext. walls

2/253	506	
ext. wall fdn. sprd.		
2/248.5	497	1.003
		6.697

	depth
	350
	200
	550

External and internal wall trenches are calculated separately in a systematic and logical way.

Excvtg.

39.69	Trs. > 0.30 wide
0.75	≤ 1.00 max. depth
0.75	
	(ext. walls
6.70	
0.45	
0.55	(int. wall

&

Disp.

Excvtd. matl.
off site

Excavation to trenches is measured in the prescribed stages of maximum depth and stating the width classification as D20:2.5–6.1–4.0, and the commencing level where >0.25 below existing ground level.

It is preferable in this example to take all the trench excavated soil as disposal off site, as D20:8.3.1.0, leaving the adjustment to be made to that backfilled on the outside of the external wall trenches later.

Foundations 3

Surf. trtmts.

Compactg. btms. of excavns.

39.69	(ext. wall
0.75	trs.
6.70	(int. wall
0.45	trs.

Compacting is measured in m² in three separate categories (D20:13.2.1–3), using the dimensions already obtained for measuring the excavation of the trenches.

mean gth.		39.688
add		
4/2/375		3.000
outer face of tr.		42.688

To arrive at the outer girth of the trench add twice times half the trench width at each of the four main corners (the other three external angles are offset by the internal ones).

E.W.S.

To faces of excavn.

≤ 1.00 max. depth; dist. betw. opp. faces ≤ 2.00

Earthwork support is measured in m² stating the maximum depth range and the distance between opposing faces (D20:7.1.1.0).

2/	39.69	(ext. wall
	0.75	trs.
2/	6.70	(int. wall
	0.55	trs.
	42.69	(ext. wall
	0.15	outer face of tr. to retain topsoil

The earthwork support to the outer face of the external wall trench must include the depth occupied by the topsoil and is shown as a separate dimension. This can be strutted from the opposing face of the trench and so the distance of ≤2.00 can apply. The provision of D20:M9a (≤0.25 high) will not apply in this case.

Ddt. ditto

2/	0.45	(tr. intersecs.
	0.55	

Finally, adjustment is needed for the trench intersections.

Disp.

Item	Surf. water

This item is needed to comply with D20:8.1.0.0.

Foundations 4

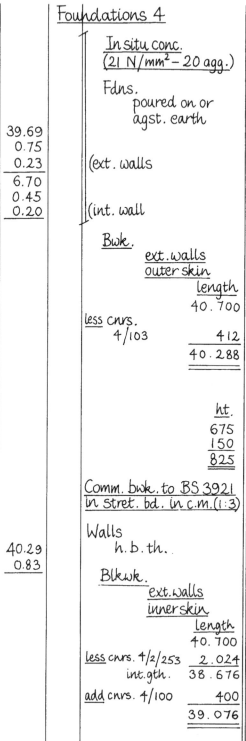

In situ conc.
(21 N/mm² – 20 agg.)

Fdns.
    poured on or
    agst. earth

39.69	
0.75	
0.23	(ext. walls
6.70	
0.45	
0.20	(int. wall

Bwk.
    ext. walls
    outer skin
      length
      40.700

less cnrs.
    4/103   412
            40.288

        ht.
        675
        150
        825

Comm. bwk. to BS 3921
in stret. bd. in c.m.(1:3)

Walls
    h. b. th.

40.29	
0.83	

Blkwk.
    ext. walls
    inner skin
      length
      40.700

less cnrs. 4/2/253 2.024
    int. gth.   38.676

add cnrs. 4/100   400
          39.076

Concrete is usually specified by strength rather than by proportions of components by volume.

Note the classification of *in-situ* concrete in foundations as E10:1 and the further requirements in E10:1.0.0.5.

The rough finish to foundation concrete to receive walling is outside the scope of surface treatments in E41:1–7.

The centre line of the outer skin is calculated by subtracting $4/2/\frac{1}{2}$ the thickness of the outer skin from the external girth of the wall.

The height of brickwork up to dpc level is calculated in waste using the dimensions given in section A–A.

The brickwork description will be obtained from the project specification. The description follows F10:1.1.1.0, except that it is unnecessary to describe the walls as vertical, as the work is deemed vertical unless otherwise described (F10:D3). The walls are described as half brick thick to overcome the difficulties resulting from bricks of varying thickness.

The part of the outer skin built in facework will be adjusted later.

Foundations 5

int. wall
<u>length</u>
7.700

less ext. walls
  2/253   <u>506</u>
      7.194

<u>ht.</u>
350
200
<u>100</u>
<u>650</u>

To obtain the length of the internal wall it is necessary to deduct the width of the external wall at each end, while the height is calculated from the dimensions given in section B–B.

<u>Blkwk. in solid, lt. wt. agg.</u>
<u>blks. w. O. P. ct. binder;</u>
<u>3.5 N/mm² crushg. strength</u>
<u>to BS 6073 : Pt.1 ; size</u>
<u>440 lg. × 215 hi. in stret.</u>
<u>bd. in c.m. (1 : 3)</u>

Walls
  100 th.
(ext. walls
 inner skin
(inner wall

39.08	
0.83	
7.19	
0.65	

The blockwork description is to include the kind, quality and size of blocks, type of bond, and composition and mix of mortar, as SMM7 supplementary information F10:S1–4. Alternatively, a proprietary product may be described such as Celcon Standard Aircrete blocks. The order of the block walling items follows the order of columns 1 and 2 in F10 (SMM7) and levels 3 and 4 in the *SMM7 Library of Standard Descriptions*. Note how the level 4 item (100 th.) is indented from the preceding items. See F10:1.1.1.0.

<u>length</u>
40.700

less cnrs.   103
  ½/50   <u>25</u>
  2/4/   <u>128</u>   <u>1.024</u>
cav. mean gth.   <u>39.676</u>

825
less dpc to grd. lev.   <u>150</u>
<u>675</u>

Sundries

39.68	
0.68	

Formg. cavs. in h.w.s
 50 wide; b.i. ends of
 s.s. butterfly wall
 ties to BS 1243 at
 the rate of 4 nr per m²

Forming cavities in hollow walls follows the requirements of F30:1.1.1.1. Under S2 and S3 the type, size and spacing of wall ties, and the type, thickness and method of fixing cavity insulation are to be given. In this example the insulation completely fills the cavity and no special fixings are required.

39.68	
0.15	

Ditto, w. Fibreglass Ltd.
 'Dritherm' 50 th. cav.
 insul. batts.

Foundations 6

	In situ conc. (1:10)
39·68	
0.68	**Fillg. h.w.s.**
0.05	≤ 150 th.

D.p.c.s of single lyr. 'Hyload' pitch polymer to BS 743, bedded in c.m. (1:3)

On surfs.
≤ 225 wide; hor.

40.29	(ext. walls
0.10	outer skin
39.08	(ext. walls
0.10	inner skin
7.19	
0.10	(int. wall

	ht.
above g.l.	150
below g.l.	75
	225

Facewk.

Ddt. comm. bwk. in stret. bd. a.b.

| 40.29 | Walls |
| 0.23 | h.b. th. |

&

Add
Fcg. bwk. in Ibstock Himley mixed russet fcg. bks. in stret. bd. in c.m. (1:3) & ptg. w. nt. weatherstruck jts. a.w.p.

Walls
h.b. th; facewk.
o.s.

Concrete filling to hollow walls is measured in m³ despite its small thickness, and is given in one of the three thickness classifications contained in E10:8.1.0.0. Alternatively, it could be considered preferable to insert the width of concrete fill in the dimensions before its depth.

Damp-proof courses are measured in m² giving the width classification and plane in which laid as F30:2.1.3.0. The description is to include the number of layers and composition and mix of bedding materials (F30:S5 and 6). No allowance is made for laps, and pointing exposed edges is deemed to be included (F30:M2 and C2). The term 'surfaces' is derived from the *SMM7 Library of Standard Descriptions.*

It is usual to allow for one course of facing bricks below ground level to counteract any irregularities in the finished ground level.

The faced brickwork, previously measured as common brickwork, is now adjusted. This is a half brick wall and finished fair on one side, measured in accordance with F10:1.2.1.0, and includes bricks, bond, mortar and pointing. The estimator needs to know the type of brick in order to assess the labour requirements, which a prime cost sum per 1000 bricks will not do.

Foundations 7

Fillg.

ext. walls <u>length</u>

      38.676

<u>less</u> cnrs. 4/248.5 994

      37.682

<u>less</u> intersec. with
  int. walls 2/100 200

      37.482

The calculation of the quantity of filling around both the external and internal walls entails extensive computation. All the figures, no matter how trivial they may seem, should be shown in waste, so that other persons can follow the dimensions with ease at a later date, as when making adjustments for variations and at the final account stage.

     <u>depth</u>

      750
<u>less</u> fdn.   225
bottom of h.c. bed 525
to top of conc. fdn.

int. wall <u>length</u>

     6.697

    <u>width</u>

     450
     100
<u>less</u> wall   350

<u>H.c. of cln. hd. broken</u>
<u>bk. or stne. graded 75</u>
<u>to dust obtd. off site</u>

An adequate description of the hardcore filling should be given or alternatively a reference made to the appropriate clause of the specification. The filling to excavations should follow the requirements of D20:9.2.3.0, distinguishing between that $\leq$ and >250 thick.

Fillg. to excvns.
  > 250 av. th.

37·48
0·25
0.53
   (ext. walls
6.68
0.35
0.35
   (int. wall·

Foundations 8

		length
ext. walls		40.700
add cnrs.	4/248.5	994
		41.694

The hardcore is followed by the filling arising from the excavations or obtained from on site spoil heaps.

	depth
	675
less topsoil	150
	525

Excvtd. matl.

Fillg. to excvns.
>250 av. th.;
arisg. from
the excvns.

41.69
0.25
0.53

The filling on the outside of the external walls consists of material arising from the excavation of the foundation trenches and is classified in the appropriate average thickness range as D20:9.1–2.

&

Ddt. Disp.

Excvtd. matl.
off site

Adjustment is then needed to the quantities of excavated material disposed off site.

Excvtd. matl.

Fillg. to excvns.
≤250 av. th.;
obtd. from
on site sp. hps.;
topsoil

41.69
0.25
0.15

The top 150 will be replaced with topsoil from on site spoil heaps in accordance with D20:9.1.2.3.

Foundations 9

Solid Flr.

	11. 000
less ext. walls 2/253	506
	10. 494
less sprd. of h.c. 2/248.5	497
	9. 997

The solid floor up to the upper surface of the concrete bed is being measured here instead of in example 6 (chapter 7), as the drawing and relevant dimensions are contained in example 1; hence it would seem illogical to measure this work elsewhere.

	7. 700
less ext. walls 2/253	506
	7. 194
less sprd. of h.c. 2/248.5	497
	6. 697

	4. 250
less ext. walls 2/253	506
	3. 744
less sprd. of h.c. 2 49	
1 75	424
	3. 320

The calculations in waste cover the dimensions for ground compaction and hardcore filling, waterproof membrane, concrete bed, insulation and screed. Figured dimensions are used wherever possible and minor variations are ignored for ground compaction where the hardcore spread varies over the length of the combined external and internal walls. Alternatively, the longer section could be split into two parts.

	4. 350
less adjust for walls 253	
153	100
	4. 250
less sprd. of h.c.	424
	3. 826

	2. 000
less ext. walls 2/253	506
	1. 494
less sprd. of h.c. 2/248.5	497
	0. 997

Surf. trtmts.

Compactg. grd.

(la. area

(smaller area

(bay

The area of ground compaction extends to the inner edge of the perimeter hardcore, as opposed to the other components which extend from wall to wall. The requirements for ground compaction are contained in D20:13.2.1.0. The compacting is deemed to include levelling (D20:C5).

10. 00
3. 32
6. 70
3. 83
1. 00
0. 75

Foundations 10

H.c.a.b.

Fillg. to make up levs.
 ≤250 av. th.

Cub. × 0.20 = _____ m³

Filling is subdivided into three categories and two average thickness ranges. Note how the superficial dimensions can be converted into cubic quantities at the taking off stage and thus reduce the number of separate dimensions entries (see D20:10.1.3.0).

10.49
3.74
7.19
4.25
1.49
0.75

(la. area

(smaller area

(bay   &

Surf. trtmts.

Compactg. fillg. &
 blindg. w. sand

&

The compacting and blinding of the filling constitutes a further item in accordance with D20:13.2.2.1.

1000 gauge polythene
lap'd. 150 at jts.

Tankg. & damp prfg.
 hor.; ld. on
 blinded h.c.

The polythene damp-proof membrane falls within the tanking and damp proofing classification in J40:1.1.0.0, and the pitch (horizontal) must be stated.

&

Conc. (21 N/mm²)

Beds
    ≤ 150 th.

Cub. × 0.10 = _____ m³

The concrete bed description must include the strength or mix and the lower thickness range, as E10:4.1.0.0. The surface on to which the concrete is poured does not have to be stated as it is blinded hardcore.

Foundations 11

la. area   10.494
                3.744
         2/ 14.238   28.476

smaller area 7.194
                4.250
         2/ 11.444   22.888

bay   2/ 0.750   1.500
                      52.864

1000 gauge polythene
a.b.

Abutments

52.86

≤ 2.00 gth.;
≤ 200 gth.

These dimensions refer to the vertical section of the damp-proof membrane on the adjoining wall faces, where it connects the horizontal area under the concrete and the damp-proof courses in both external and internal walls. Note the method of setting down the dimensions to arrive at the total length.

The polythene description has appeared before and hence the use of 'a.b.' (as before) avoids repetition. The abutment work is classified as in J40:4.2.0.0, and includes the girth in 200 stages as it does not exceed 2.00 m.

The remaining items of insulation, building paper, screed and floor tiles are taken in example 6 (chapter 7), which covers floors and their finishes, but referring back to this example for the relevant dimensions.

5 Brick and Block Walling

MEASUREMENT OF BRICK AND BLOCK WALLING

Measurement Generally

Brickwork and blockwork are measured in m² stating the nominal thickness, such as $\frac{1}{2}$B, 1B or 1$\frac{1}{2}$B, the plane where other than vertical, such as battering or tapering, and whether it has facework (measured fair) on one or both sides (F10:1.1–3.1–4.0). It must include full particulars of the bricks or blocks, type of bond, composition and mix of mortar and type of pointing (F10:S1–4), although this information is likely to be provided in preambles to the brickwork and blockwork bill or in a project specification. The provision of a prime cost sum for the supply of facing bricks is not entirely satisfactory as different types of brick can have widely differing laying costs.

With hollow walls, the skins and the forming of the cavity are each measured separately. The width of the cavity must be stated in the forming cavity item and also the type, size and spacing of wall ties, and the type, thickness and method of fixing any cavity insulation (F30:1.1.1.1 and F30:S2–3). A single omnibus item of 255 hollow wall including cavity, ties and insulation is not permissible. Walls include skins of hollow walls, hence these are not described differently from other brick and block walls (F10:D4).

When building up prices for brickwork, the estimator will need to determine the number of bricks, quantity of mortar and labour requirements for laying a typical square metre of each type of brickwork and must be supplied with the relevant particulars of the brickwork to enable this to be done. Subsequently, when erecting the building, it will be necessary to ascertain the number of bricks of each type required. Excellent guidance is given in estimating books and building price books, some of which are listed in the Bibliography at the end of the book.

The normal order of measurement is: (1) external walls, (2) internal walls and (3) chimney breasts, stacks and flues, as it is advisable to proceed in a logical and orderly sequence.

Various categories of brick and block walling are each kept separate as shown in the following lists:

(1) Walling in different bricks, blocks, mortars, bonds or types of pointing (F10:S1–4).

(2) Walls of different thicknesses (F10:1.1).
(3) Walls with facework (finished fair) on one or both sides (F10:1.2–3).
(4) Isolated piers, isolated casings and chimney stacks (F10:2–4).
(5) Battering walls (F10:1.1.2).
(6) Walls tapering one side (F10:1.1.3).
(7) Walls tapering both sides (F10:1.1.4).
(8) Walls used as formwork (F10:1.1.1.3).
(9) Boiler seatings and flue linings (F10:8–9.1).
(10) Curved work, giving the radius (F10:M4).
(11) Work built overhand (F10:1.1.1.4).
(12) Work built against or bonded to other work (F10:1.1.1.1–2).
(13) Projections (F10:5.1.1–3).
(14) Arches (F10:6.1).
(15) Isolated chimney shafts (F10:7.1.1).
(16) Closing cavities (F10:12.1.1–3).
(17) Bonding to existing (F10:25.1).
(18) Facework ornamental bands, quoins, sills, thresholds, copings, steps, tumblings to buttresses, key blocks, corbels, bases to pilasters, cappings to pilasters and cappings to isolated piers (F10:13–24).

External Walls

The walling above and the walling below the damp-proof course are often measured separately for convenience as a natural demarcation line, and they may be built in different bricks/blocks and/or mortars. It must be emphasised that facework is included in the measured items for brick and block walling. Work is deemed to be vertical unless otherwise described (F10: D3), even though it appears as a separate classification in F10: 1.1.1.

The length of external walling will be obtained by the method of girthing illustrated in chapter 3 and the height will normally be taken up to some convenient level, such as the general eaves line. Any additional areas of external wall, such as gables, parapets and walling up to higher eaves levels, will then be measured. Adjustment of walling for window and door openings will be made when measuring the windows and doors, at which time all the relevant dimensions will be to hand.

The measurement of the areas of external walls will be followed by incidental items, such as projections, facework ornamental bands and facework quoins, normally working from ground level upwards. Facing bricks are generally taken from 75 (one course) below finished ground or paving level, to allow for any irregularities, to just above soffit boarding at eaves.

Walling could be described as 102.5 or 215 thick, but is more often depicted in terms of brick thickness, such as half brick or one brick thick, as shown in the worked examples in this book. It might be considered that the

Table 5.1   *Mortar groups.*

	Group 1	Group 2	Group 3	Group 4
Cement : lime : sand	1 : 0–0.5 : 3	1 : 0.5 : 4–4.5	1 : 1 : 5–6	1 : 2 : 8–9
Cement : premixed lime and sand (proportion of lime to sand given in brackets)	1 : 3 (1 : 12)	1 : 4–4.5 (1 : 9)	1 : 5–6 (1 : 6)	1 : 8–9 (1 : 4.5)
Cement : sand and air entrainer	—	1 : 3–4	1 : 5–6	1 : 7–8
Masonry cement : sand	—	1 : 2.5–3.5	1 : 4–5	1 : 5.5–6.5

nominal thickness of a half brick wall as defined in F10: D1 should be 100 and not 102.5, which serves to illustrate the difficulties in giving precise brick thicknesses.

Expansion joints are required in long lengths of wall and are measured in metres with a dimensioned description as F30: 8.1.0.0, and the work is deemed to include preparation but the type of filler and sealant is to be stated (F30: C5&S8). A typical description would be 'Expansion joint; 20 wide in 215 facing brickwork; and filling with Fillcrete joint filler'.

Mortar groups: The mortar for use in brickwork and/or blockwork may be specified by a group number, such as by using one of the four groups listed in table 5.1, of which group 3 is the most commonly used category. Within the specified group the contractor can select one of the four types entered under the prescribed group, but must use the same mortar throughout any one type of facing brickwork.

Internal Walls

The measurement of external walls is usually followed by measurement of internal walls, which may be of bricks or blocks, but usually the latter. A careful check should be made on the type and thickness of each partition and, where there is a number of different types of partition, it is often helpful to colour each type in a different colour on the floor plans and to suitably mark each length on the floor plan as it is measured.

Chimney Breasts and Stacks

This work is described in chapter 6, although they are not used extensively in modern construction. A worked example of a concrete block flue to a gas fired appliance is included in chapter 6.

INCIDENTAL WORKS

Damp-proof Courses

These have been covered previously in chapter 4.

Rough and Fair Cutting

Rough cutting encompasses labours to common brickwork which will not be exposed, while fair cutting most commonly occurs with faced brickwork. All rough and fair cutting is deemed to be included with the brickwork or blockwork (F10: C1b). Similarly, forming rough and fair grooves, throats, mortices, chases, rebates and holes, stops and mitres, and raking out joints to form a key are also deemed to be included (F10: C1c–d), and do not therefore need to be mentioned in the brickwork/blockwork description.

Eaves Filling

Brickwork and blockwork in eaves filling is added to the general brick and block walling respectively, and no additional item is required for the extra labour involved in working between the feet of the rafters and the underside of the roof covering (F10: C1e).

Projections

Projections of attached piers, plinths, oversailing courses and the like are measured in metres, stating the width and depth of the projection, and whether horizontal, raking or vertical (F10: 5.1.1–3.0). Attached pier projections come within this category when their length on plan is ≤four times their thickness, otherwise they will be measured as walls of the overall thickness (F10: D9).

Figure 20 illustrates the projections to a 440 × 440 pier at the end of a one brick wall, the projections on each side being measured in metres, giving the width and depth in the description (440 and 112.5 respectively).

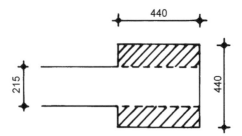

Figure 20   *Projections on pier.*

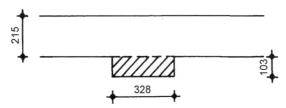

Figure 21   *Attached pier as projection.*

Figure 21 shows a half brick thick projection to a wall, where the length does not exceed four times the thickness, and hence it is measured in metres and described as a projection, 328 wide and 103 deep. Figure 22 illustrates a $4\frac{1}{2}$B long projection 1B deep on a one brick wall, where the length exceeds four times the thickness and it is therefore measured in m² as a 2B wall, and the 1B wall deducted for the area occupied by the projection.

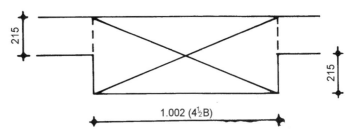

Figure 22   *Attached pier as wall.*

Deductions for String Courses and the Like

Deductions for string courses, lintels, sills, plates and the like are measured as regards height to the extent of full brick or block courses displaced and as regards depth to the extent only of full half brick beds displaced (F10:M3). Hence the deductions for a wall plate 100 × 75 and a concrete lintel

250 × 225 would be as follows. The wall plate deduction would be one course of brickwork high × a half brick bed, while the deduction for the concrete lintel in a cavity wall would be three courses of brickwork high × half a brick bed, and one course of concrete blocks high × one 100 block thick.

Facework Ornamental Bands

Facework to brick-on-edge bands, brick-on-end bands, basket pattern bands, moulded or splayed plinth cappings, moulded string courses, moulded cornices and the like are each measured separately in metres, giving the width of the band, and where sunk or projecting, the depth of set back or set forward, and are normally taken as extra over the work in which they occur (F10: 13.1–3.3.1). Labours in returns, ends and angles are deemed to be included (F10: C1f), but where such features involve the use of special bricks or blocks it is recommended that they are enumerated separately. The measurement of facework ornamental bands is illustrated in worked example 2.

Figure 23 shows a splayed capping, which is measured in metres as 'plinth cappings; 75 wide; splayed; projecting 28 from face of wall; horizontal; extra over the work in which they occur' (F10: 13.3.3.1). In practice it will be necessary to specifically identify the nature of the work on which the capping is extra over.

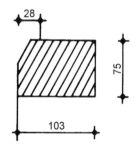

Figure 23   *Plinth capping.*

Facework Quoins

Facework quoins are formed with facing bricks which differ in kind or size from the general facings (F10: D12). They are measured in metres, stating whether flush, sunk or projecting and giving the appropriate dimensions including the mean girth, and they are normally taken as extra over the work in which they occur (F10: 14.1–3.1.1). Figure 24 shows a facework quoin which will be measured in metres and described as 'quoins; 318 mean girth; projecting 40 from face of wall; alternative groups of three quoin courses built into wall; extra over the work in which they occur'.

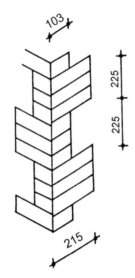

Figure 24   *Facework to quoin.*

Composite Walls

Where walls are constructed of two or more different materials, they are measured as walls of a given thickness, describing the materials and method of construction. For example, the $1\frac{1}{2}$B wall shown in figure 25 will be measured as follows:

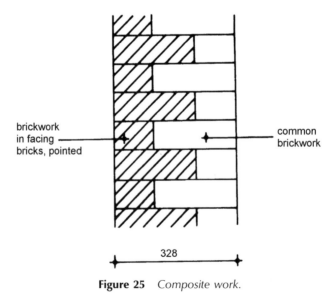

brickwork in facing bricks, pointed

common brickwork

Figure 25   *Composite work.*

			Comp. wk. of fcg bwk. in Ibstock clay Parham red stock bks.; in Eng. bond & ptg w. nt. flush jt. a.w.p.; backd. w. comm. bwk. to BS 3921; in Eng. bond in g.m. (1:1:6).
			Walls $1\frac{1}{2}$B th.; facwk. o.s.

METAL SHEET CLADDING

An alternative to external brick, block and stone walls is that of metal pro-filed steel sheeting. The most commonly used material is that produced by British Steel Strip Products with a colorcoat finish, while another type is produced by Ward Cladding Systems in a range of moduclad composite panels.

Wall claddings of this type are measured in m² as H31:2.1.0.2, with vertical angles and flashings taken in metres with a dimensioned cross section description. Wall cladding descriptions are to include minimum side and end laps, jointing or sealing, and nature, thickness and spacing of structural supports (H31: S3–5). Two examples will help to illustrate the approach.

			British Steel Strip Products: colorcoat HP200 stl. strip sheets; trapezoidal ribbed shallow profile; lapd. min. 150; 0.7 outer stl.; 80 mineral wool insul.; 0.4 inner stl.; rubber sld. jts.
	68.00 2.20		Wall claddg fxd. to hor. stl. rls. at 900 ccs. w. plastic hdd. self-tappg. scrs.
	68.00		Accessories Stand. flashgs. 0.9 th.; colorcoat silicone polyester; bent to profile 375 gth.
			Ward moduclad comp. pans.; DW 1000/B/50; 0.5 prectd. HP200 stl. weather sht. & 0.4 liner sht., complyg. w. BS EN 10147; both bonded to rigid polyurethane insul. core.
	68.00 2.20		Wall claddg. fxd. to hor. rls. w. clips at 900 ccs.; stl. batten caps to ext. jts.

Note: HP200 is a durable 200 micron high performance plastisol coating applied together with primers on to the galvanised steel substrate, to provide a leather grain finish.

WORKED EXAMPLES

Two worked examples follow covering the brickwork and facework to external walls and internal block partitions to a small building and a curved brick screen wall.

Note: Normal brick dimensions are 215 × 102.5 × 65 with mortar joints 10 thick. Hence a half brick wall is 102.5 thick although in practice 102 or 103 may be used to avoid such a small dimension as half a millimetre (1/50 of an inch). A one brick wall will be 215 thick; a hollow wall with a 100 concrete block inner skin, a 50 cavity and a half brick outer skin around 253 thick; and 255 if it has two brick skins.

This page has been left intentionally blank for make-up reasons.

Superstructure walling   DRAWING 2

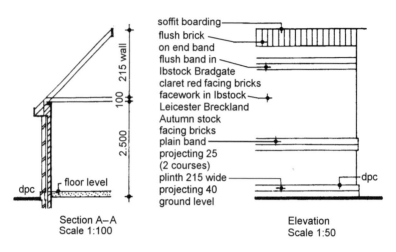

Section A–A
Scale 1:100

Elevation
Scale 1:50

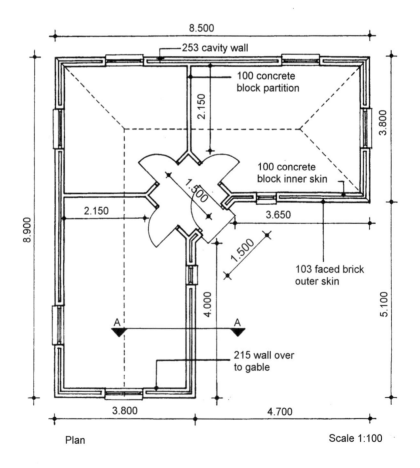

Plan   Scale 1:100

Drg.2			Example 2

Superstructure Walling 1

External Walls

It seems logical to separate external walls from internal ones and to indicate clearly the work being measured, such as outer skin, dpc to eaves level.

DPC to Eaves Level

Outer skin

len.

8.500
8.900
2/ 17.400
34.800

The mean girth of the outer skin of faced brickwork is calculated in waste, starting with the outer dimensions on the drawing and then adjusting for the splayed and square corners.

less spld. cnr.
2/1.080 2.160
         32.640
add splay   1.500
         34.140
less cnrs. 4/103 .412
m.gth. outer skin 33.728

The height of the brickwork is obtained by scaling the two deductions to be made from the total figured height.

ht.

2.500
less dpc to
  u/s flr. 150
  eaves fillg. 350 .500
         2.000

Fcg. bwk. in Ibstock Leicester Breckland Autumn Stock fcg. bks. in stret. bd. in g.m. (1:1:6) & ptg. w. nt. weather struck jts.

a.w.p.

Note the full description of the faced brickwork given in accordance with the requirements of F10:S1–5.

The wall description complies with F10:1.2.1.0, although being vertical is deemed to be included in the item by virtue of F10:D3.

Walls

| 33.73 | |
| 2.00 | |

h.b.th.; facewk. o.s.

Superstructure Walling 2

<u>Inner skin</u>

Note method of calculating the mean girth of the inner skin, starting with the external girth and deducting 4/2/ the wall thickness less half the thickness of the inner skin.

	34.140	
<u>less cnrs.</u> .253		
− .050		
4/2/ .203	1.624	
	32.516	

<u>Blkwk. in solid Thermalite 'Shield 2000' conc. blks., size 440 lg. × 215 hi. in stret. bd. in g.m.(1:1:6)</u>

In this example a proprietary product has been taken for the lightweight concrete block inner skin, as opposed to the performance description given in example 1, based on the appropriate British Standard. See F10:1.1.1.0.

32.52
2.00
———

Walls

100 th.

<u>Cavity</u>

As the width of the cavity is not stated, it has been calculated in waste.

		.253
.103		
.100	.203	
width of cav.	.050	

<u>len.</u>

Note the method of arriving at the mean girth of the cavity from the outside dimensions, by adjusting for twice the width of the outer skin and for once the width of the cavity.

	34.140	
<u>less cnrs.</u>		
2/.103	.206	
	.050	
4/	.256	1.024
	33.116	

Superstructure Walling 3

	Sundries	

33.12	Forming. cav. in holl.		Forming cavities is measured in accordance with F30:1.1.1.1, giving the width of cavity, type and thickness of cavity insulation, and type and number of wall ties and method of fixing cavity insulation (F30:S2–3). The width of the cavity behind the projecting plinth is increased and this will form a separate cavity item.
2.00	walls ,50 wide ; 0.6		
	s.s. pressed wall ties		
	w. insul. retaing. clips		
	at 5 per m²; Polyfoam		
	plus cavitybd., 25 th.,		
	1200 x 450 in size, fxd.		
	to outer skin.		

Blkwk. at eaves

		len.
		34.140
less cnrs. 4/2/253	2.024	
int. gth.	32.116	
less gable 3.800		
−2/253 506	3.294	
	28.822	
add cnrs. 2/215	430	
	29.252	

The centre line of the blockwork to eaves filling is calculated in waste. The blockwork is deemed to include labour in eaves filling between rafter feet (F10:C1e). The gable is deducted as the cavity wall is carried up to a higher level (top of ceiling joists).

Blkwk. a.b.

29.25	Closing cavs.	
	50 wide; blkwk.	
	215 th. ,500 hi.; hor.	

It was considered appropriate to measure this work as closing cavities (F10:12.1.3), although this item is more commonly encountered when closing cavities around window and door openings. An alternative would be to measure a 215 thick wall in m².

Gable wall

Fcg. bwk. a.b.

	Walls	
3.80	h.b.th.; facewk. o.s.	
0.50	&	
	Blkwk. a.b.	
	Walls	
	100 th.	
	&	
	Sundries	
	Formg. cav. in h.w.s.	
	a.b.	

These three coupled items pick up the hollow wall at the base of the triangular section to the gable. Note the extensive use of 'a.b.', to prevent the constant repetition of descriptions that have been given previously.

Superstructure Walling 4

2/   0.50

Blkwk. a.b.

Closg. cavs.

    50 wide, blkwk.
    100 th.; vert.

This item covers the closing of the cavities vertically at each end of the hollow wall to the gable.

D.p.c.s of single lyr.
Hyload pitch poly. to
BS 743 bedded in
g.m. (1:1:6).

The damp-proof course incorporated in the closing of the cavity is measured separately in m² in accordance with F30:2.1.1.0.

2/   0.50
    0.10

On surfs.
    ≤ 225 wide; vert.

          3.800
less ends covd.
by rf. slope 2/150 300
          3.500

The triangular section of faced brickwork to the gable is next measured, where the area = $\frac{1}{2}$ height × base (the height is scaled).

Comp. wk. of fcg. bwk.
a.b. backd. w. comm.
bwk. to BS 3921 in
Flem. bd.
Walls

½/   3.50
    1.75

    o.b. th.; facewk. o.s.

The gable is constructed with facing brickwork externally with a common brickwork backing and is described as composite work.

Plinth

         34.140
add cnrs. 4/2/40   .320
         34.460

The length of the plinth is taken from the outside girth with adjustment for the four corners. The plinth is below dpc but is measured here to illustrate the approach.

Superstructure Walling 5

	Fcg. bwk. a.b. but in c.m. (1:3).	A similar type of heading is retained for the ornamental bands.
34.46	E.o. facg. bwk. a.b. for bk. on end bands 215 wide; proj. 40 from face of wall; hor.	Brick on end bands are so described, giving the width, depth of set back or set forward and the plane, as F10:13.3.3.1. As they are measured extra over the work in which they occur, no adjustment is required for the facing brickwork measured previously.

Proj. courses

len.

      34.140

add cnrs. 4/2/25   .200

      34.340

	Fcg. bwk. a.b.	This feature is built in the same facing bricks as the general surfaces. The description follows the procedure prescribed in F10:13.3.3.1.
34.34	E.o. facg. bwk. a.b. for plain bands 150 wide; proj. 25 from face of wall; hor.	

len.

      33.116

add cnrs. 4/2/12.5 .100

      33.216

Build up of the slightly increased length of the wider cavity.

Sundries

33.22 0.15	Formg. cav. in h.w.s a.b. but 75 wide.	It is necessary to keep separate forming cavities of different widths as F30:1.1.1.1.
33.12 0.15	Ddt Formg. cav. in h.w.s a.b.	

Superstructure Walling 6

<u>Flush bands</u>

Fcg. bwk. a.b.

34.14

E.o. facg. bwk. a.b. for
bk. on end bands
215 wide; flush;
hor.

&

Fcg. bwk. a.b. but
in Ibstock Bradgate
Claret Red fcg. bks.

E.o. Ibstock Leicester
Breckland Autumn
facg. bwk. for plain
bands
150 wide; flush;
hor.

<u>Int. walls</u>

<u>len.</u>

2/2.150	4.300
2/1 .500	3.000
	1.700
	9.000

<u>ht.</u>

<u>less</u>	2.500
conc. flr.	.150
	2.350

Blkwk. a.b.

9.00
2.35

Walls
100 th.

Both types of band have the same length and so are grouped together.

The procedure outlined in F10:13.1.3.1 is followed for the flush brick on end band.

A description of the different type of facing brick is required (F10:S1).

No adjustment is made for the areas of faced brickwork occupied by the bands, as they have been taken as extra over the work in which they occur.

Note the build up of the lengths of internal wall, using figured dimensions on the drawing as far as possible. The length of 1.700 includes the additional walling at the two corners.

The blockwork is of similar type to that measured previously on the external walls and so does not need to be fully described.

This page has been left intentionally blank for make-up reasons.

Curved brick screen wall   **DRAWING 3**

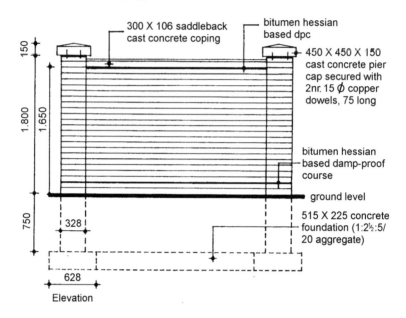

300 X 106 saddleback
cast concrete coping

bitumen hessian
based dpc

450 X 450 X 150
cast concrete pier
cap secured with
2nr. 15 Ø copper
dowels, 75 long

bitumen hessian
based damp-proof
course

ground level

515 X 225 concrete
foundation (1:2½:5/
20 aggregate)

150   1.800   1.650   750

328

628

Elevation

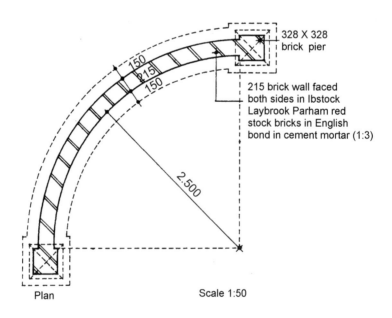

328 X 328
brick pier

215 brick wall faced
both sides in Ibstock
Laybrook Parham red
stock bricks in English
bond in cement mortar (1:3)

150   215   150

2.500

Plan   Scale 1:50

Drg.3			Example 3	

Curved Brick Screen Wall 1

Work below dpc | The measurement of the foundations has been included in this example as various points of note emanate from the curved construction.

dia.

int. rad.	2.500
½/215	107.5
2/	2.607.5
mean dia.	5.215

width

2/150	300
	215
	515

Topsoil excavation for preservation is measured in m² as D20:2.1.1.0, although a common approach in this type of situation is to commence trench excavation from ground level. The disposal of the surplus excavated material forms a separate cubic item (D20:8.3.2.1). The circumference of a quadrant ($\frac{1}{4}$ circle)$=\frac{1}{4}\pi D$.

Excvtg.

Topsoil for presvn.

(wall 150 av. depth.

¼/22/7/	5.22
	0.52

2/	0.63
	0.63

(piers   &

Disp.

Excvtd. matl.
on site; 20.00 av.
dist. in sp. hps.

Cub × 0.15 = _____ m³

Excavating topsoil from an area bounded by curved sides involves higher cost, but the SMM does not require this to be kept separate. The curved outline will be evident to the contractor from the drawings supplied with the bill of quantities.

	628
	328
2)	300
	150

An adjustment is needed for the overlapping of excavation measured at the junction of the wall and pier foundations.

2/	0.15
	0.52

Ddt. last two items

(junctn. of wall & pier fdns.

Curved Brick Screen Wall 2

Wall tr.

depth

	750
add fdn.	225
	975
less topsoil	150
tr. excvn.	825

Note the method of calculating the depth of trench excavation, using figured dimensions as far as possible.

Excvtg.

Trs. > 0.30 wide
≤ 1.00 max. depth.

The excavation is measured in m³ in the maximum depth ranges listed in D20:2.6.2.0. The trench excavation and filling includes that to the pier foundations. The commencing level of the excavation is not stated unless it exceeds 0.25 m below existing ground.

$\frac{1}{4}/\frac{22}{7}/$ 5.22
        0.52
        0.83

(wall

2/   0.63
     0.63
     0.83

&

(pier

Excvtd. matl.

Fillg. to excvns.
> 250 av. th.;
arisg. from excvns.

Filling returned to trenches is measured in m³ and classified as D20:9.2.1.0.

2/   0.15
     0.52
     0.83

Ddt. last two items

(junctn. of wall & pier fdns.

Adjustment of overlap of wall and pier foundations.

Surf. trtmts.

Compactg. btms. of
excvns.

Compacting the bottoms of excavations, under a main heading of surface treatments, in accordance with D20:13.2.3.0.

$\frac{1}{4}/\frac{22}{7}/$ 5.22
        0.52

(wall

2/   0.63
     0.63

(pier

Ddt. ditto.

(junctn. of wall & pier fdns.

Adjustment of overlap at junction of wall and pier foundations.

2/   0.15
     0.52

Curved Brick Screen Wall 3

depth	Build up of depth of earthwork support in waste. Even the most simple additions should be recorded.
750	
225	
975	

E.W.S.

To faces of excvn.
≤ 1.00 max. depth;
dist. betwn. opp.
faces ≤ 2.00; curved.
(wall

Earthwork support is measured in m² and classified in maximum depth ranges as D20:7.1–3. Curved earthwork support is so described (D20:7.1.1.1), irrespective of the radius, and is deemed to include any extra cost of curved excavation D20:C3.

2/¼/22/7/   5.22   0.98

Ddt. ditto

(junctn. of wall & pier fdns.

2/2/   0.15   0.98

Adjustment needed at junction of wall and pier foundations.

```
     628
     515
  2) 113
```
width of pier retn. 56.5

Calculation of width of pier returns.

E.W.S.

To faces of excvn.
≤ 1.00 max. depth;
dist. betwn. opp.
faces ≤ 2.00.

Earthwork support is measured around the pier foundations: 3 sides and 2 returns to each pier.

2/3/   0.63   0.98

(sides
(to piers

It could not be combined with the previous item as it is not curved work.

2/2/   0.06   0.98

(retns. to
(piers

Disp.

Item

Surf. water

This item is needed to comply with D20:8.1.0.0.

			Curved Brick Screen Wall 4

		In situ conc.
		($1:2\frac{1}{2}:5/20$ agg.)
		Fdns.
		poured on or
		agst. earth.
$\frac{1}{4}/\frac{22}{7}/$		(wall
	5.22	&
	0.52	
	0.23	*Ddt. Excvtd. matl.*
		Fillg. to excvns.
$2/$	0.63	$>$ 250 av. th.;
	0.63	(piers arisg. from
	0.23	the excvns.
		&
		Add Disp.
		Excvtd. matl.
		off site.

		Ddt. In situ conc. a.b.
		Fdns.
		poured on or
		agst. earth.
$2/$		
	0.15	(junctn. of
	0.52	(wall & pier
	0.23	(fdns.
		&
		Add Excvtd. matl.
		Fillg. to excvns.
		$>$ 250 av. th.;
		arisg. from
		the excvns.
		&
		Ddt. Disp.
		Excvtd. matl.
		off site.

In-situ concrete foundations are measured in accordance with E10:1.0.0.5.

The concrete can be described by volume and size of aggregate as in this example or can be given a performance rating as in example 1.

Note the method of adjusting the filling and disposal of excavated material for the volume occupied by the concrete.

This item relating to the concrete deduction for the junction of the wall and pier foundations, entails the reverse process to the previous one.

			Curved Brick Screen Wall 5	

Bwk. below dpc

<u>ht.</u>

	750
g.l. to dpc	150
	900

<u>Comm. bwk. to BS 3921</u>
<u>in Engl. bd. in c.m. (1:3)</u>.

The kind, quality and size of bricks are defined in BS 3921, and the composition and mix of mortar and bond of brickwork are required by F10:S2–3.

Walls

o.b. th.; curved
to 2.61 rad.

	150
	75
	225

$\frac{1}{4}/\frac{22}{7}$ 5.22
     0.90

The radius is to be stated in the description of curved brickwork (F10:M4), and General Rules 14.2 prescribes that the radius is the mean radius unless otherwise stated (i.e. $2.500 + \frac{1}{2}/215 = 2.608$).

<u>Ddt. comm. bwk. in</u>
<u>Engl. bd. a.b.</u>

Walls

o.b.th.; curved
to 2.61 rad.

$\frac{1}{4}/\frac{22}{7}$ 5.22
     0.23

It is better to describe brick walls as half brick and one brick thick rather than giving their actual thickness, which can result in fractions of a millimetre.

&

<u>Add</u>

<u>Fcg. bwk. in Ibstock</u>
<u>Laybrook Parham Red</u>
<u>Stock bks. in Engl. bd.</u>
<u>in c.m. (1:3) & ptg.</u>
<u>w. nt. flush jts. a.w.p.</u>

Facings are normally taken to one course of bricks below ground, to allow for possible irregularities in the final ground surface. Hence an adjustment is needed to cater for the facework up to damp-proof course level.

Walls

o.b. th.; facewk.
b.s.; curved to
2.61 rad.

Curved Brick Screen Wall 6

Comm. bwk. in Engl. bd. a.b.

Walls
    o.b.th.

(piers

2/	0.33
	0.90

The wall is measured through the pier and the projections beyond the wall faces are measured in metres, stating the width and depth of the projection (F10:5.1.1.0), as illustrated in figure 20.

Ddt. comm. bwk. in Engl. bd. a.b.

Walls
    o.b.th.

(piers   &

Add

Fcg. bwk. a.b.

Walls
    o.b.th.; facewk. o.s.

2/	0.33
	0.23

This is followed by the adjustment of facings for commons in the piers.

```
    328
    215
  2)113
    56.5
```

Note method of calculating the depth of brick projections to piers in waste.

Comm. bwk. a.b.

Projs.
    328 x 57 proj.; vert.
(piers

2/2/	0.90

Ddt. comm. bwk. a.b.
Projs. a.b.

        &

(piers

2/2/	0.23

Add fcg. bwk. a.b.
Projs. a.b.; facewk. o.s.

It is necessary to adjust the projections for the lengths in faced brickwork. No additional facework is measured to the ends of the piers as the facing bricks have already been taken and the additional labour is deemed to have been included (F10:C1f).

Curved Brick Screen Wall 7

D.p.c.s of single lyr.
hessian based bit. to
BS 743 bedded in c.m.(1:3)

On surfs.
  ≤ 225 wide; hor.;
  curved.

| $\frac{1}{4}/\frac{22}{7}/$ | 5.22 | (wall |
| | 0.22 | |

Note the inclusion of curved in the description of the dpc in accordance with F30:M1.

D.p.c.s a.b.

On surfs.
  > 225 wide; hor.

| 2/ | 0.33 | (piers |
| | 0.33 | |

The remainder of the description follows F30:2.1.3.0, including the gauge, thickness or substance of sheet metal, number of layers, and composition and mix of bedding material measured in m², and distinguishing between those not exceeding and exceeding 225 wide.

Adjust. of excvn.

depth of bwk. below grd.	750
less topsoil	150
	600

Adjustment of excavated material filling and disposal for the volume occupied by brickwork in the depth of the trench below topsoil.

Ddt. Excvtd. matl.

Fillg. to excvns.
  a.b.d.

$\frac{1}{4}/\frac{22}{7}/$	5.22	(wall
	0.22	
	0.60	&
2/	0.33	(piers
	0.33	
	0.60	

Note the use of 'a.b.d.' (as before described) to reduce the length of the filling description by reference to a previous similar item.

Add Disp.
Excvtd. matl.
  off site.

2/628	1.256
2/328	656
	1.912
less 215	
2/150 300	515
	1.397

Note build up of girth of topsoil adjustment around the two piers.

Curved Brick Screen Wall 8

	Excvtd. matl.	Adjustment of topsoil around the outside of the wall and piers, measured in m³ in accordance with D20:9.1.2.3.
	Fillg. to excvns. ≤ 250 av. th.; topsoil, obtd. from on site sp. hps.	

$\frac{2}{\frac{1}{4}\frac{22}{7}}$
| 5.22 |
| 0.15 |
| 0.15 | (wall

2/
| 1.40 |
| 0.15 |
| 0.15 | (piers

Wk. above dpc

Note build up of height of faced brickwork to wall, by deducting the height of the dpc above ground level from the overall figured dimension of 1.800 on the drawing.

wall ht.

```
               1.650
less dpc to g.l.    150
               1.500
```

Fcg. bwk. a.b.

The same bricks, mortar and pointing as taken before for work below dpc, and so the words 'as before' (a.b.) can be used to reduce the length of the description.

$\frac{1}{4}\frac{22}{7}$
| 5.22 |
| 1.50 |

Walls

o.b.th.; facewk. b.s.; curved to 2.61 rad.

pier ht.

```
               1.800
less dpc to g.l.    150
               1.650
```

Measurement of piers above dpc level – using the same procedure as before – one brick wall and two projections, all in facing bricks.

Fcg. bwk. a.b.

2/
| 0.33 |
| 1.65 |

Walls

o.b. th.; facewk. o.s.

	Curved Brick Screen Wall 9	
	Fcg. bwk. a.b.	Projections measured in accordance with F10:5.1.1.0, with the inclusion of facework.
²/₂/ 1.65	**Projs.** 328 × 57 proj.; vert.; facewk. o.s.	An alternative approach would be to terminate the projections at the same level as the top of the curved wall and measure the top 150 of the pier as an isolated pier F10:2.3.1.0.
	<u>D.p.c.s below copgs</u>	As the damp-proof courses below the wall coping and pier caps are similar to those 150 above ground level, the description heading merely refers back to the earlier entry.
	<u>D.p.c.s a.b.</u>	
	On surfs. ≤ 225 wide; hor.; curved.	
¼/²²/₇/ 5.22 0.22	(wall	
	On surfs. >225 wide; hor.	
²/ 0.33 0.33	(piers	Sufficient information about the coping must be given to enable the estimator to insert a realistic price against the item in the bill of quantities. Alternatively, a proprietary item may be used when a catalogue reference and the manufacturer's name may be sufficient to identify the product and price it. This is a linear item in accordance with F31:1.2.0.0, and including the number of units in the description.
	<u>Copg.</u>	
	Precast conc.	
¼/²²/₇/ 5.22	Copg., saddleback, 2ce thro., 300 × 106; curved to rad. of 2.61; b. & p. in c.m. (1:3) (In 5 nr. units).	
	<u>Pier caps</u>	
	Precast conc.	
²/ 1	Pier cap, 450 × 450 × 150, 4 times wethd. & thro. a/r to match copg.; b. & p. in c.m. (1:3); fxd. w. cop. dowel, 15 dia. & 75 lg.	Pier caps are enumerated with a dimensioned description as F31:1.1.0.0, and including the method of bedding and fixing (F31:S4).

6 Fires, Flues, Vents and Stone Walling

CHIMNEY BREASTS AND STACKS

Brickwork in Breasts and Stacks

Open fireplaces are now becoming much less common than hitherto, being largely replaced by gas fires with appropriate flues and terminals, whose measurement is described and illustrated later in the chapter.

The projecting chimney breasts and chimney stacks are not usually measured when the general brickwork to a building is being taken off. Where a chimney breast is located on an external wall, the chances are that the wall at the back of the fireplace with its external facework will be measured with the general brickwork, and the projecting breasts and chimney stack will be left to be measured later. With fireplaces on internal walls, it is probable that the whole of the enclosing brickwork will be measured together following the taking off of the general brickwork.

The brickwork projecting from the face of the wall on which the chimney breast is located, will be measured as projections as defined by F10:D9 and in accordance with F10.5.1.1.0 in metres, stating the width and depth of projection, where the length on plan does not exceed four times the thickness, otherwise it will be measured as a wall of the combined thickness. Brickwork in chimney stacks is measured separately in m^2 (F10:4.1–3.1.0).

No brickwork will be taken for the void occupied by the fireplace opening, but the brickwork will be measured solid above the fireplace opening where the extra labour in 'gathering over' is offset by the saving in brickwork.

Flues

In the measurement of chimney flues, no deduction of brickwork for voids will be made when the void is $\leq 0.25\,m^2$ in cross sectional area (F10:M2b). The normal domestic flue measures 225×225 ($0.051\,m^2$), so that it would have to be a very large flue for the deduction provision to operate. However, when dealing with a stack containing several flues, their combined total area has to be assessed against the $0.25\,m^2$ cross sectional area requirement.

Clay and precast concrete flue linings are measured in metres, giving a dimensioned description, and with cutting to form easings and bends and cutting to walls around linings deemed to be included (F30:11.1.0.0 and F30:C7). Brick flue linings are measured in square metres stating the thickness in accordance with F10:9.1.0.0.

Chimney pots are best enumerated, with a dimensioned description and manufacturer's reference, under the classification of proprietary items (F30:16.1.1.0). For example, 'Hepworth terracotta pocket beehive chimney pot YQ48R, 450 high'. Alternatively, a general description referring to the appropriate British Standard can be used, such as 'Chimney pot to BS 1181: 225 diameter × 450 high, type E' (General Rules 6.1).

FIREPLACES

The measurement of fireplaces, surrounds, hearths and stoves may be either regarded as proprietary items (F10:16.1.1.0) or as fixtures, furnishings and equipment (N10:1.1–2.0.0 and appendix A). The supply of stoves or grates and slabbed tile surrounds and hearths is frequently covered by prime cost sums or basic prices, but the work in fixing and supply of the incidental materials has also to be taken into account.

Consideration should also be given at this stage to any air ducts laid under solid floors and connected to fireplaces. The ducts would normally be measured in metres with a full description and any fittings enumerated.

FLUES TO GAS FIRED APPLIANCES

Gas fired appliances may be connected to precast concrete or clayware gas flue blocks, bonded into the surrounding brickwork or blockwork, or to stainless steel insulated pipes and fittings or other metal gas flue systems as produced by Marflex.

Gas flue blocks are enumerated (F30:15.1.0.0) giving the relevant details as provided in example 4, while gas flue pipes are measured in metres, giving the type, nominal size and method of jointing and support (Y10:1.1.1.1), with bends and fittings enumerated as extra over the pipes in which they occur (Y10:2.1–4.1–6.1). Example 4 shows how a gas flue can be carried over from the eaves of the building, or some intermediate point, to the apex of the building at its ridge, where the gas terminal is positioned.

VENTS

Vents are often required to provide ventilation under hollow boarded floors and to toilets and food stores where provided. The formation of the opening and the building in of the air brick or ventilating grating are deemed to be included in the enumerated component item. Air bricks, ventilating gratings and soot doors are enumerated, giving the size of opening and nature and thickness of the wall and including lintels and archs where required (F30:12–14.1.0.1–2). Care must be taken not to miss the provision of ventilating gratings, often of metal or fibrous plaster, on the inside face of the wall. Working wall plaster around them is deemed to be included in the wall plaster items.

The following example illustrates the method of measuring a typical vent:

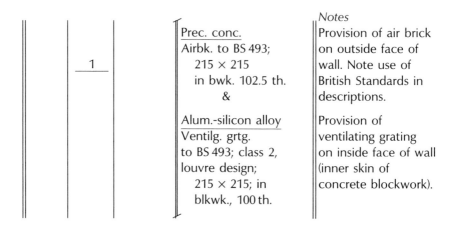

				Notes

	1		Prec. conc.
			Airbk. to BS 493;
			    215 × 215
			    in bwk. 102.5 th.
			        &
			
			Alum.-silicon alloy
			Ventilg. grtg.
			to BS 493; class 2,
			louvre design;
			    215 × 215; in
			    blkwk., 100 th.
				Provision of air brick on outside face of wall. Note use of British Standards in descriptions.
				
				Provision of ventilating grating on inside face of wall (inner skin of concrete blockwork).

RUBBLE WALLING

It is possible that the measurement of rubble walling may appear in some examination syllabuses, and so a brief summary is given of the main provisions of the *Standard Method* relating to this section.

Rubble work is generally defined as natural stones either irregular in shape or roughly dressed and laid dry or in mortar with comparatively thick joints. Dry rubble walling, with no mortar in the joints, is frequently used in boundary walls.

The measurement of this class of work follows the rules prescribed in work section F20 (Natural stone rubble walling), with work measured in m² giving the thickness, plane (other than vertical) and any facework. Full particulars of the stone, type of walling, heights of courses (where applicable), mortar and pointing are to be given (F20:S1–10). Rough and fair square cutting are deemed to be included (F20:C1l), while rough or fair raking or circular cutting are measured in metres stating the thickness (F20:26–27.1.0.0). Levelling uncoursed rubble work for damp-proof courses, copings and the like, labours in returns, ends and angles and dressed margins to rubble walling are deemed to be included (F20:C1g–j).

Natural Stone Dressings

Natural stone dressings involve the use of stones which are carefully cut and dressed to shape and laid with fine joints. Work of this type is measured in the window surrounds in example 20. Quoin stones and jamb stones are measured in metres with a dimensioned description, which includes the various labours (F21:10–11.1.1–3.0). Definitions of the principal stonework labours are given in chapter 10.

WORKED EXAMPLES

Worked examples covering a flue and terminal to a gas fired appliance, and a random rubble boundary wall follow.

Gas flue blocks   **DRAWING 4**

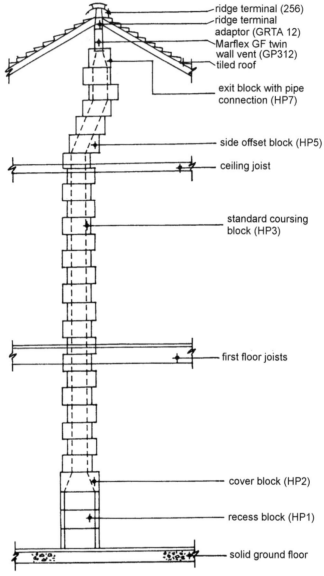

ridge terminal (256)
ridge terminal adaptor (GRTA 12)
Marflex GF twin wall vent (GP312)
tiled roof

exit block with pipe connection (HP7)

side offset block (HP5)

ceiling joist

standard coursing block (HP3)

first floor joists

cover block (HP2)

recess block (HP1)

solid ground floor

Note: Marflex gas flue blocks to BS 1289 built into concrete block inner skin of cavity wall

Elevation   Scale 1:50

Drg.4	Gas	Flue	Blocks 1	Example 4

Marflex HP precast
conc. flue blks.
containg. single flue
jtd. w. bondg. nibs
in sealg. compd. &
b.i. conc. blkwk.
inner skin of cav. wall.

Starting with a general description of the gas flue blocks in a heading, eliminates the need to repeat these details in each and every block item.

$\underline{3}$ Recess blk. HP1;

405 × 147 × 222.

$\underline{1}$ Cover blk. HP2;

385 × 140 × 222.

Gas flue blocks are enumerated and the size of block and number of flues in each block are to be stated (F30:15.1.0).

$\underline{21}$ Stand. csg. blk. HP3;

355 × 140 × 222.

$\underline{3}$ Side offset blk. HP5;

400 × 140 × 222.

Alternatively, they can be measured under F30:16.1.1 as proprietary items, when they are enumerated with a dimensioned description and the manufacturer's reference, as has been done in this example. F30:S12–13 requires the method of building or fixing to be stated.

$\underline{1}$ Vert. exit blk. HP7;

280 × 181 × 222.

No adjustment of the concrete block inner skin is required following displacement by the flue block, as all blocks are $\leq 0.25\,\mathrm{m}^2$ (F10:M1b).

Gas Flue Blocks 2

Marflex GF twin
wall gas vent pipe
of alum. & zalutite
w. twist lock s & s jts.

1

Pipe len. 300; 125
  nom. dia.; GP 312.

The short length of pipe connecting
the flue blocks to the adaptor is
enumerated as a pipework ancillary
and stating the method of jointing
(Y11:8.1.1.0).

Marflex GF
fabricated alum.
ridge tile adaptor.

1

Ridge tile adaptor;
  GRTA 12;
  360 x 79 x 237.

A ridge tile adaptor is required to
connect the gas vent pipe to the
ridge terminal. This is taken as an
enumerated item, giving a
dimensioned description and the
manufacturer's reference. H60:10
would seem to be the most
appropriate SMM7 section.

Ridge Terminal

1

Marley gas vent
  ridge terminal;
  conc. compg. ridge
  tile & hood to
  match roof tiles;
  w. s.s. mesh grille;
  code 256; bolted
  to GRTA 12 adaptor;
  450 x 300 in size.

The ridge terminal is dealt with in a
similar manner, giving a brief
description, manufacturer's
reference, method of fixing and
overall dimensions as H60:10.3.1.0.

This page has been left intentionally blank for make-up reasons.

Stone rubble walling   DRAWING 5

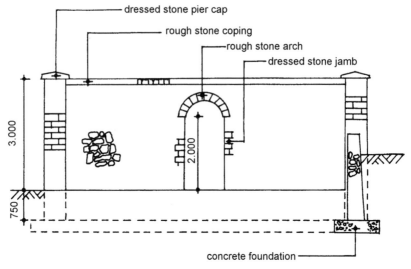

dressed stone pier cap

rough stone coping

rough stone arch

dressed stone jamb

concrete foundation

Elevation and Section A–A

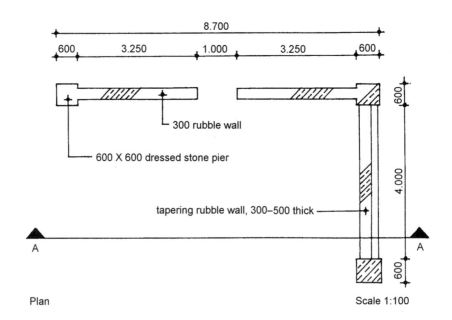

8.700

600   3.250   1.000   3.250   600

300 rubble wall

600 X 600 dressed stone pier

tapering rubble wall, 300–500 thick

600

4.000

600

A   A

Plan   Scale 1:100

| Drg.5 | | | | Example 5 | |

Stone Rubble Walling 1

The measurement of the foundations and excavation has been omitted as this type of work has already been covered in examples 1 and 3.

len. of wall

2/3.250	6.500
	1.000
	7.500

ht. of wall

	3.000
	750
less	3.750
st. copg. 150	
below grd.	
750-150 600	750
fcd. wk.	3.000

A start is make with the taller stone wall, distinguishing between faced work above and below ground and unfaced work below ground, with the calculations entered in waste.

Nat. st. random rubble in Cotswold Guiting limestone, uncsd., in c.l.m.(1:2:9).

The description of the materials used in the rubble stone walling must give the information detailed in F20:S1.

7.50	Walls
0.60	300 th.
	(unfcd.
	(below grd

The walls are measured in accordance with F20:1.1.1.5, although work is deemed to be vertical unless otherwise stated.

| 7.50 | Walls |
| 3.00 | 300 th.; fcd.b.s. |

	Copgs.
	300 wide; av. 150 hi;
7.50	hor.; ro. dressed;
	fcd. on exp. faces.

Copings are measured in accordance with F20:16.0.3.0. Levelling uncoursed work to receive the coping is deemed to be included (F20:C1j).

Stone Rubble Walling 2

			Tapg. wall
			ht.
	fcd. b.s.		700
	fcd. o.s.		950
	unfcd.		650
			2.300

It is necessary to subdivide the tapered wall into three separate sections: upper section – faced both sides; middle section – faced one side; and lower section – unfaced, using scaled dimensions.

thickness

$$\frac{0.200 \times 0.700}{2.300} + 0.300$$

Th. at btm. of wall fcd. b.s. $= 361$

The thickness to be stated for tapering walls of diminishing thickness is to be the mean thickness (F20:D7). The thickness of each separately described wall must be calculated as shown.

$$\frac{0.200 \times 1.650}{2.300} + 0.300$$

Th. at btm. of wall fcd. o.s. $= 443$

```
          443
          500
     2) 943
        471.5
```

av. th. of unfcd. wall

```
          361
          443
     2) 804
        402
```

av. th. of wall fcd. o.s.

```
          300
          361
     2) 661
        330.5
```

av. th. of wall fcd. b.s.

Nat st. a.b.

	Walls	
4.00	472 th.; tapg. o.s.	
0.65		

Tapering walls are measured in accordance with F20:1.1.3.4–5, stating whether tapering on one or both sides and whether faced on one or both sides

	Walls	
4.00	402 th.; tapg. o.s.;	
0.95	fcd. o.s.	

	Walls	
4.00	331 th.; tapg. o.s.;	
0.70	fcd. b.s.	

	Jamb sts.	
2/ 2.00	attchd.; av. 175 x 150 on face; lapd. 50; 300 th.; ro. dressed.	

Jamb stones are measured in metres as F20:11.2.1.8. The finish, being different from the random rubble walling, must be stated. Bonding to other work will not apply as it is of the same material as the adjoining walling.

Stone Rubble Walling 3

Nat. st. a. b.

Jamb sts.
   attchd.; ro. dressed;
   as dimsnd. diagrm.

2/ | 2.00

dia.

	1.000
add	
arch sts.	
2/ ½ /200	200
mean dia.	1.200
of arch	

An alternative approach to that shown on sheet 2 of this example. It incorporates a dimensioned diagram in place of the dimensioned description as advocated in General Rules 5.3. An isometric diagram is needed in this case to show clearly the construction required.

The mean diameter of the arch is calculated in waste as this is needed to determine its mean length on face as F20:M8.

½/22/7/ | 1.20

Arch
   ro. dressed; 200 hi.
   on face; 300 wide
   exp. soff.; semi-circ.
   (In 1 nr.).

dia.

1.200
200
1.400

The arch is measured in metres, giving the number, height on face, width of soffit and shape of arch as F20:24.1.0. The circumference of a semi-circle is $\frac{1}{2}\pi D$. Alternatively, arches may be enumerated.

½/22/7/ | 1.40

Fair rakg. or circ.
   cuttg., 300 th.

Fair raking or circular cutting to stonework is measured in metres stating the thickness to the upper surface of the arch as F20:27.1.0.

1

Centrg.
   st. arch; semi-circ.;
   300 wide soff.; 1.00
   span.

Centring to arches is enumerated, giving the details listed in F20:36.1.1.0 and F20:M12.

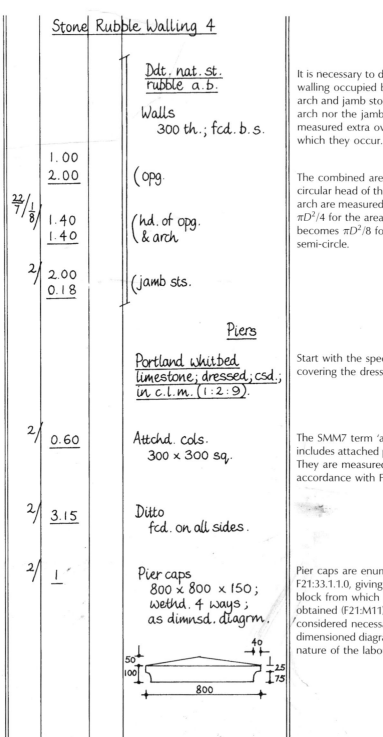

Stone Rubble Walling 4

<u>Ddt. nat. st.</u>
<u>rubble a.b.</u>

Walls
    300 th.; fcd. b.s.

It is necessary to deduct the areas of walling occupied by the opening, arch and jamb stones, as neither the arch nor the jamb stones are measured extra over the work in which they occur.

1.00	
2.00	(opg.

The combined area of the semi-circular head of the opening and the arch are measured using the formula $\pi D^2/4$ for the area of a circle, which becomes $\pi D^2/8$ for the area of a semi-circle.

$\frac{22}{7} / \frac{1}{8}$ 1.40   (hd. of opg.
        1.40   (& arch

2/   2.00   (jamb sts.
    0.18

Piers

<u>Portland whitbed</u>
<u>limestone; dressed; csd.;</u>
<u>in c.l.m. (1:2:9).</u>

Start with the specification heading covering the dressed stonework.

2/   0.60   Attchd. cols.
           300 × 300 sq.

The SMM7 term 'attached columns' includes attached piers and pilasters. They are measured in metres in accordance with F21:4.1.0.0.

2/   3.15   Ditto
           fcd. on all sides.

2/   1   Pier caps
           800 × 800 × 150;
           wethd. 4 ways;
           as dimnsd. diagrm.

Pier caps are enumerated as F21:33.1.1.0, giving the smallest block from which the stone can be obtained (F21:M11). It was considered necessary to add a dimensioned diagram because of the nature of the labours entailed.

7   Floors and Partitions

SEQUENCE OF MEASUREMENT

It is essential to adopt a logical order in taking off this class of work to reduce the risk of omission of any items. The best procedure is probably to take each floor complete, starting with the highest floor in the building and working downwards.

The work on each floor can conveniently be subdivided into: (1) construction and (2) finishes, and they are best taken in this sequence, following the order of construction on the site. In the case of solid floors, the finishes may be measured in a finishes section which picks up all the items covered in the Surface Finishes Sections of the *Standard Method*.

SUSPENDED TIMBER FLOORS

Suspended timber floors consist of boarding nailed to timber joists. On a ground floor the joints will generally be supported on timber plates bedded on brick sleeper walls built off the concrete oversite. Honeycomb brick sleeper walls have been largely superseded by solid walls incorporating airbricks, as illustrated in example 7. With upper floors the joists will be deeper and will usually be supported at their ends on walls or loadbearing partitions, and some form of strutting will normally be incorporated at about 2.000 to 2.500 centres, where the clear span of the joist exceeds 3.000.

In 1997, European strength classes were incorporated into BS 5268: Structural use of timber, thereby substituting C16 for SC3, C24 for SC4 and C27 for SC5. The prefix C denotes coniferous for softwoods and D denotes deciduous for hardwoods.

The sequence of taking off should preferably follow the order of construction on site:

(1) plates and bedding and possibly adjustment of brickwork (any supporting beams would also be taken at this stage;
(2) floor joists;
(3) strutting, if required;
(4) boarding.

117

Table 7.1 *Basic sawn sizes of softwood (from BS 4471: 1987) measured at a moisture content of 20 per cent.*

Thickness (mm)	Width (mm)								
	75	100	125	150	175	200	225	250	300
16	X	X	X	X					
19	X	X	X	X					
22	X	X	X	X	X	X	X		
25	X	X	X	X	X	X	X	X	X
32	X	X	X	X	X	X	X	X	X
36	X	X	X	X					
38	X	X	X	X	X	X	X		
44	X	X	X	X	X	X	X	X	X
47	X	X	X	X	X	X	X	X	X
50	X	X	X	X	X	X	X	X	X
63		X	X	X	X	X	X		
75		X	X	X	X	X	X	X	X
100		X		X		X	X	X	X
150				X		X			X
200						X			
250								X	
300									X

Basic sawn softwood timber sizes are shown in table 7.1 and basic lengths in table 7.2.

Table 7.2 *Basic metric lengths of sawn softwood from BS 4471: 1987.*

Length (m)						
1.80	2.10	3.00	4.20	5.10	6.00	7.20
	2.40	3.30	4.50	5.40	6.30	
	2.70	3.60	4.80	5.70	6.60	
		3.90			6.90	

Note: Lengths of 6 m and over may not be readily available.

Plates

Plates are measured in metres classified as plates with a dimensioned description and, where the length is >6.000 in a continuous length, the length is stated (G20:8.0.1.1). Where brickwork has been initially measured over the plate and it measures 100 × 75 or more, then brickwork will be deducted in accordance with F10: M3.

Floor Joists

In simple circumstances the number of floor joists is obtained by taking the length of the room, from which 100 is deducted to allow for spaces between the end joists and walls, and dividing the adjusted length by the spacing of the joists, usually 350 to 450. This gives the number of spaces to which one is added to give the number of joists. In other cases it will be necessary first to position members of a different size, such as trimmer and trimming joists and those which are required to carry the weight of partitions above. Only then can the positions of the normal bridging joists be determined at the required spacing. The joist layout for upper floors must always be done by reference to supporting walls and partitions.

Floor joists and beams are classified as floor members and measured in metres giving a dimensioned description in nominal sizes (G20:6.0.1.0 and G20:D1). Where joists are >6.000 in one continuous length, the length is stated, as these become progressively more expensive. The building in of ends of timbers into brickwork is deemed to be included. Metal hangers (G20:21), as described in example 9, avoid the need for bedding ends of timbers into external walls with the possibility of exposure to dampness and consequent risk of timber decay. It is necessary to state whether timbers are sawn or wrought (G20:S1).

A worked example of the measurement of trimming floor joists around an opening is given in this chapter. The tusk and tenon joints and dovetail notches have been largely superseded by metal hangers of the type shown in example 9.

Joist Strutting

The most usual form of strutting of floor joists is herringbone strutting which, like solid or block strutting, is measured in metres over the joists, stating the depth of the joists and giving a dimensioned description (G20:10.1–2.1.0), as illustrated in example 9.

Floor Boarding

Floor boarding takes various forms, such as square edged or tongued and grooved timber boarding in hardwood or softwood, parquet flooring, chipboard and plywood.

Floor boarding is measured in m^2, giving a dimensioned description as K20:2.1.1.0 (pages 73–4 of SMM7). Work sections K11 and K21 apply to rigid sheet flooring and timber strip and board fine flooring respectively. In the case of the former it is usually necessary to measure battens beneath the edges of sheet flooring running at right angles to the line of the joists. The battens should be described as individual supports (G20:13.0.1.0). Floorings in openings, although often $\leq 1.00\,\text{m}^2$ in area or ≤ 300 in width are not usually separated from the general flooring and enumerated or measured in metres respectively as K20:2.1–3.1.0, since they are regarded as a natural extension of the general flooring. Bearers under flooring in openings are classified as floor members.

Doors are usually hung to open into rooms and so will normally be located on the room side of the wall. The floor finish in the door opening will accordingly generally be the same as that in the hall, passage or landing from which the room is entered.

Skirtings are usually taken at the same time as measuring the internal finishes, although they can be taken with the floors.

SOLID FLOORS

There are two main types of concrete floor construction:

(1) ground floors consisting of a concrete bed usually laid on a bed of hardcore;
(2) upper floors consisting of suspended concrete slabs, which may be of *in-situ* or precast concrete.

In each case the floor finish can be measured at the same time as taking off the floor construction or be left to be measured with the other finishes. Concrete beds to ground floors are often included with the foundations/ substructure measurement.

Concrete Beds

In-situ concrete beds are measured in m^3, stating the appropriate thickness range as E10:4.1–3.0.5 and including in the description where poured on or against earth or unblinded hardcore, because of the additional concrete which fills the interstices on the upper surface of the earth or unblinded hardcore. The prescribed thickness ranges are ≤ 150, 150–450 and >450. Treating the surface of *in-situ* concrete is classified and given in m^2 as E41:1–7. Common surface treatments include mechanical tamping, power floating and trowelling, but a non-mechanical tamped finish is deemed to be included (E10:C1).

Hardcore and similar beds are measured in m^3, classified as to whether the average thickness is $\leq$ or >250, the nature of the filling material and its source and/or treatment (D20:10.1–2.1–3.1–4). The measurement of filling is illustrated in example 1. Where the surface of hardcore filling requires blinding, this is included in surface treatments for compacting (D20:13.2.2.1), but where specific blinding beds are required which increase the overall thickness of filling, these are measured as separate items of filling (D20:M18).

Suspended Concrete Slabs

Suspended *in-situ* concrete floor slabs are also measured in m^3 and classified in one of the three thickness stages given in E10:5.1–3.0.1.

Chases formed in new brickwork to receive the edges of floor slabs are deemed to be included in the brickwork rates (F10:C1c), and hence are not measured.

Bar reinforcement is billed in tonnes but entered on the dimensions paper in metres, keeping each nominal size, and straight, bent and curved bars, and links separate (E30: 1.1.1–4.1–4). Hooks and tying wire, and spacers and chairs where at the discretion of the contractor, are deemed to be included (E30:C1). Horizontal bars, including those sloping $\leq$308, with a length of 12.000 or more, and vertical bars, including those sloping >308, with a length of 6.000 or more, are each kept separate in 3.000 stages (E30:1.1.1–3.1–4 and E30:D1–2).

Fabric reinforcement is measured in m^2, stating the mesh reference and weight per m^2 and minimum laps (E30:4.1.0.0 and E30:S4). A commonly used fabric is ref. A193, weighing 3.02 kg/m^2, with 100 minimum laps.

Formwork to the soffits of floor slabs is measured in m^2 and the slab is classified as to thickness $\leq$200 and thereafter in 100 stages (E20:8.1–2.1.0). The heights to soffits are classified as $\leq$1.500 and thereafter in 1.500 stages (E20:8.1–2.1.1–2). Formwork to edges of suspended slabs $\leq$1.000 high is measured in metres, classified in the three height stages given in E20:3.1.2–4.0, whereas that to edges >1.000 in height is measured in m^2.

Where beams are attached to slabs, their *in-situ* concrete volume is measured as part of the floor slab where their depth is $\leq$ three times their width (depth being measured below the slab) in accordance with E10:D4a. Formwork to beams attached to slabs is measured in m^2, stating the number of beams in each item, the shape of the beam and the height to soffit in 1.500 stages (E20:13.1.1.1–2). The measurement of concrete and formwork to beams is illustrated in example 10.

Precast Concrete Floors

Precast concrete beam and block floors are now being used increasingly, particularly for the ground floors of domestic buildings, as they provide a soundly constructed floor, free from dry rot and other forms of insect and fungal attack to which timber floors are susceptible, with a good platform to receive the insulation and floor finish. A typical form of construction is illustrated in figure 26. Such work is measured in m^2 in accordance with E60.1.1.0.0 and E60:S1-6.

Another alternative to reinforced *in-situ* concrete slabs on upper floors, as shown in example 10, is to use precast concrete hollow core slabs grouted after laying. These are measured in the same way as the precast concrete beam and block floor in example 8. This type of floor has the advantage of

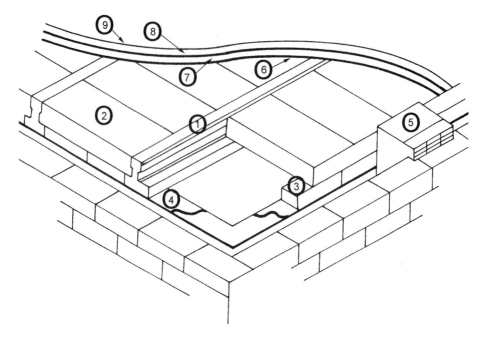

Key

1 Prestressed concrete beam, 150 deep (Jetfloor standard)
2 Concrete building block infill, 100 deep
3 Split course block
4 Bitumen damp-proof course
5 Void ventilator and air brick
6 Grouting coat of cement and sand (1:6)
7 EPS insulation board, grade SD to BS 3837, Part 1, 40 thick
8 1000 g polythene sheet vapour check
9 Chipboard flooring grade type C to BS 5669, 18 thick

Figure 26 *Jetfloor beam and panel construction.*

faster erection times and the virtual elimination of propping, shuttering and pouring of concrete.

Floor Finishes

The majority of floor finishes are measured in accordance with the numerous work sections contained under M Surfaces finishes, of which M10, M40–42,

M50 and M51 find the most frequent application. Where the finishes vary from one room to another it is usual to insert a dividing strip between the different floor finishings which is measured in metres with a dimensioned description in accordance with M40:16.4.1.0.

Variations in the thickness of different floor finishings are generally overcome by varying the thickness of the screed, to maintain a uniform finished floor level throughout. Screeds are measured in m^2, giving the plane, thickness and number of coats (M10:5.1.1.0). Details must also be given of mix composition, surface finish and the nature of the base (M10:S1, 3&5). An explanation of the term 'falls' in relation to screeds and floor finishes is given in chapter 2. Typical specification clauses for screeds are given in Appendix 4 at the back of the book.

PARTITIONS

A timber stud partition is measured in example 11 and figures 27 and 28 give details of the construction at right angled intersections and corners of stud partitions to help the student in calculating the required number of studs.

An understanding of the details of construction of stud partitions is necessary, as the drawings from which they are measured do not usually include a detailed layout of the members involved.

Block partitions have been covered in example 2 (chapter 5) and the timber stud partition in example 11 (in this chapter), and panel and dry partitions will now be considered.

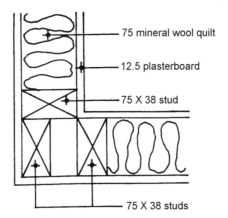

75 mineral wool quilt

12.5 plasterboard

75 X 38 stud

75 X 38 studs

Figure 27 *Corner to stud partition.*

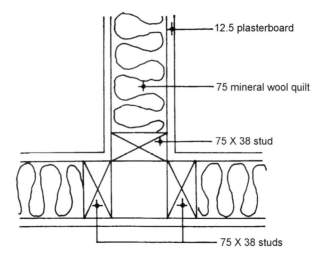

12.5 plasterboard

75 mineral wool quilt

75 X 38 stud

75 X 38 studs

Figure 28   *Intersection of stud partitions.*

Panel Partitions

Partitions which are of panel, as distinct from sheet and stud construction, are measured in metres, stating the height and thickness of the partition, and whether a factory or site applied finish is required, where not at the discretion of the contractor (K30:1.1.1–2.0). Trims, being items fixed on site, as cover pieces to edges or panel joints, are measured in metres with a dimensioned description (K30:2.1.0.0 and K30:D2). Openings are enumerated as extra over the partitions in which they occur with a dimensioned description and they include the components filling the openings (K30:3.1–5.1.0 and K30:D3).

The description of panel partitions is to include the kind and quality of materials; method of construction; layout of joints; method of fixing; complex integral services; method of bedding, jointing and pointing; and details of ironmongery, glass, linings and the like (K30:S1–7). Panel partitions can take a variety of forms, including frame with infill panel partitions, self finished prefabricated panel to panel partitions with featured joints, plasterboard panel to panel partitions and laminated plasterboard partitions comprising sheets bonded together on site. The last two types normally have a seamless finish ready to receive decoration. Panels may incorporate glazing, doors and windows in their construction.

Dry Partitions

Dry partitions are usually proprietary partitions and are measured in metres, stating the thickness and the height in 300 stages, and whether boarded on one or both sides (K10:1.1.1–2.0). These partitions are deemed to include such items as head and sole plates, studs, jointing battens and insulation, where part of the proprietary system. Angles, junctions and fair ends to partitions are measured in metres, stating the partition thickness (K10:3–5&8).

A typical example of proprietary dry plasterboard partitioning is Gyproc metal stud partitions which are 75 thick and comprise studs 48 thick, 12.5 taper edge wallboard each side; joints filled, taped and finished flush; holes filled with joint filler; and surface finished with one coat of Gyproc drywall top coat. Non-proprietary dry partitions may also be constructed using dry linings on timber studs.

Where plasterboard dry linings to walls are required, they are measured in accordance with K10:2, 6&7.

WORKED EXAMPLES

Worked examples follow covering a variety of different forms of floor construction to ground and upper floors. Example 8 covers precast concrete beam and block construction to the ground floor, while in example 10, brief details are included of the measurement of reinforced *in-situ* concrete and structural steelwork which it is hoped will be of value to the student.

The last worked example in this chapter covers the measurement of a timber partition, since this work is of similar character to that dealt with in timber floors.

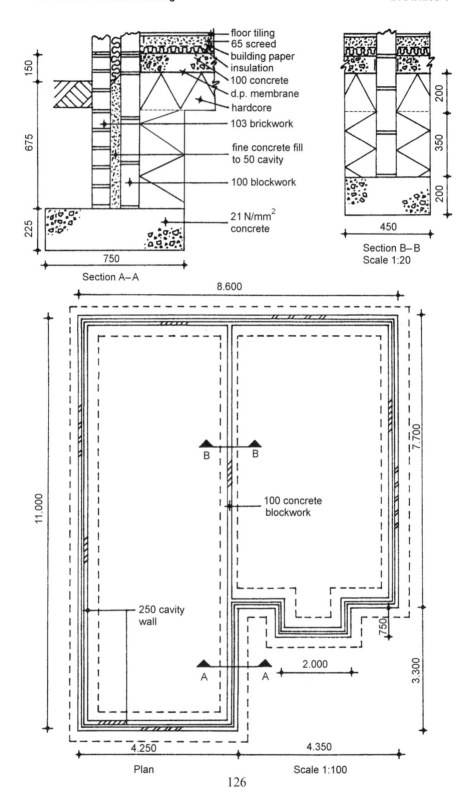

floor tiling
65 screed
building paper
insulation
100 concrete
d.p. membrane
hardcore

103 brickwork

fine concrete fill
to 50 cavity

100 blockwork

21 N/mm^2
concrete

150

675

225

750

Section A–A

200

350

200

450

Section B–B
Scale 1:20

8.600

B   B

100 concrete
blockwork

7.700

11.000

250 cavity
wall

750

2.000

3.300

A   A

4.250   4.350

Plan   Scale 1:100

126

Drg.1			Solid Ground Floor 1 Example 6

<u>Grd. Flr.</u>
<u>Insul.</u>

The dimensions are extracted from example 1 and hence do not require recalculating. The insulation board, whose provision is now advisable to meet the energy conservation requirements of the Building Regulations, is measured under P10. See P10:D1a which defines the range of work covered by this section.

<u>Expd. polystyrene EPS</u>
<u>insul. bd.; grade SD to</u>
<u>BS 3837 pt. 1; 40 th.</u>

10.49	(la. area
3.74	Plain areas
7.19	(smaller hor.; on conc.
4.25	(area
1.49	(bay
0.75	&

<u>Bldg. paper sheets to</u>
<u>BS 1521, class A</u>

The building paper is also measured in accordance with P10.

Plain areas
    hor.; on polystyrene

&

<u>Ct. & sd. screed (1:3-4.5)</u>
<u>stl. trowld.; one ct.; 65th.</u>
Flrs.
    lev. or to falls only
    ≤ 15° from hor.;
    on bldg. paper

The cement and sand screed is covered by M10:5.1 and it is necessary to state the composition and mix of the materials and the nature of the base.

&

<u>Marley flex. vinyl tiles</u>
<u>to BS 3261, pt. A; 300</u>
<u>× 300 × 2; fixd. w.</u>
<u>Marley Embond adhesive;</u>
<u>strt. butt jtd. both ways</u>
Flrs.
    >300 wide; lev. or
    to falls only ≤ 15°
    from hor.; on screed

Plastics tiling is dealt with in M50:5.1.1, giving a description of the tiling, nature of base and method of fixing and treatment of joints (M50:S1,5&8).

<u>Bldg. paper a.b.</u>

		65
52.86	Upstands	19
	84 gth.	84

An upstand to the building paper is taken as a linear item stating the girth, which is assumed to extend past the edges of the tiles to separate materials with different rates of expansion.

Suspended floors   DRAWING

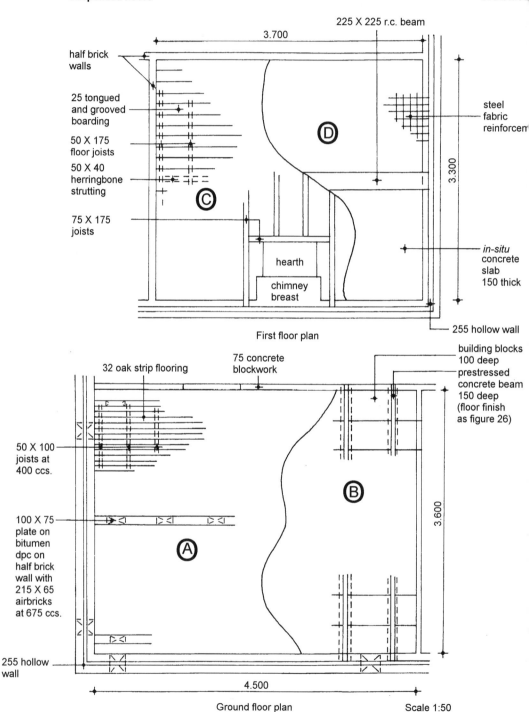

225 X 225 r.c. beam

3.700

half brick
walls

25 tongued
and grooved
boarding

50 X 175
floor joists

50 X 40
herringbone
strutting

75 X 175
joists

steel
fabric
reinforcem

3.300

in-situ
concrete
slab
150 thick

hearth

chimney
breast

First floor plan

255 hollow wall

75 concrete
blockwork

building blocks
100 deep

32 oak strip flooring

prestressed
concrete beam
150 deep
(floor finish
as figure 26)

50 X 100
joists at
400 ccs.

3.600

100 X 75
plate on
bitumen
dpc on
half brick
wall with
215 X 65
airbricks
at 675 ccs.

255 hollow
wall

4.500

Ground floor plan   Scale 1:50

Drg. 6	Timber Suspended Ground Floor 1

Example 7

Comm. bwk. to BS 3921
in stret. bd. in c.m. (1 : 3)

3/	4.50	Walls
	0.30	h. b. th.

```
                4.500
less 2/330       660
        675 )3.840
                6 + 1
```

Terra cotta, sq. hole
patt. to BS 493

| 3/ | 7 | Air bks.
215 × 65; in
h.b. wall |
|---|---|---|

D.p.c.s of single lyr.
'Hyload' pitch polymer
to BS 743, bedded in
c.m. (1 : 3)

3/	4.50	On surfs.
	0.10	≤ 225 wide; hor.

Sn. swd. grade C16
to BS 5268 ; vacuum
impregnated w.
preservative

| 3/ | 4.50 | Plates
100 × 75 |
|---|---|---|

The flooring examples cover the measurement of two separate constructional methods for each floor (A & B and C & D respectively). It is assumed that the concrete oversite and associated work will have been measured previously as part of the substructure.

The brickwork is measured up to the underside of the plate, using the classification given in F10:1.1.1.0.

The number of air bricks in each sleeper wall is found by dividing the length between the centres of the end air bricks by their spacing and adding one to the total to convert the intervening spaces to the number of air bricks. Air bricks are enumerated and described in accordance with F30:12.1 and are deemed to include bedding and pointing (F30:C1b) and any necessary forming of openings (F30:C8). No deductions are made to the brickwork for these voids as each is $\leq 0.10\,\text{m}^2$ (F10:M2a).

Damp-proof courses on sleeper walls are measured in m^2 and classified as ≤ 225 wide as F30:2.1.3.

The kind and quality of timber is to be given and whether it is sawn or wrought (G20:S1). European strength classes were incorporated into BS 5268 in 1997, whereby C16 replaced SC3 and C24 replaced SC4.

Plates are separately classified as G20:8.0.1 with a dimensioned description.

Timber Suspended Ground Floor 2

		4.500
less ends	2/75	150
400	)4.350	
	11 + 1	

The procedure for calculating the number of joists is similar to that adopted for air bricks.

Sn. swd. a.b.

Joists are classified as floor members in G20:6.0.1.

12/ 3.60

Flr. membrs.

50 × 100

Wrot European oak to class 1 of BS 1186 pt. 1

This main heading refers to the quality and kind of timber to be used in the strip flooring.

4.50
3.60

Strip flrg.
32 th.; t. & g. in 75 widths; secret nailed to swd. jsts.; spld. headg. jts.; sanded to a fin.; trtd. w. 2cts. 'Bournseal' on completn.; > 300 wide.

This heading is followed by a detailed description of the strip board flooring and its final treatment. Specifying the correct type of hardwood is always difficult and a variety of approaches is used in practice.

The description takes account of the requirements of K21:S1–3, 5, 9&12.

0.75
0.08

(dr.
(opg.

This item also picks up the small area of flooring in the door opening. The appropriate reference is K21:2.1.1.

Sn. swd. a.b.

3/ 0.08

(brrs. Flr. membrs.
in dr.
(opg.   50 × 75

Finally, the short bearers needed to support the flooring in the door opening are measured.

Drg. 6 and Fig. 26		Example 8	
Precast Concrete Suspended Ground Floor 1			

		4.500 3.600	The area of the decking must include the bearings on the perimeter which vary between 75 and 100 deep.
	add bearys.		
	2/100 200 200		
	4.700 3.800		
4.70	Jetfloor standard precast composite conc. deck'g. comprs'g. beams 150 dp. at 675 ccs. and infill blks. 100 dp.		The T beams and infill blocks are included in a single item in m² in accordance with E60:1.1. The thickness stated is the overall thickness (E60:M1).
3.80			
	Slabs		
		150 th.	
		4.700	
		3.800	
		8.500	
	add cnrs. 2/75		
	2/100 350		
	8.850		
	D.p.c.s of bit. to BS 6398 ; bedded in c.m. (1:3)		The damp-proof course follows as a linear item in accordance with F30:2.1.3.
8.85	On surfs. ≤ 225 wide; hor.		
	Proprietary items		These ventilator/air bricks are one-off specialist features and are best taken in accordance with F30:16.1.1. The manufacturer's description is sufficient to identify them.
4	Void ventilators & air bks.; supplied by Marshalls Flooring Ltd. Hoveringham, Nottingham		

Precast Concrete Suspended Ground Floor 2

Grouting

Grout of ct. & sand
(1:6) brushed on to
upper surf. of precast
conc. beams & infill
panels

The grout brushed on to the upper surface of the beams and blocks will fill the interstices and provide a reasonably smooth surface to receive the polystyrene board. The grout has been measured as a screed under M10:5.1.0.0. Alternatively, it may be included as part of the concrete decking item.

| 4.70 |
| 3.80 |

Flrs.
lev. & to falls only
≤ 15° from hor.

Insul.

Expd. polystyrene
EPS insul. bd.; grade
SD to BS 3837,
pt. 1; 40 th.

The polystyrene board is measured as the area between the enclosing walls. It is covered under P10 as an insulating component.

| 4.50 |
| 3.60 |

Plain areas
hor.; on grouted
surf. to prec.
conc.
&
1000 gauge polythene
sheets lap'd 150 at jts.

The polythene sheeting which acts as a vapour barrier is measured in m^2 in accordance with P10:1.1.1.0, giving particulars as required by P10:S1&2.

Plain areas
hor.; on
polystyrene

Chipbd. moisture
resistant flrg. grade
type C4 to BS 5669;
18 th.

The chipboard is classified as rigid sheet flooring and hence is covered by K11:2.1.1 and giving the type, quality and thickness of material as K11:S1. It is good practice to leave a 10 mm space around the periphery of the chipboard, but this is insufficient to warrant any adjustment of the floor area.

| 4.50 |
| 3.60 |
| (dr. 0.75 |
| opg. 0.08 |

Flrs.
>300 wide;
(dr. appld. to polythene
opg. sheet'g. w.
adhesive.

Drg. 6	**Timber Upper Floor 1**	**Example 9**

<table>
<tr><td></td><td></td><td>

nr. of jsts.

	3.700
less end spaces	
to ¢ of jsts. 2/75	150
400)3.550
9 + 1

len.

	3.300
add beargs. 2/100	200
	3.500

Sn. swd. grade C16
to BS 5268 ; vacuum
impreg. w. preservative

Flr. membrs.

</td><td>

The number of floor members is found by taking the length of the room less the 50 space at each end and half the thickness of the end joists. This dimension is divided by the spacing of the joists, to which one is added to convert spaces between joists to the number of joists.

The first heading specifies the kind and quality of timber and that it is sawn as G20:S1, and using the European strength class of C16 (formerly SC3).

Floor members are measured in accordance with G20:6.0.1.

</td></tr>
</table>

10/	3.50	(jsts 50 × 175

Ddt. ditto

2/	1.00	(trimd. jsts.
2/	3.50	(trimg. jsts.

Add ditto

2/	3.50	(trimg. jsts. 75 × 175
	1.13	(trimmer

Adjustments follow for the trimming work around the hearth and chimney breast.

Jst. struttg.
50 × 40; herringbone;
jsts. 175 dp.

3.70	

The joist strutting measured over the joists accords to G20:10.1, giving the depth of joists served by the strutting.

Flr. membrs.
50 × 75

2/	0.50	(cradlg. (pieces (to hth.

Finally, the cradling pieces on each side of the hearth are measured and these are also classified as floor members.

Timber Upper Floor 2

Galvd. mild steel

SPH jst. hangers to
type 'S' stand.; supplied
by Expanded Metal
Company

2	(trimmg. (jsts. to (trimmer	to fit jsts., 70 th. & 175 dp.

Ditto

2	(trimmer (to trimmd. (jsts.	to fit jsts., 50 th. & 175 dp.

Wrot swd.

3.70
3.30
	Bd. flrg.
	    25 th.; t. & g.
	    fxd. w. flr. brads
	    in 125 widths
	    to swd. jsts.; spld.
	    headg. jts.;
	    >300 wide.

0.90 0.34	(chy. (breast Ddt. ditto.
0.85 0.53	(hth.

```
              850
   2/530   1.060
           1.910
```

Wrot European oak
a.b.

Abutments

1.91	50 x 25 margins

Joist hangers are enumerated with a dimensioned description or dimensioned diagram (G20:21.1–2). Alternatively, a specific product of a manufacturer can be given as has been done in this example (General Rules 6.1). It is necessary to distinguish between joists to be supported where the thickness or depth varies as this affects the price.

Floorboarding follows the procedure outlined in K20:2.1.1, and stating the type, quality and thickness of timber (K20:S1), method of jointing (K20:S2), constraints on width of boards (K20:S9) and method of fixing where not at the contractor's discretion (K20:S12).

The margins around the hearth are taken as abutments (K20:8.1&K20:D8). The type of abutment must be stated.

Drg. 6   Concrete Upper Floor 1   Example 10

		len.	width
		3.700	3.300
	add beargs.		
	2/100	200	200
		3.900	3.500

Conc. (21 N/mm²)

3.90	Slabs
3.50	reinfd.; ≤ 150 th.
0.15	

0.90	Ddt. ditto.
0.34	(chy.
0.15	(breast

Fwk.
Soffits of slabs

3.70	slab ≤ 200 th.; hor.;
3.30	ht. to soff. 1.500–3.000

	3.900	3.500
less cover 2/40	80	80
	3.820	3.420

Reinft.

Fabric

3.82	stl. to BS 4483 ref.
3.42	A193; weighg.
	3.02 kg/m²; 100
	min. laps

	900
add cover 2/40	80
	980

0.98	(chy.
0.34	(breast Ddt. ditto.

The overall dimensions of the concrete slab are calculated in waste, making allowance for the bearings on all four edges.

In-situ concrete slabs are measured in m³ stating the thickness range and reinforced classifications as E10:5.1.0.1. The required strength of the concrete is given as an alternative to the mix.

Formwork is classified as E20:8.1.1.2, according to the slab thickness and height to soffit. Where the thickness of the slab is >200, the formwork is given separately in 100 stages of slab thickness, while the height to soffit of slab is given in 1.500 stages. Voids ≤5.00 m² irrespective of location are not deducted (E20:M4), hence there is no deduction of the chimney breast. It is assumed that the finish produced by the formwork is at the contractor's discretion and that no further details are required in accordance with E20:S2.

It is usual to allow about 40 cover to all reinforcement to prevent the possible onset of corrosion.

Fabric reinforcement is measured in m² giving the particulars listed in E30:4.1.0.0 and E30:S4 (minimum laps). Tying wire, cutting, bending, and spacers and chairs which are at the contractor's discretion are deemed to be included (E30:C2).

Where chases are necessary to receive the concrete slabs, they are deemed to be included with the brickwork or blockwork (F10:C1c).

Concrete Upper Floor 2

Worked finishes

3.70	Trowellg. surf. of conc.
3.30	

Worked finishes is the main heading to E41.

	Ddt. ditto
0.90	(chy.
0.34	(breast

Trowelling the surface of concrete is so described and measured in m² (E41:3.0.0.0). It is assumed that the floor finishes will be measured elsewhere.

R.C. Beam

len.

$$
\begin{array}{rr}
 & 3.700 \\
\text{add beargs. } 2/100 & 200 \\
\hline
 & 3.900 \\
\end{array}
$$

Conc. (21 N/mm²)

Slabs

	reinfd.; ≤ 150 th.
3.90	(beam under slab
0.23	
0.23	

In-situ concrete attached beams to suspended slabs are included with the slabs, except where they are deep beams with a depth/width ratio exceeding 3 : 1, measured below the slab (E10:D4a). Although this results in a thicker slab over a small area, it is considered permissible to include the beam concrete in with the slab concrete, without any change of thickness classification. E10:M2 appears to support this approach. Furthermore, it would be unrealistic to measure a small section of slab in the combined thickness range of 150–450.

gth. of beam

$$
\begin{array}{lr}
\text{sides } 2/225 & 450 \\
\text{Soff.} & 225 \\
\hline
 & 675 \\
\end{array}
$$

Fwk.

	Beams attchd. to slabs
3.70	reg. sq. shape;
0.68	ht. to soff. 1.500–3,000
	(In 1 nr.)

Formwork to attached beams of regular shape is measured in m², stating the number of members, and shape and height to soffit (E20:13.1.1.2).

	Ddt. Fwk.
3.70	Soff. of slab a.b.
0.23	

Deduction of formwork to soffit of slab for area occupied by beam.

Concrete Upper Floor 3

<u>R.C. Beam</u>

<u>len.</u>
3.900
less cover ²/40   80
3.820
add hkd. ends ²/300   600
4.420

It is assumed that the beam is reinforced with 3 nr. 25 diameter mild steel bars. The length of the bars is obtained by taking the full length of the beam, deducting the 40 cover at each end and adding twelve times the diameter for each hooked end. The reinforcement will subsequently be reduced to tonnes. It is measured in accordance with E30:1.1.1.0, with each diameter kept separate, The bars are described as straight, as hooks, tying wire and spacers and chairs which are at the discretion of the contractor are deemed to be included (E30:C1).

³/ | 4.42

<u>Reinft.</u>
M.S. bars to BS 4449
25 ⌀ nom. size;
strt.

Alternative Approach substituting a steel joist encased in concrete for the reinforced concrete beam.

3.90

<u>Isolated struct. steel members to BS 4</u>
Plain beams
≤ 40 kg/m
× 0.03 =ㅤㅤtonnes
(203 × 133 × 30 kg beam

Isolated structural members are described in accordance with the classifications in G12:5.1.1.0, including the use to which they are put and the weight range. They are deemed to include fabrication and erection (G12:C2). Any fittings are grouped together irrespective of the member to which they are attached (G12:M2). The unit of billing is tonnes, which is achieved by multiplying the length by the weight per metre beneath the description.

<u>width</u>	<u>depth</u>
133	203
add cover ㅤ 75	40
208	243

<u>Conc. (21 N/mm²)</u>

Build up of dimensions of concrete casing to steel beam below slab in waste.

3.90
0.21
0.24

Slabs
reinfd.; ≤ 150 th.

The concrete in the suspended slab is deemed to include concrete casings to steel beams (E10:D4), but giving the slab thickness range classification, as in the case of the reinforced concrete beam.

<u>gth. of beam</u>	
sides ²/243	486
soff.	208
	694

<u>Fwk.</u>

Formwork to attached beams is measured in accordance with E20:14.1.1.2. This will be followed by the adjustment of the formwork to the soffit of the slab as before.

3.70
0.69

Beam casgs. attchd to slabs
reg. rect. shape;
ht. to soff. 1.500–3.000
(In 1 nr.)

DRAWING 7

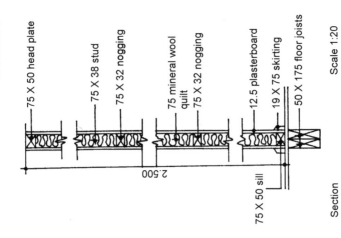

75 X 50 head plate

75 X 38 stud

75 X 32 nogging

75 mineral wool quilt

75 X 32 nogging

12.5 plasterboard

19 X 75 skirting

50 X 175 floor joists

2.500

75 X 50 sill

Section

Scale 1:20

Stud partition

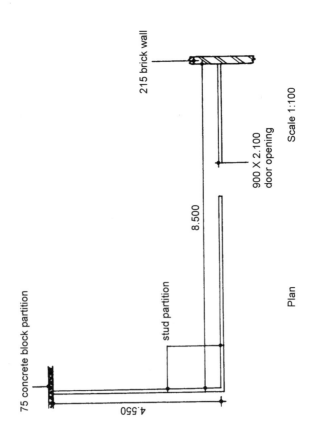

75 concrete block partition

215 brick wall

stud partition

8.500

900 X 2.100 door opening

4.550

Plan

Scale 1:100

Drg. 7	Partitions 1 Example 11

STUD PARTITION

Construction lens.

$$2.500$$
less 2/50 $$\quad 100$$
studs $$\quad 2.400$$

$$2.500$$
less 2/50 $$100$$
$$2.100 \quad 2.200$$
stud over dr. $$\quad 300$$
$$\quad 900$$

add
jbs. 2/50 $$100$$
t. tenons 2/100 $$200 \quad 300$$
dr. hd. $$1.200$$
$$8.575$$
$$4.550$$
hd. plate $$13.125$$
less dr. opg. $$\quad 900$$
sill $$12.225$$

nr. of studs

less ends $$.050 \quad 4.550$$
2/½/.038 $$.038 \quad .088$$
$$450\,)\ 4.462$$
$$10 + 1\,(11)$$

$$5.100$$
less ends 2/½/.038 $$\quad .038$$
$$450\,)\ 5.062$$
+1 at cnr. −1 dr. fr. $$12 + 1 + 1 − 1\,(13)$$

less ends $$.050 \quad 2.500$$
2/½/.038 $$.038 \quad .088$$
$$450\,)\ 2.412$$
$$6 + 1 − 1\,(6)$$

len. of noggings

$$8.500$$
$$4.550$$
$$13.050$$
less 30/38 $$1.140$$
2/50 $$100$$
$$900 \quad 2.140$$
$$10.910$$

The lengths of partition members and numbers of studs are calculated in waste. The lengths of studs comprise the room height (floor to ceiling) less the thickness of both sill and head plate, while those over the door head are obtained by subtracting the thickness of door frame and head plate plus the height of the door opening from the room height.

The door head length consists of the width of the door opening plus the door jambs and framed joints, although the latter are not always used in practice. The sill length is the same as that of the head plate less the width of the door opening.

To calculate the number of studs, the length of each section between centres of end studs is taken and divided by the spacing of the studs, adding one to convert spaces to studs and adjusting for door frames and an additional stud at the corner as shown on figure 27.

The lengths of the two sets of noggins shown on the section are obtained from the total lengths of partition less the thickness of studs and frames and the width of the door opening. The spacing of 450 is determined by the width of the sheets of plasterboard (900 in this case).

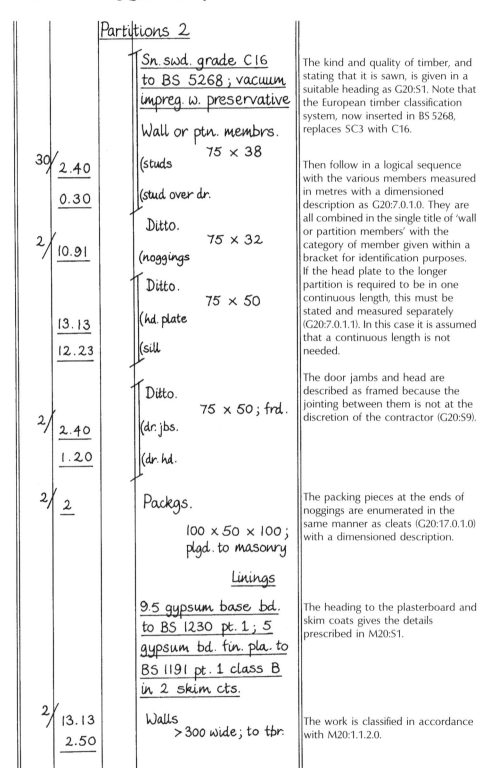

Partitions 2

Sn. swd. grade C16
to BS 5268 ; vacuum
impreg. w. preservative

The kind and quality of timber, and stating that it is sawn, is given in a suitable heading as G20:S1. Note that the European timber classification system, now inserted in BS 5268, replaces SC3 with C16.

Wall or ptn. membrs.

75 × 38

30/ 2.40   (studs

0.30   (stud over dr.

Ditto.
75 × 32
2/ 10.91   (noggings

Then follow in a logical sequence with the various members measured in metres with a dimensioned description as G20:7.0.1.0. They are all combined in the single title of 'wall or partition members' with the category of member given within a bracket for identification purposes. If the head plate to the longer partition is required to be in one continuous length, this must be stated and measured separately (G20:7.0.1.1). In this case it is assumed that a continuous length is not needed.

Ditto.
75 × 50
13.13   (hd. plate

12.23   (sill

Ditto.
75 × 50 ; frd.
2/ 2.40   (dr. jbs.

1.20   (dr. hd.

The door jambs and head are described as framed because the jointing between them is not at the discretion of the contractor (G20:S9).

2/ 2   Packgs.

100 × 50 × 100 ;
plgd. to masonry

The packing pieces at the ends of noggings are enumerated in the same manner as cleats (G20:17.0.1.0) with a dimensioned description.

Linings

9.5 gypsum base bd.
to BS 1230 pt. 1 ; 5
gypsum bd. fin. pla. to
BS 1191 pt. 1 class B
in 2 skim cts.

The heading to the plasterboard and skim coats gives the details prescribed in M20:S1.

2/ 13.13   Walls
2.50   > 300 wide ; to tbr.

The work is classified in accordance with M20:1.1.2.0.

<u>Partitions 3</u>

2/	0.90	
	2.10	

Ddt. 9.5 gyp. pla. bd. a.b.
Walls
>300 wide; to tbr.
(dr. opg.

	len.
	8.500
	4.550
	13.050

less cov. fill. 4/19   .076

inner side   12.974

	13.050
add cnr. 2/75	.150
	13.200

less cov. fills.
3/19   57
     38   .095

outer side   13.105

	ht.
less	2.500
cov. fill 19	
sktg. 75	94
	2.406

	12.97	
	2.41	
	13.11	
	2.41	

<u>Emulsn. ptg.; pla.;</u>
<u>prep. & 2 cts.</u>

Gen. surfs.
>300 gth.

	width
	900
add archve. 2/50	100
	1.000

	ht.
	2.100
add archve.	50
	2.150
less sktg.	75
	2.075

2/	1.00	
	2.08	

Ddt. ditto.
(dr. opg.

The main areas of plasterboard are followed by the deduction on both sides for the door opening. The door linings are assumed to be within the door opening sizes. Calculations in waste are required to determine the areas to be painted with emulsion paint. The adjustments include the area occupied by cover fillets, skirting and door opening.

Emulsion paint is measured in m² to general surfaces in accordance with M60:1.0.1.0. It is necessary to give the kind and quality of painting materials, nature of base and number of coats (M60:S1–6).

The architraves surrounding the door opening are added to the areas of emulsion painting to be deducted as they will be covered in a separate painting item.

<u>Part|itions 4</u>

<u>sktgs. & cov. fillets</u>
The kind and quality of timber and whether sawn or wrought must be given (P20:S1). Skirtings, picture rails, architraves and the like are taken as linear items with a dimensioned overall cross section description as P20:1.1.0.0.

	13.125
<u>less</u>	
archves. ²/50 100	
dr. opg. 900	1.000
sktg.	12.125

2/ 12.13

Wrot. swd.

sktgs. pict. rls., archves. & the like
      19 × 75 chfd.
      &

<u>K.p.s., 2 u/c, 1 f/c &</u>
<u>alkyd gloss; wd.</u>

Gen. surfs.
      isol.; ≤ 300 gth.
      &

<u>Primg.; wd.</u>

Gen. surfs.
      isol.; ≤ 300 gth.; on
      site prior to fxg.

The painting item heading to the skirtings, etc. is to include the kind and quality of materials and nature of base, preparatory work where necessary, undercoats (number), finishing coats (number) and surface finish (M60:S1–6). The painting follows the procedure prescribed in M60:1.0.2.4. Priming on site prior to fixing applies to backs of skirtings only.

	2.500
<u>less</u> sktg.	75
	<u>2.425</u>

Wrot. swd.
Cov. fillets, stops, trims, beads, nosgs. & the like
(inside   19 quadrant
(cnr.   &
(ends   <u>K.p.s., 2 u/c, 1 f/c</u>
(top   <u>alkyd silk; wd.</u>

Gen. surfs.
      isol.; ≤ 300 gth.

Cover fillets and the like are measured as P20:2.1.0.0, followed by their appropriate painting items. These are followed by the fillets which encompass the outside corner.

2/2/ 2.43
      2.43
2/ 13.13

<u>less</u>	2.425
cov. fill.	19
	<u>2.406</u>

Wrot. swd.
Cov. fillets, etc.; a.b.
(ext.   38 × 19 chfd.
(cnr.
            &

<u>K.p.s., etc. a.b.</u>
Gen. surfs.
      isol.; ≤ 300 gth.

The door, door lining, architraves and painting to them will be measured elsewhere.

2/ 2.41

Partitions 5

Insul.

len.

as noggings   10.910

ht.

2.500

less
head & sill   100
2 noggings
   2/32   64   164
          2.336

The internal spaces of the stud partitions are filled with insulating quilt to reduce the amount of sound transmitted between adjoining rooms. The areas to be measured are obtained by taking the overall dimensions and deducting the thickness of the intervening members, as the area measured is that covered (P10:M1).

over dr. opg.

len.

300

less stud   38

262

The quilt is described as between members, stating the centres of members. The measurement rules are contained in P10:2.3.2.0.

10.91	Paper fcd. min. wool insul. quilt ; 75 th.
2.34	Betw. membrs. membrs. at 450 ccs.; vert.

The type, quality and thickness of material should be given along with the extent of laps, if any, and method of fixing where it is not at the discretion of the contractor (P10:S1–3).

0.26	Cover dr.
0.30	opg.

A similar approach is adopted for thermal insulation work.

8 Pitched and Flat Roofs

Introduction

Both flat and pitched roofs can conveniently be subdivided into two main sections for purposes of measurement, that is, construction and coverings. The order of measurement of these two sections varies in practice, but on balance it is probably better to take the construction first since this follows the order of erection on the site.

The general rules for the measurement of the lengths of rafters, hips and valleys were described in chapter 3. The work to the eaves of pitched roofs and rainwater goods is normally measured with the roofs.

With flat roofs the taker-off must ensure that all rolls, drips, gutters and associated work are drawn on the roof plan before starting measurement, otherwise dimensions may be woefully deficient.

PITCHED ROOFS

Roof Timbers

The order of items in a traditional cut roof should follow a logical sequence such as plates, rafters, ceiling joists, collars, purlins, struts, ridge boards and hip and valley rafters. The European strength classes inserted in BS 5268 should be noted, i.e. C16 replacing SC3 and C24 in place of SC4.

When determining the length of the plate, it is customary to allow for laps at corners equal to the width of the plate, but not for laps in running lengths. Where joints are not described they are at the discretion of the contractor (G20:S9). Deduction of brickwork for the plate will be needed where the brick walling has been previously measured over the plate and the plate has a minimum size of 100 × 75 (F10:M3).

The roof timbers are measured in metres as pitched roof members, giving a dimensioned description, and these include struts, purlins, rafters, hip and valley rafters, ridge boards, ceiling joists, binders and bracing (G20:9.2.1.0 and G20:D6). Where the length of a member is >6.000 in one continuous length, the length is stated as it involves an increase in cost (G20:9.2.1.1).

The size and spacing of rafters can be determined by reference to Appendix A of the Building Regulations Approved Document A1/2 (1992).

For example, 50 × 100 softwood rafters (class C16 and C24), at 400 centres laid to a pitch >308 but ≤458 with a dead load of >0.50 but ≤0.75 kN/m², will have a maximum clear span of 2.41 m (eaves or ridge to purlin or between purlins).

The number of rafters is found by taking the total length of roof between the end rafters and dividing it by the spacing of the rafters to give the number of spaces and then adding one to give the number of pairs of rafters on the two roof slopes. Some adjustment of the spacing of rafters may be necessary to fit the space available, but if this results in an increase in the spacing of the rafters, then the data must be checked to ensure that the requirements of Appendix A of Approved Document A1/2 are met. Where a roof has a hipped end, an additional rafter must be added for each hipped end, but apart from this the number of rafters will be the same as if the roof were gabled.

It will also be necessary to measure to the extremities of the splayed ends of the rafters when calculating their lengths as shown in figure 11. This will also apply to ceiling joists and collars as illustrated in example 12.

The adjustment of roof timbers around chimney stacks is usually taken after the measurement of the main roof timbers and coverings and a worked example is provided in example 13.

Roof trusses, trussed rafters and trussed beams are enumerated and provided with a dimensioned description as G20:1–3.1.0.0. The work is deemed to include webs, gussets and the like (G20:C2). The measurement of roof trusses is illustrated in example 15.

Covering Materials

Coverings to pitched roofs can take a variety of forms including plain tiling, interlocking tiling, slating, profiled protected metal sheeting, shingles and thatch. Limitations of space will not permit the measurement of all of these different coverings to be included in this chapter, although as many as possible are covered. It was felt desirable to assist the reader by giving concise descriptions of the main characteristics of some of the more important coverings to pitched roofs.

With *roof tiles* it is necessary to distinguish between single lap and double lap tiling. For example, double lap plain tiles in clay or concrete normally measure 265 × 165, may be cross cambered and have two nibs and two holes for nailing near the head of the tile. They are normally nailed at each fourth or fifth course and at top course and eaves, preferably with two nails per tile, to a minimum pitch of 358. The head lap of 65 minimum occurs where there are three thicknesses of tile and two thicknesses elsewhere, while the gauge measured from tail to tail is 88 minimum. Concrete tiles last longer than clay tiles but soon lose their colour. Both concrete and clay tiles should conform to BS 5534: Part 1. A plain clay tiled roof is measured in example 12.

Single lap tiles take various forms such as Redland Norfolk pantiles, 381 × 227, in eight colours with a minimum head lap of 75 and a minimum pitch of 22.58 or 100 lap for lower pitches, with a minimum gauge of 256 and maximum gauge of 306. By contrast, the double Roman tile is considerably larger at 418 × 330 requiring fewer tiles and has proved very popular. The recommended batten size is 38 × 25 for rafters at 600 maximum centres for all tiles, fixed through an approved reinforced roofing underlay, such as 1F reinforced bitumen felt to BS 747, to rafters with wire, cut or improved nails as detailed in BS 5534: Part 1. Nailing of tiles is usually carried out with 38 × 3.35 mm aluminium clout headed nails.

Slates made from natural slate should meet the requirements of BS 680: Parts 1 and 2, and are made in a variety of sizes ranging from 610 × 355 down to 305 × 205 or even smaller, and are frequently blue-grey or Westmorland green in colour. They are durable and suitable for low pitches and complicated roof details. Slated roofs at a pitch of 308 have a head lap of 75 and can be either twice head nailed or centre nailed with copper nails to BS 1202: Part 2, measuring 20–25 longer than two thicknesses of slate, as they are single nailed to softwood battens or boards. A slated roof is measured in example 15.

Another alternative is man made roofing slates such as Euroslates and Eternit 2000 slates. Euroslates are manufactured from polyester resin and resemble natural slate in appearance, while Eternit slates consist of a high quality resin painted outer coat, densely pigment impregnated cement intermediate layer and pigmented fibre cement base. Fixing arrangements for both products are similar to those for natural slates but their lives are likely to be shorter.

Wood shingles provide a very attractive silvery grey roof covering and are usually made from Western Red Cedar. Standard shingles are 400 × 75–355 and 10–2 tapered thickness and weigh as little as one-tenth of the weight of concrete tiles and are normally guaranteed for 25 years. A minimum of three thicknesses of shingles is normal increased to four where the roof is very exposed or has a shallow pitch. They are nailed to battens with stainless steel nails to a 125 gauge where 22.58 or steeper pitch, but do not require a felt underlay.

Thatching with reed from the Norfolk Broads provides an attractive, albeit expensive, form of roof covering to steep roofs, with a life of 50–75 years, excluding ridges which are probably limited to 20 years. It is usually laid about 300 thick on 25 × 19 battens at 225 centres, while eaves and verges have tilting boards.

Measurement of Roof Coverings

The area of roof covering is measured first in m² stating the pitch, and is deemed to include underlay and battens (H60:1.1.0.0 and H60:C1). Note the various particulars that are required in the billed description of slating or tiling, including the kind, quality and size of slates or tiles, method of fixing, minimum laps and spacing of battens (H60:S1–4).

These items are followed by linear roofing boundary items, such as work at abutments, eaves and verges. Also measured linearly are ridges, hips, vertical angles and valleys (H60:3–9). In each case the method of forming must be given. Boundary work to roofs is deemed to include undercloaks, cutting, bedding, pointing, ends, angles and intersections (H60:C2).

These measurements will be followed by the adjustment of roof coverings for chimney stacks or other openings through the roof. Where the opening is $\leq 1.00\,m^2$, no deduction will be made from the area of roof covering (H60:M1), and no boundary work will be measured to the void (H60:M2).

Measurement of various types of slating and tiling is covered in work sections H60 (plain roof tiling), H61 (fibre cement slating), H62 (natural slating), H63 (reconstructed stone slating and tiling), H64 (timber shingling), H65 (single lap roof tiling) and H66 (bituminous felt shingling), all of which share the same rules of measurement.

Profiled Metal Sheeting

Profiled protected metal sheeting is now used extensively, especially for commercial and industrial buildings. For example, European Profiles Ltd manufacture EP Insulite, a fully bonded composite roof panel which is considered to be attractive and environmentally friendly. EP Insulite uses EP1000/32 coated steel roof sheeting, a heavy density expanded polystyrene insulation core and a flat steel white enamel coated liner, with an overall nominal depth of 93. The core insulation is moulded to match exactly the shape of the steel inner and outer skins. The weatherside surface is coated in colorcoat HP200 plastisol leather grain PVC. Alternatively, a PVf2 coating system can be specified. It can be used on pitches as low as 48, is available in lengths up to 8 m and is fixed with carbon steel self drilling/tapping fasteners.

Plannja have introduced Scan roof – a versatile roof tile system which combines traditional appearance with a range of polyester coated finishes with light weight, ease of installation and good weather tightness properties. It is equally effective for residential, commercial and industrial uses with integral battens, providing a self supporting, loadbearing roof fixed directly to rafters and trusses. Another Plannja product is Royale sheets with a deep Dutch pantile appearance. The minimum roof pitch is 128.

The measurement of profiled sheet roofing is covered in work sections H30 (fibre cement), H31 (metal), H32 (plastics) and H33 (bitumen and fibre).

The billed description for EP Insulite panels, measured in m^2, in accordance with H31:1.1.0.0 could appear as follows:

EP Insulite composite panel, comprising EP1000/32 coated steel sheet, in colorcoat HP200 plastisol leather grain PVC; heavy density expanded polystyrene core and white steel liner, 0.4 mm thick; overall nominal depth 93; 150 end laps bedded in mastic and 20 rebated insulation side laps; fixing to supporting steelwork at 600 centres with carbon steel self drilling/tapping fasteners, in accordance with manufacturer's instructions.

Roof coverings
  158 pitch

This item would be followed by linear items in metres with a dimensioned cross section description covering eaves, verges, ridges, etc. as H31:4–6. EPI rooflights which are factory assembled double skin glass fibre reinforced plastics (GRP) are enumerated as extra over the roof covering with a dimensioned description as H31:18.3.1.0.

Roof Void Ventilation

In recent years the ventilation of roof voids, lofts or spaces has become a very important aspect of pitched roof design, to prevent condensation in the roof which could reduce the thermal performance of the insulating materials and the structural performance of the roof construction. Hence it is felt desirable to show how this can be achieved in a variety of ways, to provide the student with a better understanding of these developments as an aid to their measurement.

The Building Regulations, Approved Document F2, section 1 (1995) requires that pitched roof spaces should have ventilation openings at eaves level to promote cross ventilation. The openings shall have an area on opposite sides at least equal to continuous ventilation running the full length of the eaves and 10 wide, increasing to 25 wide for roofs of less than 158 pitch or where the ceiling follows the pitch of the roof. It further describes how purpose made components are available to ensure that quilt and loose fill insulation will not obstruct the flow of air where the insulation and the roof meet.

It further prescribes that a pitched roof that has a single slope (monopitch) and abuts a wall should have ventilation openings at eaves level and at high level. The area of the high level vent should be at least equal to continuous ventilation running the full length of the junction and 5 wide. In practice, some designers prefer to provide both eaves and high level ventilation openings in normal type pitched roofs, especially on those exceeding 10 m span, and these are illustrated in figures 29, 30 and 31.

Figure 29 illustrates the Redland DryVent ridge and RedVent eaves ventilation to a plain tiled pitched roof space. A timber ridge batten is fixed to the apex of the timber trussed rafters with a stainless steel fixing strap fixed to each rafter, and there is a polypropylene ridge to ridge seal and a PVC-U profile filler unit. The eaves ventilation comprises a PVC-U fascia grille unit

Redland DryVent Ridge

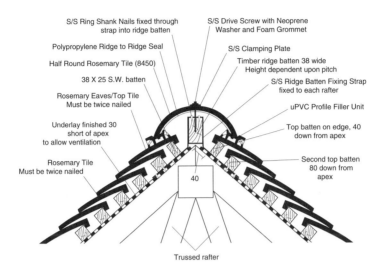

S/S Ring Shank Nails fixed through strap into ridge batten

Polypropylene Ridge to Ridge Seal

Half Round Rosemary Tile (8450)

38 X 25 S.W. batten

Rosemary Eaves/Top Tile Must be twice nailed

Underlay finished 30 short of apex to allow ventilation

Rosemary Tile Must be twice nailed

S/S Drive Screw with Neoprene Washer and Foam Grommet

S/S Clamping Plate

Timber ridge batten 38 wide Height dependent upon pitch

S/S Ridge Batten Fixing Strap fixed to each rafter

uPVC Profile Filler Unit

Top batten on edge, 40 down from apex

Second top batten 80 down from apex

40

Trussed rafter

RedVent Eaves Ventilation

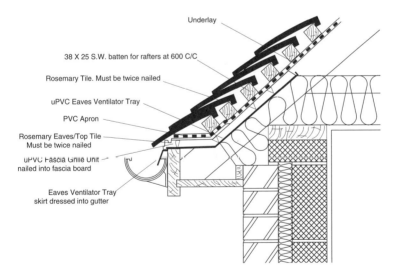

Underlay

38 X 25 S.W. batten for rafters at 600 C/C

Rosemary Tile. Must be twice nailed

uPVC Eaves Ventilator Tray

PVC Apron

Rosemary Eaves/Top Tile Must be twice nailed

uPVC Fascia Grille Unit nailed into fascia board

Eaves Ventilator Tray skirt dressed into gutter

Figure 29 *Redland ridge and eaves ventilation (plain tiles).*

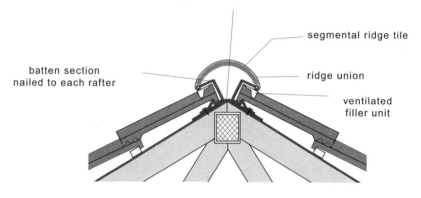

VENTILATED DRY RIDGE
(with interlocking tile)
underlay cut to
provide ventilation

segmental ridge tile

batten section
nailed to each rafter

ridge union

ventilated
filler unit

Eaves Detail
(Interlocking tile)

ventilation
ducts nailed
to each other

underlay support nailed
to each rafter

comb eaves
filler &
fixing clip
(when required)

10mm continuous
strip ventilator
nailed to fascia board

Figure 30   *Marley ridge and eaves ventilation (interlocking tiles).*

nailed to the fascia board and a PVC-U eaves ventilator tray supporting a PVC apron and with an eaves ventilator tray skirt below dressed into the eaves gutter. These items are measured in example 12. Individual vent tiles will be enumerated with a dimensioned description as H60:10.1.1.0.

Figure 30 illustrates a Marley ventilated dry ridge and a ventilating eaves detail and the billed descriptions for these linear items in accordance with H65:4.0.0.0 and H65:6.0.0.0 follow.

Ridge
Marley ventilated dry ridge, comprising extruded black PVC-U batten sections, 3.000 long, nailing to each side of ridge apex with ventilation slots in top flanges of batten sections; fitting high impact polystyrene profiled filler units to top course of roofing tiles; segmental clay ridge tiles 300 diameter with injection moulded PVC-U brown ridge unions and attached neoprene rubber seals; fixing all in accordance with manufacturer's instructions and BS 5534: Part 1.

Eaves
Marley ventilating eaves comprising moulded black high density polypropylene strip ventilators, 500 long, nailed to top of fascia; PVC-U comb fillers for profiled tiles clipped into strip ventilator; continuous roll duct trays nailed to each rafter; fixing all in accordance with manufacturer's instructions and BS 5534: Part 1.

Figure 31 illustrates a Burlington natural slate vent and a Burlington slate ridge vent to ventilate a roof void and a billed description of each of these items follows, being numbered items in accordance with H62: 10.1.1.0.

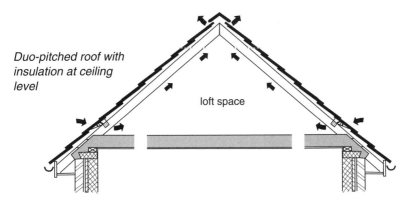

Figure 31 *Burlington roof void ventilation to slated roof.*
(figure continued on page 152)

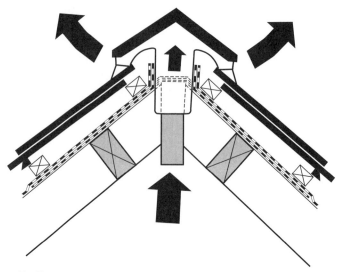

Burlington slate ridge vent used to ventilate a roof void.

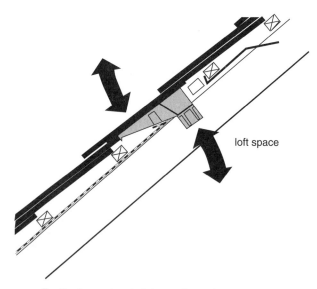

Burlington natural slate vent used to allow ventilation of the loft space.

Figure 31   *Burlington roof void ventilation to slated roof (continued).*

Ridge

Burlington ridge vent, comprising blue grey angular slate ridge 460 long; PVC-U vent body; PVC-U extension piece fitted to underside of ventilator with wing nuts; bedding base of slate vent in cement mortar (1 : 3); fixing vent by nailing to ridge or rafters; fixing all in accordance with manufacturer's instructions.

Eaves

Burlington natural slate vent, comprising blue grey 510 × 355 slate with rectangular cut-out and bonded colour matched polypropylene tray with grille to underside of slate vent; PVC-U pipe adaptor fitted with wing nuts; fixing all in accordance with manufacturer's instructions in positions.

Figure 32 illustrates eaves ventilation to a Euroslate pitched roof measured as a linear item with the following billed description in accordance with H63:4.0.0.0.

Eaves

Euroslate PVC-U grilled soffit ventilator 200 wide; ERV 2000 PVC-U roll panel 250 long, nailed to softwood rafters and 250 wide strip of Hi-ten 125 sarking felt, nailed to softwood rafters and bottom edge dressed into gutter; fixing all in accordance with manufacturer's instructions.

Some good working details of Cavity Tray polypropylene eaves ventilators, strip soffit and circular soffit ventilators are contained in *Building Technology, fifth edition* by the present author and there are measured items of these components in example 15.

Eaves and Rainwater Goods

Rainwater downpipes and gutters are available in PVC-U, cast iron, aluminium and glass fibre reinforced plastics (GRP).

PVC-U pipes and gutters are used extensively on grounds of lower cost, self-decoration (black, grey, brown and white) and less maintenance. However, colours fade and a quick thaw following a heavy snowfall on a pitched, well insulated roof can result in deformation of PVC-U gutters.

Cast iron rainwater goods need regular painting to avoid corrosion but could be a good choice in the long term.

Aluminium: the main advantages claimed for these products are long life, low maintenance, corrosion resistance, simple and quick assembly, extensive range of shapes and colours and cost effectiveness.

GRP rainwater goods are especially suitable as a replacement for cast iron or lead gutters to match the existing.

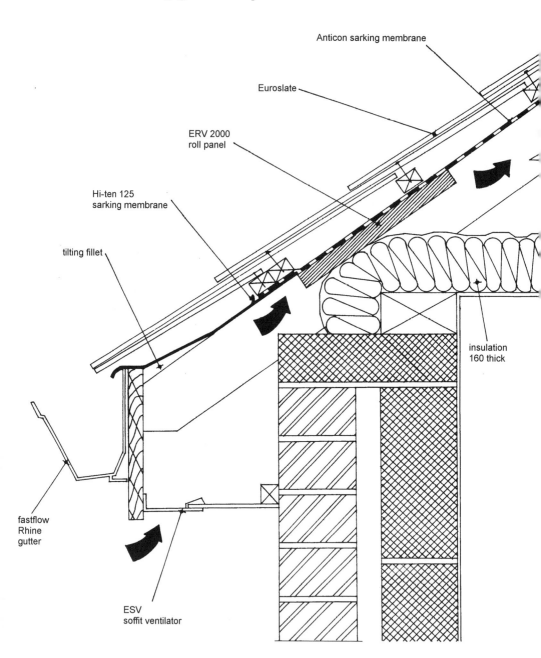

Figure 32   *Eaves ventilation to Euroslate roof.*

Table 8.1 *Schedule of rainwater goods.*

Material	Sizes (mm)	Lengths (m)	Spacing of brackets (m)	Joints (for gutters) or fixing to walls (for downpipes)
Gutters				
PVC-U	h/r 112/150/170 ogee 120 × 75	2/4	1 maximum (2/25 × 5 sherardised round headed screws) 150 from joints	Fixing clip/union/ synthetic rubber seals
Cast iron	h/r 102/127 ogee 115 × 60 153 × 102	1.83	0.9 ogee screwed direct to fascia	Sealant and bolted together with galvanised gutter bolt and nuts M6 ×20 or 50
Aluminium	h/r 110/113/125 ogee 155 × 100	1.83	0.6	Snap fit joints (extruded) spigot and socket bolted and silicone sealant
GRP	h/r 110/125/135 ogee 160 × 90	1.80	0.9	Spigot and socket, sealant and bolted
Downpipes				
PVC-U	circ. 68 sq. 65 × 65	2.5/3/5.5	Spigot and socket unsealed	Pipe clips maximum 1.83 apart
Cast iron	circ. 64–102 rect. 76 × 76 to 153 × 102	1.83	Unsealed	Standard brackets or ears to sockets screwed to wall, 1.83 apart
Aluminium	circ. 63/76/102 sq. 72 × 72 to 102 × 102	1/2/3	Caulking foam silicone and sealant	Standard brackets or eared sockets Nr. 16 × 76 dome headed pipe screws
GRP	circ. 50/75/100 rect. 65 × 65 to 150 × 100	1.80	Pointed with mastic	Eared sockets with screws, seals, washers and covers

Key: h/r = half round; circ. = circular; rect. = rectangular; sq. = square.

The schedule given in table 8.1 lists the essential details of downpipes and gutters in four selected materials.

The design of rainwater systems can be either based on the requirements of Building Regulations, Approved Document H3 (1990) or BS 6367: 1983,

Code of Practice for Drainage of Roofs and Paved Areas, based on gutters laid level. PVC-U, cast iron and aluminium rainwater goods are measured in the roof examples 12, 15 and 16 respectively.

Eaves or verge soffit boards and fascia boards (including barge boards), and gutter boards (including sides) are each measured separately in metres where the width is ≤300, with a dimensioned overall cross section description. Where the width is >300, there is an option to measure either in m^2 or in metres. If measured in m^2 a dimensioned description is required (G20:14–16.1–3.1–2.0). Adjustments will have to be made to the length measured on the outside face of the external walls for corners to arrive at the correct girth of fascia board. The student must also be sure to deduct the width of gables, where no fascia is required.

The supporting bearers to eaves and soffit boarding are measured by length as individual supports, giving a dimensioned overall cross section description (G20:13.0.1.0). Remember to take the painting of the exposed timber surfaces as general surfaces (M60:1.0.1–2.0).

Both downpipes and eaves gutters are measured in metres over all fittings and branches, with the fittings, made bends and special joints and connections enumerated as extra over the pipe or gutter in which they occur (R10:1.1.1.1, R10:10.1.1.1, R10:2.1–4, R10:11.1–2.1.1 and R10:M1 and M6). Pipe and gutter types, nominal sizes, method of jointing, and type, spacing and method of fixing supports shall be included in the descriptions of pipes and gutters. The pipe and gutter items will be followed by their painting (M60:8.2.2.0 and M60:9.0.2.0). Painting work on surfaces >300 girth is given in m^2, with that to narrower isolated surfaces in metres.

FLAT ROOF COVERINGS

Asphalt

The rules for the measurement of asphalt coverings to flat roofs are detailed in work section J21. The main areas of asphalt roofing are measured in m^2, stating the pitch. In the case of flat roofs, the pitch is usually described as the ratio of one vertical unit to a number of horizontal units, such as 1 in 40. The particulars listed in J21:S1–4 must be included and these cover the kind, quality and size of materials, thickness and number of coats, nature of base on which applied and surface treatments. Furthermore, the asphalt item is deemed to include cutting, notching, bending and extra material for lapping the underlay and reinforcement, and working the asphalt into recesses (J21:C1b and c). Linear items will follow the coverings, such as skirtings, fascias and aprons, linings to gutters, channels and valleys, coverings to kerbs, internal angle fillets and appropriate labours (J21:5–17), followed by any enumerated items such as collars around pipes, linings to cesspools and sumps, and roof ventilators (J21.18–20 and 23).

Built up Felt

Built up felt roof coverings are measured in m², and stating the pitch as J41:2.1.0.0. Full particulars of the felt are to be given including the kind and quality of felt, nature of base on which applied and method of jointing (J41:S1–3). The measurement of the main areas of roof covering will be followed by such linear items as are appropriate, giving the girth in 200 mm stages (J41:4–15.2.0.0), finishing with any enumerated items (J41:16–19.1.0.0).

Sheet Metal

The measurement of sheet metal roofing in lead, aluminium, copper, zinc and stainless steel is dealt with in work sections H71–75. The main areas of sheet metal are taken first, to be billed in m², stating the pitch, and allowing for the additional material at drips, welts, rolls, seams, laps and upstands/downstands, in accordance with the allowances given in H71–75:M2. Full particulars of the sheet metal must be given as required by H71:–75:S1–6, including the underlay, thickness, weight and temper grade, method of fixing, details of laps, drips, welts, beads, rolls, upstands and downstands and the like, and type of support materials and special finishes.

Flashings, aprons and cappings and the like are measured in metres, usually with a dimensioned description (although a dimensioned diagram may be used as an alternative), and stating whether horizontal, sloping, vertical, stepped or preformed (H71–75:10–18.1.0.1–5). Gutters and welted, beaded and shaped edges are also measured as linear items. Saddles, soakers and slates, hatch covers and ventilators are each enumerated with a dimensioned description (H71–75:25–28.1.0.0).

WORKED EXAMPLES

Worked examples follow covering the measurement of a tiled traditional cut pitched roof, adjustment of roofing for a chimney stack, lead covered flat roof to bay, slated trussed rafter pitched roof, felt covered timber flat roof and asphalt covered concrete flat roof, together with the associated rainwater goods.

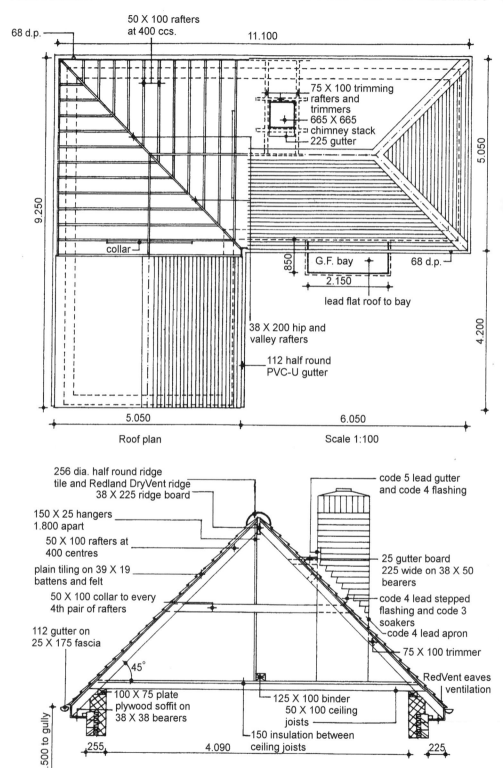

50 X 100 rafters
at 400 ccs.

68 d.p.

11.100

75 X 100 trimming
rafters and
trimmers
665 X 665
chimney stack
225 gutter

9.250

5.050

collar

850

G.F. bay

68 d.p.

2.150

lead flat roof to bay

38 X 200 hip and
valley rafters

4.200

112 half round
PVC-U gutter

5.050

6.050

Roof plan   Scale 1:100

256 dia. half round ridge
tile and Redland DryVent ridge
38 X 225 ridge board

code 5 lead gutter
and code 4 flashing

150 X 25 hangers
1.800 apart

50 X 100 rafters at
400 centres

plain tiling on 39 X 19
battens and felt

50 X 100 collar to every
4th pair of rafters

112 gutter on
25 X 175 fascia

45°

25 gutter board
225 wide on 38 X 50
bearers

code 4 lead stepped
flashing and code 3
soakers

code 4 lead apron

75 X 100 trimmer

RedVent eaves
ventilation

4.500 to gully

100 X 75 plate
plywood soffit on
38 X 38 bearers

255

4.090

125 X 100 binder
50 X 100 ceiling
joists

150 insulation between
ceiling joists

225

Section   Scale 1:50

158

Drg. 8	Pitched Tiled Roof 1 Example 12	The order of measurement adopted is: (1) construction; (2) coverings; and (3) eaves and rainwater goods.

Pitched Roof
Constn.

Plate

	11.100
less o'hg at eaves 2/225	450
	10.650
	9.250
less eaves & verge	300
	8.950
	10.650
2/	19.600
len. on ext. face	39.200
less crnrs. 4/2/255	2.040
len. on int. face	37.160
less gable	4.090
	33.070
add laps at crnrs. 3/2/100 600	
	33.670

With the roof timbers adopt a logical sequence such as plates (separate SMM7 classification), rafters, ceiling joists, collars, purlins, ceiling beams, struts, ridge board and hip and valley rafters in a traditional cut roof. Note the detailed build up of length of plate, making allowance for the gable and laps at corners, measuring to the extremities of the timber. No allowance is made for laps in running lengths. The timber now carries the European classification of C16 replacing the original SC3. In the case of plates, external angles do not cancel out internal angles and the lapped length at the external angle needs to be added.

Sn. swd. grade C16 to
BS 5268; vacuum
impregntd. w. preservative

| 33.67 | Plates
100 × 75 | Plates are classified as G20:8.0.1.0 with a dimensioned description. |
|---|---|---|

Common rafters

len. – ½ o'll span × sec. 15° +
       spld. ends

= 2.525 × 1.414 + 0.075

= 3.645

number	11.100
less o'hg 2/225 450	
walls 2/255 510	
end spaces 2/50 100	1.060
400)	10.040
	25 + 1
	4.200

less end space 50		
gable wall 215		
verge	75	340
26	400)	3.860
11		10 + 1
37		

Note the method used for calculating the lengths of rafters to their extremities. The number of pairs of rafters in the main length of the roof is obtained from the number of spaces, with one added to convert spaces into pairs of rafters. Next proceed to calculate the number of rafters in the bottom leg of the roof.

Pitched Tiled Roof 2

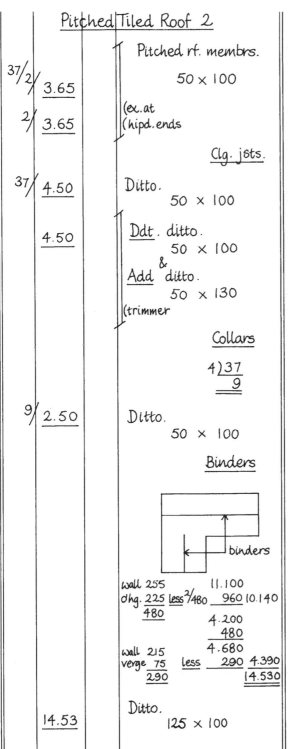

Pitched rf. membrs.

50 × 100

³⁷/₂/ 3.65

²/ 3.65

(ex.at
(hipd.ends

Two additional rafters are added for the extra ones required at the hips.

All roof timbers, except plates, are classified as 'pitched roof members', giving a dimensioned description as G20:9.2.1.0. The wording of the description follows that contained in SMM7 and the *SMM7 Library of Standard Descriptions*, based on the format of the *Common Arrangement* (CPI).

Clg. jsts.

³⁷/ 4.50

Ditto.
50 × 100

Ddt. ditto.
50 × 100
&
Add ditto.
50 × 130

4.50

(trimmer

The lengths of ceiling joists are scaled off the 1:50 section. A larger member is needed to support the ends of ceiling joists across the intersection of the two roofs. All labours on members are deemed to be included (G20:C1).

Collars

4)37
  9

The collar calculation is rather approximate as the presence of the hipped corner prevents the use of a collar for some distance from the corner. One collar is provided to each fourth pair of rafters.

⁹/ 2.50

Ditto.
50 × 100

Binders

Note the method of setting out the calculations for the length of binders in waste.

```
wall  255          11.100
dhg.  225  less ²/480   960  10.140
      480
               4.200
                480
wall  215      4.680
verge  75   less    290   4.390
      290          14.530
```

It is assumed that no lengths exceeding 6.000 will be required. No distinction is made between members of different types in bill descriptions.

Ditto.
125 × 100

14.53

Trimming of roof timbers around the chimney stack will be taken later in roofwork adjustment for the chimney stack in example 13.

Pitched Tiled Roof 3

Ridge bd.

	11.100
less 2/½/5.050	5.050
	6.050

	9.250
less ½/5.050	2.525
	6.725
less	290
	6.435
add bearg.	100
	6.535

6.05	**Pitched rf. membrs.**
	38 × 225 ; in one
	continuous len.: 6.05 m

6.54	**Ditto**
	38 × 225; in one
	continuous len.: 6.54m

Calculation of the lengths of ridge board are shown in waste. Cutting and fitting ends of rafters against the ridge boards and over wall plates are deemed to be included, as they are covered by G20:C1.

Ridge boards are dealt with in the same way as other roof members as G20:9.2.1.1, keeping members >6.000 in one continuous length separate with the length stated.

Hangers

	6.050
	6.435
1.800)	12.485
	7 + 1

8/ 2.08	**Ditto**
	150 × 25

The hangers are spaced 1.800 apart. Their length is scaled from the 1:50 section. The adjusted ridge board length is used to calculate the number of hangers.

Hip & valley rafters

	5.000
add tapd. end	75
	5.075

4/ 5.08	**Ditto**
	38 × 200

The length of hip rafter is found by plotting the height of the roof on plan at the hipped end and drawing the hip to the slope of the roof, as illustrated in figure 12, and adding an allowance for a tapered end. Alternatively, the length can be calculated.

Pitched Tiled Roof 4

Insulation

		len. of side rf.	
	255	4.200	
add intersectn.	225	480	
		4.680	
less gable	215		
jsts. ¹¹/50	75		
	550	840	
		3.840	

	len. of main rf.	width	
	11.100	5.050	
less ²/480 960		960	4.090
jsts.²⁶/50 1.300	2.260		
	8.840		
add			
²/o'lap at			
eaves ²/350	700		700
	9.540		4.790

Build up of dimensions of insulation above the ceiling in the roof void. In arriving at the length of the insulation it is necessary to deduct the thickness of the ceiling joists, as the insulating quilt is to be laid between the ceiling joists and the area measured is to be the area covered by insulation (P10:M1). The overlap at eaves is that shown in figure 29.

9.54	Mineral fibre quilt insul.
4.79	150 th., to BS 5803, Pt. 1,
3.84	Ritemk. certified
4.79	Betw. membrs.
	membrs. at 400 ccs.;
	hor.; w. 100 laps;
	installed to BS 5803, Pt. 5.

The quilt insulation laid between members is measured in m² in accordance with P10:2.3.1.0, and stating the type, quality and thickness of insulating material as P10:S1.

Roof Coverings

	11.100
add proj. of tilg.	
over gutter ²/40	80
	11.180
	4.200
less proj. of tilg.	40
	4.160

Redland rosemary clay plain brindled tiles, 265 × 165; double lapd. 65 min. hd. lap. nailg. w. two 40 mm × 11 g alum. nls. per tile every 4ᵗʰ cos. & at eaves, 38 × 25 sn. swd. battens vac. impregntd. w. preservative to BS 4072 at 100 ccs.; fxd. w. galv. nls.; reinfd. bit. felt underlay to BS 747 type 1F, fxd. w. galvd. clout nls.

Roof tiling is measured in m² for the surface area of the roof, including the projections over eaves gutters. The length on slope has been obtained from the 1:50 section, but it could be calculated in the same way as for the rafters. The roof tiling description encompasses the pitch (H60:1.1.0.0) and the particulars listed in H60:S1–4, including the kind, quality and size of materials, method of fixing, minimum laps and spacing of battens. Coverings are deemed to include underlay and battens (H60:C1), but the materials involved must be mentioned in the description (General Rules 2.11).

Pitched Tiled Roof 5

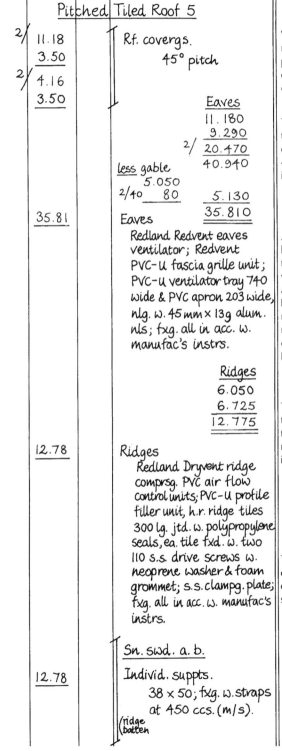

2/	11.18	
	3.50	
2/	4.16	
	3.50	

Rf. covergs.
45° pitch

With the detailed specification requirements contained in the previous heading, the item description is confined to roof covering and pitch.

Eaves
11.180
9.290
2/ 20.470
40.940

less gable
5.050
2/40   80   5.130
35.810

35.81 Eaves

The eaves lengths are built up from the figured dimensions on the drawing and include an additional 40 mm for the projection of roof tiles into the eaves gutters.

Redland Redvent eaves ventilator; Redvent PVC-U fascia grille unit; PVC-U ventilator tray 740 wide & PVC apron 203 wide, nlg. w. 45 mm × 13g alum. nls; fxg. all in acc. w. manufac's instrs.

A Redland dry tech arrangement has been adopted for the eaves to show the method of measurement, as the ventilation of roof voids has become a statutory requirement under the Building Regulations, to eliminate the risk of excessive condensation. The relevant measurement rules are contained in H60:4.0.0.0 and H60:6.0.0.0.

Ridges
6.050
6.725
12.775

12.78 Ridges
Redland Dryvent ridge comprsg. PVC air flow control units; PVC-U profile filler unit, h.r. ridge tiles 300 lg. jtd. w. polypropylene seals, ea. tile fxd. w. two 110 s.s. drive screws w. neoprene washer & foam grommet; s.s. clampg. plate; fxg. all in acc. w. manufac's instrs.

The ridge lengths are derived from the previously calculated lengths for the ridge board. A traditional form of mortar bedded ridge tiling is covered in example 15.

The Redland DryVent ridge system entails the inclusion of a ridge board extension supported by stainless steel straps as illustrated in figure 29.

Sn. swd. a. b.

12.78 Individ. suppts.
38 × 50; fxg. w. straps at 450 ccs. (m/s).
(ridge
(batten

Pitched Tiled Roof 6

		450) 12.775
		28 + 1

Stainless steel

29	Straps
	Redland ridge batten
	straps; bendg. ard.
	ridge batten & nlg.
	w. s.s. clout nls.

The number of straps is found by dividing the total length of ridge by the spacing of the straps (450 centres) and adding one to convert spaces to straps.

The straps are measured in accordance with G20:20.1.0.0, although the reference to a manufactured product eliminates the need for a dimensioned description.

Hips

	add tile proj.	5.000
		40
		5.040

Redland rf. tilg. a. b.

3/ 5.04	Hips
	Redland h.r. ridge
	tiles; 256 dia. ×
	300 lg.; b & p
	in c.m. (1 : 3).

The hips are measured as H60:7 in metres, giving the appropriate particulars with regard to kind, quality and size of tiles and method of fixing (H60:S1–2).

3	Galv. steel
	Hip irons to BS 5534, pt. 1
	scrg. to tbr.

Hip irons are enumerated as H60:10.4.1.0.

Redland rf. tilg. a. b.

5.04	Valleys
	Redland p.m. plain
	valley tiles; nld.
	to battens.

Purpose made plain valley tiles are taken as covering to the valley, using tiles that match the plain tiles.

2/ 3.50	Verges
	Tile-&-a-half width
	tiles & ex. underclk.
	cos.; b & p & fin.
	w. fillet in c.m.(1 :3)

A description of the work required in forming the verges having regard to the tiling manufacturer's recommendations. H60:C2 states that undercloaks and bedding and pointing are deemed to be included, but details must be given of the method of forming (H60:S5).

Pitched Tiled Roof 7

<table>
<tr><td></td><td colspan="2" align="right">Fascia & Eaves
Boards</td></tr>
<tr><td></td><td colspan="2">Wrot. swd.</td></tr>
<tr><td>35.81</td><td colspan="2">Fascia bds.
25 x 175; grvd.</td></tr>
<tr><td></td><td>len. on ext. face</td><td>39.200</td></tr>
<tr><td></td><td>less gable 4.090</td><td></td></tr>
<tr><td></td><td>2/255   510</td><td>4.600</td></tr>
<tr><td></td><td></td><td>34.600</td></tr>
<tr><td></td><td>add cnrs. 3/2/210</td><td>1.260</td></tr>
<tr><td></td><td>extreme len.</td><td>35.860</td></tr>
</table>

Plywd. to BS 6566
w. WBP bondg.

35.86	Eaves soff. bds. 15 x 210

Sn. swd. a.b.

34.60	Individ. suppts. 38 x 38; plugd. to bwk.

Wrot. swd.

<table>
<tr><td>2/ 1</td><td>Eaves gusset ends
225 x 250 (av.) x 25;
irreg. shape</td><td></td></tr>
<tr><td></td><td>2/175</td><td>350</td></tr>
<tr><td></td><td>2/25</td><td>50</td></tr>
<tr><td></td><td></td><td>400</td></tr>
</table>

Knot & prime on wd.

<table>
<tr><td>35.81
0.40</td><td>Gen. surfs.
>300 gth.; on site
prior to fxg.</td><td></td></tr>
<tr><td></td><td>(fascia   2/210</td><td>410</td></tr>
<tr><td></td><td>2/15</td><td>30</td></tr>
<tr><td></td><td></td><td>450</td></tr>
<tr><td></td><td></td><td>34.600</td></tr>
<tr><td></td><td>add cnrs. 2/210</td><td>420</td></tr>
<tr><td></td><td>mean len.</td><td>35.020</td></tr>
</table>

The length of the eaves has been used for the fascia boards, as the difference is minimal. Avoid splitting hairs when measuring building work and working to unreasonably precise dimensions.

For the soffit boarding, the length on the external face of the brickwork has been taken as the starting point, with subsequent adjustments for the gable and the width of the soffit boarding, which is let into a groove in the fascia. The relevant measurement requirements are contained in G20:15.3.2.0 and G20:16.3.2.0.

The bearer supporting the eaves soffit is classified as 'individual supports' and measured in accordance with G20:13.0.1.0.

Eaves gusset ends are enumerated with a dimensioned description (G20:18.0.1.0).

Priming on all surfaces of exterior timber members on site prior to fixing is advisable. The build up of the exterior girth is entered in waste. The measurement requirements for this work are contained in M60:1.0.1.4, as the girth exceeds 300.

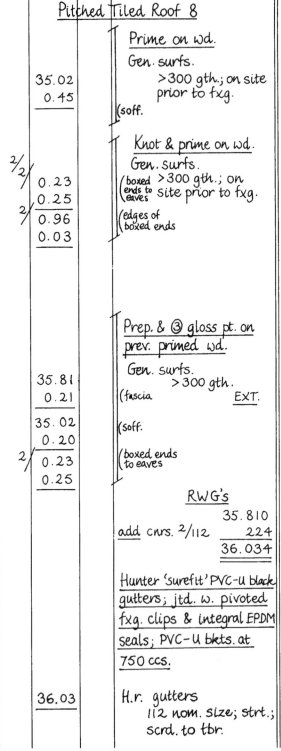

Pitched Tiled Roof 8

Prime on wd.
Gen. surfs.
>300 gth.; on site
prior to fxg.
(soff.

35.02	
0.45	

The preparatory painting work to the soffit is kept separate from the fascia as no knotting is required on the plywood surface.

Knot & prime on wd.
Gen. surfs.
(boxed >300 gth.; on
ends to site prior to fxg.
eaves
(edges of
boxed ends

²/₂/ 0.23
0.25
²/ 0.96
0.03

The painting of the boxed or spandrel ends to the eaves is classified in the >300 girth range, as the total dimension measured on all faces exceeds this figure. Even if this had not been the case, it should still be in the higher classification as it can be painted together with the fascia board and soffit and is not an isolated item. The same approach applies to the gloss painting.

Prep. & ③ gloss pt. on
prev. primed wd.
Gen. surfs.
>300 gth.
(fascia   EXT.
(soff.
(boxed ends
to eaves

35.81
0.21

35.02
0.20
²/ 0.23
0.25

The final preparation of the wood surfaces and application of three coats of gloss paint are taken to all the exposed surfaces and classified as M60:1.0.1.0. The work must be described as external as otherwise it will be deemed to be internal and could generate a lower price (M60:D1).

RWG's

	35.810
add cnrs. ²/112	224
	36.034

Hunter 'surefit' PVC-U black
gutters; jtd. w. pivoted
fxg. clips & integral EPDM
seals; PVC-U bkts. at
750 ccs.

36.03

H.r. gutters
112 nom. size; strt.;
scrd. to tbr.

Calculation of gutter length from length of fascia board. Gutters are classified as R10:10.1.1.1, stating the type, nominal size, method of jointing, type, spacing and method of fixing supports. A particular product has been specified in this case, but alternatively a gutter to BS 4576, Part 1 could be specified. The black coloured gutter will not require painting.

Pitched Tiled Roof 9

PVC-U gutters a.b.

These fittings are enumerated in accordance with R10:11.2.1.0, stating the type of fitting as extra over the gutter in which they occur. Balloon gratings can be included in the outlets item if required.

4 | E.O. for
angles

2 | Ditto. for
outlets

2 | Ditto. for
s.e.'s

	4.500
add for	
offset bend	150
	4.650

Downpipes are measured in metres over all fittings and descriptions are to include the type, nominal size, method of jointing, type, spacing and method of fixing supports, as R10:1.1.1.1. The fixing to backgrounds is described in accordance with General Rule 8.3.

Hunter PVC-U black pipewk. w. pushfit jts.; PVC-U bkts. at 1.800 ccs.

2/ 4.65 | Pipes
68 nom. size; strt.; plug'g. & scrg. to masonry

2/ 2 | E.O. for
offset bends

&

Ditto. for
connectn. to clayware gully w.c.m. (1:3)

Offset bends, two needed to form a swanneck, are enumerated as extra over the pipe in which they occur as R10:2.4.5.0. The connection of the bottom end of downpipes to gullies is measured as R10:2.2.1.0.

Drg. 8   Example 13
Roofwork Adjustment for Chimney Stack

Comm. rafters | Softwood heading as before, with grade C16 superseding SC3.

Ddt.
Sn. swd. grade C16
to BS 5268; vac.
impregntd. w.
preservative

Pitched rf. membrs.
50 × 100

| 1.17 | |

Start with adjustment of structural timbers for the chimney stack opening. A single rafter is deducted across the chimney stack with the length scaled between the centre lines of the trimmers. This is followed by the substitution of the thicker trimming timbers for those previously measured. There is no requirement to describe them as trimming members.

Ddt ditto.
50 × 100

2/ 3.65

&

Add ditto.
75 × 100
(trimmg.
(rafters

	len.
	665
add spaces 2/50	100
trimmg. rafters 2/75	150
t. tenons 2/150	300
	1.215

The build up of the length of trimmers including tusk tenon joints. The alternative is to use joist hangers as illustrated in example 9.

2/ 1.22

Add ditto.
75 × 100
(trimmers

Deduction of tiling and the measurement of extra work to the boundaries around the stack (abutments and eaves) only operate when the void is >1.00 m² (H60:M1–2). In this case the area is $1.25 \times 0.67 = 0.84\,m^2$, and hence no adjustments are made. Coverings are deemed to include work in forming voids ≤1.00 m², other than holes (H60:C1b).

Note: An identical procedure will follow for the adjustment of ceiling joists; the dimensions have not been included in this example as it is largely repetitive.

Wrot. swd.

Gutter bds.
225 × 25

0.67

Sn. swd.

Individ. suppts.
38 × 50
(gutter bd. brrs.

3/ 0.37

Gutter boarding is so classified giving an overall dimensioned cross section (G20:14.3.2.0); when >300 wide it may be measured in m² giving a dimensioned description, or in metres giving an overall dimensioned cross section.

The gutter board bearers are classified as 'individual supports' as G20:13.0.1.0.

Roofwork Adjustment 2

	<u>Wrot. swd.</u> Gutter bds. 300 × 25 (lier bd.	Length of lier board provided up the roof slope to support the lead, where a 930 length is needed to bridge the trimming rafters. Included in gutter board classifications as side to gutter (G20:M3&D9).
0.93		

2/ 1	Gusset ends to chy. gutter
	   irregular shape;
	  225 × 225 × 25 o/a.
		Gusset end is an enumerated item as G20:17.0.1.0, giving a dimensioned description.

<u>Lead sheet covergs./flashgs.</u>

                      <u>gutter len.</u>

add passgs.(ends)   665
           2/150   <u>300</u>
                   <u>965</u>

Note build up of dimensions of lead lining to gutter, calculated where appropriate in accordance with the table of allowances in H71:M2. The 500 allowance for upstands in SMM7 seems high as a more usual dimension would be 150.

                    <u>width</u>

upstd. to wall   500
gutter bott.   225
up rf. slope   <u>300</u>
                <u>1.025</u>

	<u>Lead code 5 to BS 1178</u>
	Gutters
	  1.025 gth.; irreg.
	  shape w. slopg. side;
	  cop. nld. at 50 ccs. to
	  swd. at top edge.
		The gutter is measured as a linear item in accordance with H71:19.1.0.1–4. The description could conceivably be reduced in length by the inclusion of 'tapering'. The type and quality of material and method of fixing are to be stated (H71:S1&3).

0.97		

                 <u>flashg. len.</u>

add passgs.(ends)   665
         2/100   <u>200</u>
                 <u>865</u>

The substance of the material required by H71:S2 is provided by the code reference in the description.

	<u>Lead code 4 to BS 1178</u>
	Flashgs.
	  150 gth.; hor.; fxg. w.
	  ld. tacks & ld. wedg'g.
	  into bk. jt.
	        &
	  <u>Ct. mtr. (1 : 3)</u>
		Flashing is measured as a linear item as H71:10.1.0.1, giving a dimensioned description and the plane in which the lead is laid.

0.87		
	<u>Ptg. in flashgs.</u>	The pointing in flashings measurement procedure is prescribed in F30:6.0.0.0. This is deemed to include cutting or forming grooves or chases (F30:C4). Passings of flashings are added in arriving at the length.

                  <u>apron len.</u>

add passgs.(ends)   665
         2/100   <u>200</u>
                 <u>865</u>

Roofwork | Adjustment 3

Lead code 4 to BS 1178

0.87	Aprons 300 gth.; hor.; fxg. w. ld. tacks & ld. wedg'g. into bk. jt.; dressg. over slating & tilg.

&

Ct. mtr. (1:3)
Ptg. in flashgs.

| The lead apron is measured in accordance with H71:11.1.0.1&7, giving a dimensioned description. Fixing is included in the description in accordance with H71:S3. The lead codes specified are the minimum lead thicknesses recommended in BS 5534 and by the Lead Development Association. |

The length on slope is scaled from the section.

stepped flashgs.

add passgs.(ends) 2/100

| 1.000 |
| 200 |
| 1.200 |

Lead code 4 to BS 1178

2/ 1.20	Flashgs. 200gth.; stepped; fxg. w. ld. tacks & ld. wedg'g. into bk. jt.

Stepped flashings are measured in accordance with H71:10.1.0.4.

&

Ct. mtr. (1:3)
Ptg. in flashgs.

Pointing forms a linear item as F30:6.

soakers

100)1.250
   13

len.
lap+ gauge +25
= 65 + 100+25
= 190

The length of slope is divided by the gauge of the tiling to give the number of soakers on each side of the stack. The girth of the soakers is normally 150 (100 horizontally and 50 vertically).

Lead code 3 to BS 1178

2/13/ 1	Soakers 190 x 150 o/a; in slatg. or pl. tilg.; hand to others for fxg.

An enumerated item for soakers with a dimensioned description as H71:26.1.1.1.

&

Soakers
190 x 150 o/a;
fxg. only

Separate enumerated fixing item by a different trade (roof tiler) as H60:10.5.1.1.

Drg. 8			**Example 14**

Lead Covering to Flat Roof to Bay 1

Lead sheet coverg.

			len.
			2.150
		add rolls 2/250	500
		turn dn. (gutter)	
		2/50	100
			2.750
			width
			850
		add upstand	500
		turn dn.	50
		(gutter)	
			1.400

Lead code 6 to BS 1178

2.75		Rf. covergs.	
1.40		pitch : 1 in 54;	
		cop. nld. to swd. bdg.	
		on 2 edges of ea.	
		sheet.	
			2.150
		add passgs. (ends)	
		2/100	200
			2.350

Lead code 4 to BS 1178

		Flashgs.	
2.35		150 gth.; hor.; fxg.	
		w. ld. tacks & ld.	
		wedg'g. into bk. jt.	
		&	
		Ct. mtr. (1:3)	
		Ptg. in flashgs.	

The lead sheet covering to the ground floor bay has been measured to show the approach to this class of work. The measurement of the roof timbers, insulation and rainwater goods have been omitted as they are adequately covered elsewhere and there are insufficient details provided. Allowances for rolls and upstands are made in accordance with the table of allowances in H71:M2, despite the excessive amount provided for upstands. Assuming that rolls are provided at 750 centres, there will be two rolls in the length of the roof. Drips are usually provided in the opposite direction at 2.000 centres; hence no drips are required in this example.

Lead sheet flat roof coverings are measured in m², giving the pitch, type and quality of materials and method of fixing and type of support materials (H71:1.1.0.0).

A flashing will be required to ensure a watertight joint between the lead upstand at the back of the roof and the adjoining brickwork. The flashing is measured as a linear item, giving the particulars required in H71:10.1.0.1 and the pointing follows as F30:6.

Pitched trussed rafter slated roof   **DRAWING 9**

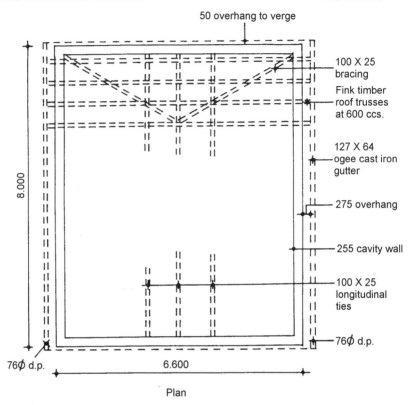

50 overhang to verge

100 X 25 bracing

Fink timber roof trusses at 600 ccs.

127 X 64 ogee cast iron gutter

275 overhang

255 cavity wall

100 X 25 longitudinal ties

76⌀ d.p.

8.000

76⌀ d.p.

6.600

Plan

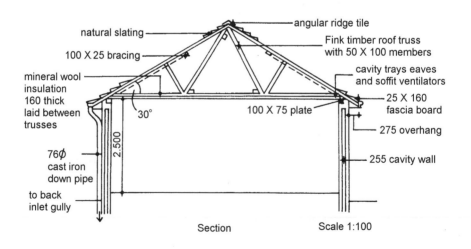

angular ridge tile

natural slating

Fink timber roof truss with 50 X 100 members

100 X 25 bracing

cavity trays eaves and soffit ventilators

mineral wool insulation 160 thick laid between trusses

25 X 160 fascia board

100 X 75 plate

275 overhang

30°

255 cavity wall

76⌀ cast iron down pipe

2.500

to back inlet gully

Section   Scale 1:100

Drg. 9
Pitched Trussed Rafter Slated Roof 1

Example 15

As the building has gables at both ends, plates are only required on the longer walls and will terminate on the inner face of the gable walls.

Construction

Plates

```
                    8.000
less gable walls
         2/255      510
              2/  7.490
                  14.980
```

The usual heading for softwood is provided, with which the reader will now be coming familiar.

Sn. swd. grade C16
to BS 5268; vac.
impregntd. w. preservative

14.98	Plates

Plates
   100 × 75

Plates are measured in accordance with G20:8.0.1.0, including a dimensioned description. It is assumed that continuous lengths >6.00 are not required.

```
                    7.490
less end spaces
         2/50       100
              600) 7.390
                   13 + 1
less cav. walls  6.600
         2/255     510
clear span       6.090
```

G20:2.1.0.0 requires a dimensioned description to be provided for trussed rafters, but General Rules 5.3 permits dimensioned diagrams showing the shape and dimensions of the work to be used in place of dimensioned descriptions. Moreover, the *SMM7 Measurement Code* (4.4, page 9) states that there may be occasions where it is more appropriate to issue architect's or engineer's drawings with the bill of quantities, rather than producing dimensioned diagrams, subject to identifying the drawings in the bill description.

Dimensioned
Description Approach

Trussed rafters

14	

Fink pattern;
50 × 100 membrs.
comprisg. 2nr. rftrs.,
2 nr. ties, 2nr. struts,
1 nr. clg. jst.; 6.090
clear span; 275 eaves
projn.; 30° pitch; jtd.
w. 18g s.s. nl. plates

Alternative forms of approach are provided for the benefit of the reader. The dimensioned description includes the pattern, size of members, number and type of members, clear span, eaves projection, pitch and form of jointing.

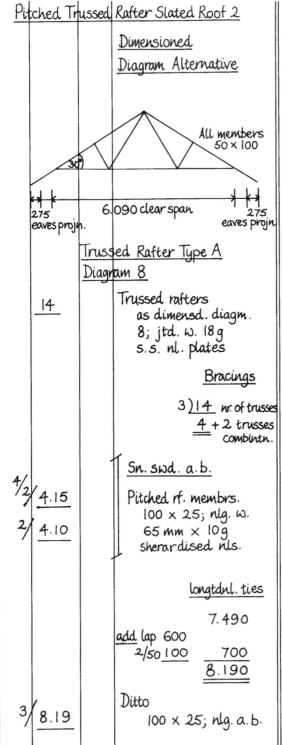

Pitched Trussed Rafter Slated Roof 2

Dimensioned
Diagram Alternative

All members 50 × 100

275
eaves projn.

6.090 clear span

275
eaves projn.

30°

The dimensioned diagram of a trussed rafter is kept quite simple, showing the disposition and sizes of members, clear span and eaves projection.

Trussed Rafter Type A
Diagram 8

14

Trussed rafters
as dimensd. diagm.
8; jtd. w. 18 g
s.s. nl. plates

The trussed rafters are enumerated with reference to the dimensioned diagram and stating the method of jointing members.

Bracings

3)14 nr. of trusses
4 + 2 trusses
combintn.

Sn. swd. a.b.

⁴⁄₂/ 4.15

2/ 4.10

Pitched rf. membrs.
100 × 25; nlg. w.
65 mm × 10 g
sherardised nls.

The diagonal bracings to withstand excessive wind pressures, as outlined in BS 5268, Part 3 are measured in accordance with G20:9.2.1.0. The lengths are calculated by drawing a right angled triangle with the base representing bracing on plan and the vertical side the rise of the roof. The hypotenuse then represents the length of bracing. The bracings connect each set of three trussed rafters leaving two rafters with slightly shorter bracings.

longtdnl. ties

7.490
add lap 600
2/50 100   700
8.190

The longitudinal ties are taken under the same classification, with a substantial lap added to each length (spacing of rafters + thickness of two rafters).

3/ 8.19

Ditto
100 × 25; nlg. a.b.

Pitched Trussed Rafter Slated Roof 3

14/2/ 1

<u>Galvd. steel</u>

I.G. Metal Fixings
truss clips, reference
GTS 50
    fxg. w. 30 × 3.75
    sq. twisted
    sherardised nls.
(trussed rafters
(to wall plate

There is one truss clip to each connection between a trussed rafter and the wall plate to give increased rigidity. The measurement provision is in G20:24.1.0.0, which requires a dimensioned description to be given. It is considered advisable to specify a named product.

2.000)7.490
4+1

5/2/ 1

I.G. Metal Fixings
vert. restraint straps
    30 × 5 × 800 gth.;
    nlg. to tbr. & plg'g.
    & scrg. to blkwk.
(wall plate
(to wall

Vertical galvanised steel restraint straps are provided at 2.000 centres to prevent movement of the wall plate by rigidly fixing it to the blockwork below. G20:20.1.0.0 requires the provision of a dimensioned description or dimensioned diagram and G20:S2 refers to the method of fixing.

<u>Insulation</u>

<u>Rockwall mineral wool
roll insulatn.; 600 wide
& 100 th.</u>

7.49
6.60

Betwn. membrs.
    membrs. at 600 ccs.;
    hor.; installed to
    BS 5803, Pt.5

The insulation is provided in two layers: the bottom layer between the ceiling joists and the upper layer continuous over the joists. A total insulation thickness of 160 is deemed necessary to satisfy the Building Regulations, Approved Document L (1995).

<u>Rockwall mineral wool
mat. insulatn., 60th.</u>

7.49
6.09

Across membrs.
    membrs. at 600 ccs.;
    hor.; installed to
    BS 5803, Pt.5

No deductions are made for the thickness of the joists when measuring the area of the bottom layer, as the 600 wide rolls can be squeezed in between the joists. It also entails a larger area being taken over the top of the external wall in the manner shown in figure 30. The bottom insulation is classified as P10:2.3.1.0 and the top insulation as P10:2.2.1.0. The area measured is that covered (P10:M1).

Pitched Trussed Rafter Slated Roof 4

			Roof Coverings

There is an extensive build up of dimensions in waste to arrive at the area of slating required on each roof slope. The recommended projection of slates into gutters is 50 and a maximum unsupported overhang at verges of 50.

```
                        slope length
                          6.600
     add eaves
         projns.
              2/275        550
                      2) 7.150
                         3.575
        sec. 30°   1.1547
                   4.128
   add eaves slate    50
   proj. over gutter  ____
                   4.178


                     rf. length
                      8.000
   add verge o'hg 2/50  100
                      _____
                      8.100
```

510 × 255 Burlington
blue/grey natural slates;
75 min. hd. lap; nlg.
ea. slate w. two 50 mm
cop. nls. as BS 1202, pt. 2;
50 × 25 sn. swd. battens,
vac. impreg. w. preservative
at 220 max. ccs., fxd. w.
60 mm × 2.64g sherardised
wire nls. to BS 1202; reinfd.
bit. felt to BS 747, type IF,
fxg. w. galv. clout nls.; all
in accordance w. BS 5534, pt. 1

A full description of the slates, battens, felt underlay and method of fixing is required in accordance with H62:1.1.0.0 and H62:S1–4. Other requirements include the roof pitch, minimum laps and spacing of battens. Slate sizes vary from 610 × 305 to 305 × 205; 510 × 255 has been selected as a commonly used size. Batten sizes vary from 38 × 25 with rafter spacings ≤450 centres to 50 × 25 for rafter spacings ≤600 centres.

			Rf. covergs.
2/	8.10		30° pitch
	4.18		
			Eaves
2/	8.10		

Eaves
Cavity Trays type V 600
eaves ventilator in
galv. stl., 580 × 180,
fxd. w. nls. to tbr. at
600 ccs.; all in accord.
w. manufac's. instrs.

In this example the roof void is ventilated using a galvanised steel continuous eaves ventilator supplied by Cavity Trays, together with circular soffit ventilators in the soffit boarding. The latter are measured with the soffit boarding. The SMM7 reference for eaves to slated roofs is H62:4.0.0.0.

Pitched Trussed Rafter Slated Roof 5

Ridge

Burlington rf. slating a.b.

8.10	Ridges Angular blue clay ridge tiles; ea. 450 lg.; bedded & ptd. in c.m. (1:3).

A common finish to the ridges of slated roofs is blue clay ridge tiles to match the blue/grey slates. The relevant measurement provision is H62:6.0.0.0, and the usual requirements as to kind, quality and size of material and method of fixing apply (H62:S1–2). There are similar requirements for verges (H62:5.0.0.0), noting that the method of forming must be given (H62:S5).

Verges

2/2/ 4.18	Verges Slate-&-a-half width slates & ex. underclk. cos.; b. & p. & fin. w. fillet in c.m. (1:3)

Fascias & Eaves Boards

2/ 8.00	Wrot. swd. Fascia bds. 25 × 160; grvd. & Eaves soffit bds. 13 × 240; v-jtd. matchbdg. & Sn. swd. a.b. Individ. suppts. 38 × 38; plugd. to bwk.

The fascia boards are measured in metres as ≤300 wide with a dimensioned overall cross section description (G20:15.3.2.0). The fascia boards are grooved to receive one side edge of the soffit boards.

The eaves soffit boards are formed of V-jointed matchboarding and similar measurement rules apply as for fascia boards (G20:16.3.2.0).

The bearer supporting the eaves soffit is classified as 'individual supports' (G20:13.0.1.0).

2/ 2	Wrot. swd. Eaves gusset ends 275 × 275 (av.) × 25; irreg. shape.
	2/160 320 2/25 50 370

The gusset ends to the eaves are enumerated, giving the overall sizes (G20:18.0.1.0).

Dimensions in waste (sidecasts) cover the overall girth of the fascia for priming purposes.

Pitched Trussed Rafter Slated Roof 6

2/ 8.00
0.37
	Knot & prime on wd.
	
	Gen. surfs.
	  >300 gth.; on site
	  prior to fxg.
	(fascia

2/ 8.00 0.51	2/240 480 2/13 26 (soff. 506
2/2/2/ 0.28 0.28	(boxed ends (to eaves
2/2/ 1.12 0.03	(edges to (boxed ends

All the eaves timbers require the preparatory treatment of knotting and priming on all surfaces on the site prior to fixing; this is measured in accordance with M60:1.0.1.4. Alternatively, the combined girth of the fascia and soffit could be calculated in waste and then entered as a single item.

2/ 8.00 0.20	**Prep. & ③ gloss pt. on** **prev. primed wd.** Gen. surfs. >300 gth. (fascia EXT.
2/ 8.00 0.24	(soff.
2/2/ 0.28 0.28	(boxed ends (to eaves

The final preparation and painting, usually three coats in number comprising two undercoats and a finishing coat, is now taken in accordance with M60:1.0.1.0. All external painting work must be clearly indicated (M60:D1). Isolated paintwork ≤300 girth is measured as linear items.

2/ 41	**Soff. ventilators**
	
	200)8.000
	Plastic 40+1
	Vents
	  Cavity Trays white
	  injectn. mo. polypropylene
	  twist & lock circ.
	  soff. ventilator (csv);
	  79/70 × 50 w. 4 mm
	  fly screeng. grid,
	  inc. formg. hole in
	  tbr. 13 th.

Soffit ventilators are enumerated with a full description including forming the hole in the soffit board. The 70 diameter hole is formed with a hole saw and fixing tool provided by the supplier. When the soffit ventilators are fixed at 200 centres this is equivalent to a gap of 10 000 mm^2 per metre and satisfies the requirements of Building Regulations, Approved Document F2 (1995).

		Pitched Trussed Rafter Slated Roof 7
		RWG's
		Alumasc Apex Heritage
		range c.i. gutters to BS 460;
		jtd. w. high qual. sealant
		& bolted tog. w. galv. gutter
		bolts & nuts size M6 x 25 mm
2/	8.10	Ogee gutters
		127 x 64 nom. size;
		strt.; scrd. to tbr.
		w. 30 mm x 16 g bright
		zinc gutter scrs. at
		900 ccs.
2/	2	E.O. for
		s.e.'s
2/	1	Ditto for
		outlets
		Prime & ③ alkyd gloss
		paint on met. surfs.
2/	8.10	Eaves gutters
	0.36	>300 gth.

EXT.

2.500
add to g.l.   200
offset   275
2.975

		Alumasc Apex Heritage
		range, c.i. pipewk. to
		BS 460; jts. wedged w. ld.
		slips; ears cast on.
2/	2.98	Pipes
		76 nom. size, strt.;
		plug'g. & scrg. to masonry
		thro. c.i. bobbins to
		give 32 paintg. gap
	2	E.O. for
		offsets 300 proj.
		&
		Ditto. for
		connectns. to clayware
		gullies w. c.m. (1:3)
		Prime & ③ alkyd gloss paint
		on met. surfs.
2/	2.98	Services
		isolated surfs. ≤ 300 gth.

EXT.

Gutters are measured in metres, stating whether straight or curved, and giving the type, nominal size, method of jointing and type, spacing and method of fixing supports, including the fixing background (R10:10.1.1.1). The length is measured over all fittings and branches (R10:M6) and is deemed to include joints in the running length (R10:C9). Ogee gutters require no brackets as they are screwed through the vertical back surface to the fascia.

Fittings to the gutter are enumerated as extra over the gutter in which they occur, stating the type of fitting (R10:11.2.1.0).

Painting to the cast iron eaves gutters is measured in m^2 where the girth is >300 and in metres when ≤300 girth (M60:8.2.1–2.0). The girth measurement embraces all inside and outside exposed surfaces.

A suitable heading is produced for the rainwater pipes, followed by the measurement of the pipes as R10:1.1.1.1, giving the type, nominal size, method of jointing, type, spacing and method of fixing supports.

Offsets and connections are enumerated as extra over the pipes in which they occur (R10:2.2&4.1&5.0).

The painting of the pipes is taken under the heading of services and is a linear item as the girth is ≤300 ($76 \times 22/7 = 239$), and is described as to isolated surfaces (M60:9.0.2.0) and external (M60:D1).

Asphalt and felt covered flat roofs   **DRAWING 10**

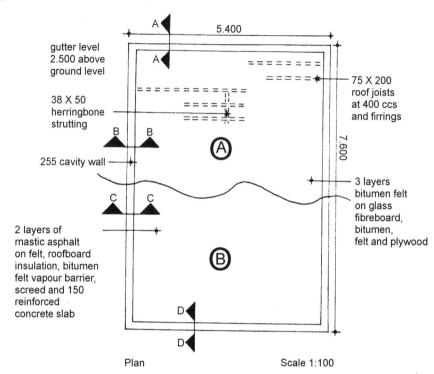

gutter level
2.500 above
ground level

38 X 50
herringbone
strutting

255 cavity wall

2 layers of
mastic asphalt
on felt, roofboard
insulation, bitumen
felt vapour barrier,
screed and 150
reinforced
concrete slab

5.400

75 X 200
roof joists
at 400 ccs
and firrings

7.600

3 layers
bitumen felt
on glass
fibreboard,
bitumen,
felt and plywood

Plan   Scale 1:100

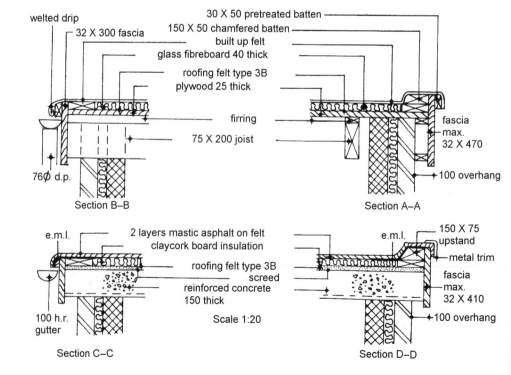

welted drip
32 X 300 fascia

30 X 50 pretreated batten
150 X 50 chamfered batten
built up felt
glass fibreboard 40 thick
roofing felt type 3B
plywood 25 thick

firring

75 X 200 joist

76Ø d.p.

Section B–B

fascia
max.
32 X 470

100 overhang

Section A–A

e.m.l.

2 layers mastic asphalt on felt
claycork board insulation
roofing felt type 3B
screed
reinforced concrete
150 thick

Scale 1:20

100 h.r.
gutter

Section C–C

e.m.l.

150 X 75
upstand
metal trim
fascia
max.
32 X 410

100 overhang

Section D–D

Drg. 10

Example 16

<u>Built up felt and Timber Flat Roof 1</u>

Two examples are illustrated on the same drawing, with alternative forms of construction and covering, annotated A and B respectively. The number of roof joists is found by taking the internal length of the building, less a space 87 wide at each end to include half the thickness of the end joist. This dimension is divided by the spacing of the joists, to which one is added to convert the spaces between joists to the number of joists. As there is a small surplus this could result in a slightly closer spacing of joists.

<u>Construction</u>

<u>jsts.</u>
7.600

<u>less</u>
walls 2/255 510
end spaces
to ₵ jsts.2/87 174   684
                400)6.916
                      18 + 1

<u>len.</u>
5.400

<u>add</u>
o'hg. 2/100 − 32
      = 2/68   136
              5.536

<u>Sn. swd. grade C24</u>
<u>to BS 5268; vac.</u>
<u>impreg. w. preservative</u>

Joists are measured as G20:9.1.1.0 and where >6.000 in one continuous length, this has to be stated and the actual length given. Grade C24 strength timber is needed to ensure joists of adequate strength to carry the substantial unsupported span.

Flat rf. membrs.
(jsts.   75 × 200
                7.600
         less   510
               7.090

19/	5.54

	7.09

Jst. struttg.
38 × 50; herringbone;
jsts. 200 dp.

Joist strutting is measured over the joists as G20:10.1.1.0, giving the depth of joists served by the strutting.

<u>firrgs.</u>
<u>depth</u>

gutter end   10
fall to rf. 1 in 40
40) 5.536   138
      max. depth   148

Firrings are required to provide the necessary fall to the roof and 1 in 40 is considered advisable on timber flat roofs to counteract sagging roof joists and avoid leaking roofs. The dimensions in waste show how deep the largest member must be to achieve this. The firrings are classified as individual supports (G20:13.0.1.1). Two firrings could be cut diagonally from a length of timber measuring 5.540 × 75 × 158.

19/	5.54

Individ. suppts.
75 wide × 148 max. depth, taperg. to 10 min. depth.

7.600
add o'hg. 2/68   136
              7.736

Built up Felt and Timber Flat Roof 2

	Plywd.; marine WBP grade to BS 1088; 25 th.
7.74 5.54	Rfs. >300 wide; nld. to tbr.
	&
	Bit. rfg. felt, type 3B to BS 747; lapd. 150 at jts.

	Plain areas
	    hor.; partially
	    bonded in hot bit.
	    on tbr.; in one layer
	(vapour control
	(layer

Plywood avoids the problems of uneven boards but must be adequately moisture resistant. It is classified as roof boarding in K11:4.1.1.0.

A single layer of type 3B bitumen roofing felt provides a sound vapour barrier to prevent condensation in a warm deck flat roof if partially bonded in hot bitumen. It is classified as P10:1.1.1.0, with the appropriate heading, stating the particulars required by P10:S1&2. The remainder of the description includes the method of fixing (P10:S3). A batten 150 wide is fixed around the perimeter of the roof and will be measured subsequently.

	7.736	5.536
less battens 2/150	300	300
	7.436	5.236

Insulation

'Foamglas' T4 cellular glass insulatn. in slabs 600 × 450 × 40 th.

7.44
5.24
	Plain areas
	    hor.; bonded in hot
	    bit. on bit. rfg. felt.

Insulation is covered by P10:3.1.1.0. A specific product has been listed in this case but a more general description could be used, such as 'foamed glass slabs: density 125–135 kg/m³; 40 thick'. It is possible to omit the vapour barrier with Foamglas slabs.

	7.736	5.536
less kerb 2/150	300	150
	7.436	5.386

Built up felt rfg. to BS 747 in 3 layers; bottom layer type 3B glass fibre base; second layer type 5U polyester base; top layer type 5E polyester based mineral surfaced; hot bit. bonded betw. layers

7.44
5.39
	Rf. covergs.
	    ld. on insul. bd.;
	    pitch 1 in 40.

Built up felt roofing is normally laid in three layers, although there are a number of single layer products on the market with long guarantee periods. J41:2.1.0.0 and J41:S1–3 require the kind, quality and size of materials to be stated, together with the nature of the base on which it is applied and the method of jointing. Light coloured chippings may be used in lieu of the mineral surfaced top layer where little foot traffic is expected, and this can be included in the main heading.

Built up Felt and Timber Flat Roof 3

kerb

2/ 5.400   10.800

add   7.600

o'hg. 3/2/100   600

19.000

eaves

7.600

add o'hg. 2/100   200

7.800

19.00

Built up felt rfg. a.b.
Covergs. to kerbs
200 – 400 gth.;
on tbr.; nlg. on
one edge.

7.80

Eaves
≤ 200 gth.; on tbr.;
nlg. on one edge.

7.736

5.536

2/ 13.272

26.544

less cnrs. 4/150   600

25.944

25.94

Sn. swd. a.b.
Individ. suppts.
150 × 40, nlg. to tbr.
(edge of insul. bd.

7.736

2/5.536   11.072

18.808

less cnrs. 2/150   300

18.508

Ditto
150 × 50; chmfd.;
nld. to tbr.

18.51

7.600

add o'hang 2/100 200   7.800

5.400

2/100 200   5.600

2/ 13.400

26.800

add cnrs. 4/30   120

26.920

J41:5.1–2.0.0 provides for treatment to eaves and J41:15.1–2.0.0 for coverings to kerbs. Such items are categorised as boundary work and are deemed to include all cutting, ends, angles, bending etc. (J41:C2).

The supporting batten enclosing the insulating board is taken as an individual support (G20:13.0.1.0). A similar approach is adopted for other battens in the roof.

Built up Felt and Timber Flat Roof 4

Sn. swd. a. b.

Individ. suppts.

26.92	30 × 50; nld. to tbr. (outside fascia

Ditto.

2/ 5.40	68 × 30; plugd. & scrd. to masonry (fascia suppts.

All the supporting timbers are taken at this stage, all in accordance with G20:13.0.1.0. The fascia supports fall into the same category, except that they are plugged and screwed to masonry.

Fascias

av. depth at sides of rf.

	depth
high side	470
gutter side	300
	2)770
av. depth	385

long side	len.
	7.600
add o'hgs. 2/100	200
	7.800

short side	5.400
add o'hgs. 2/100	200
	5.600

The depth of the fascia to the high side of the roof is calculated from the thickness of the adjoining members and because of the deep firrings has a depth of 470. This will entail cross-tonguing to connect the separate boards but this has not been included in the description, as G20:C1 states that the work is deemed to include labours on items of timber and, unlike a groove to receive the edge of soffit boarding, is a matter which the contractor will need to decide upon and is at his discretion.

Wrot. swd.

7.80	Fascia bds. 32 × 470; x. tgd. & Ditto. 32 × 300; x. tgd.

2/ 5.60	Ditto. 32 × 470 taper'g. to 32 × 300 on top edge; av. 385 dp.; x. tgd.

The fascia boards are measured in accordance with G20:15.2.2.0, giving a dimensioned overall cross section description. It is unnecessary to describe fascia boards as >300 wide as this is apparent from the dimensions given in the item description in the bill. The lengths are measured on the outside of the fascia to allow for mitred angles. Another alternative would be to use exterior quality plywood for the fascias, thereby eliminating horizontal joints.

Built up felt and Timber Flat Roof 5

Painting

total gth. of fascias

high side	2/470	940
	2/32	64
		1.004
short sides	2/385	770
	2/32	64
		834
gutter side	2/300	600
	2/32	64
		664

The calculation of the girths of the painted surfaces on the fascia boards becomes quite prolonged because of their varying heights.

7.80	Knot & prime on wd.
1.00	Gen. surfs.
2/ 5.60	> 300 gth.; on site
0.83	prior to fixg.
7.80	
0.66	

Knotting and priming is taken to all surfaces of the fascia board, made up of twice the depth of the fascia plus twice its thickness. The work is measured in accordance with M60:1.0.1.4.

expsd. gth. of fascias

high side: 470-50	420
	32
	30
	482
short sides: 385-50	335
	32
	30
	397
gutter side: 300-50	250
	32
	30
	312

To arrive at the areas of exposed surfaces after erection which are to be painted, it is necessary to adjust the outer face of the fascia board for the 30 × 50 batten and add the thickness of the fascia and depth of the exposed portion on the rear face of the fascia. The varying heights of fascia give rise to three separate but similarly described items for entry on the dimensions paper.

7.80	Prep. & ③ gloss pt. on
0.48	prev. primed wd.
2/ 5.60	Gen. surfs.
0.40	> 300 gth.
7.80	EXT.
0.31	

The paintwork is measured in accordance with M60:1.0.1.0, and the fact that it is external needs to be identified (M60:D1). Care must also be taken to include the appropriate items of supplementary information from M60:S1–8.

<u>Built up Felt and Timber Flat Roof 6</u>

RWG's

<u>Marley Alutec alum.
gutters to BS 1474; supptd
by cast alum. bkts. at 915
ccs.; s & s bolted jts. sealed
w. apprvd. silicone sealant</u>

7.80	H.r. gutters
	    100 nom. size;
	    strt.; scrd. to tbr.

1
─
2
	E.O. for
	    outlets
	Ditto
	    s.e.'s

<u>Prime & ③ alkyd gloss
pt. on met. surfs.</u>

7.80 0.32	Eaves gutters > 300 gth.

EXT.
2.500
add offset   100
2.600

<u>Marley Alutec alum.
pipewk. to BS 1474;
ears cast on; jts. sealed
w. polyurethane filler rod
& apprvd. silicone sealant</u>

2.60	Pipes
	    76 nom. size; strt.;
	    plugd. & scrd. to masonry
	    w. nr. 16 × 50 mm pipe
	    scrs.

1
─
	E.O. for
	    offsets 135 proj.
	    &
	Ditto. for
	    connectns. to clayware
	    gullies w. c.m. (1:3)

<u>Prime & ③ alkyd gloss
paint on met. surfs.</u>

2.60	Services isoltd. surfs. ≤ 300 gth.

EXT.

RWG's is an abbreviation for rainwater goods. The use of easily recognisable abbreviations throughout the descriptions saves considerable time in entering them on dimensions paper. Many of the more common abbreviations have become well established by constant use over the years (see also Appendix 1).

Gutters are measured in metres, stating whether straight or curved and giving the type, nominal size, method of jointing and type, spacing and method of fixing supports, including the fixing background (R10:10.1.1.1).

Fittings are enumerated as extra over the gutter, stating the type of fitting (R10:11.2.1.0).

Painting to gutters is measured in m^2 where the girth is >300 (inside and out) and in metres when ≤300 (M60:8.2.1–2.0).

The pipework is measured in accordance with R10:1.1.1.1, giving the type, nominal size, method of jointing, type, spacing and method of fixing supports.

Offsets and connections are enumerated as extra over the pipes (R10:2.2.1.0&R10:2.4.5.0).

Painting of the pipes is taken as a linear item as the girth is ≤300 ($76 × 22/7 = 239$), and described as to isolated surfaces (M60:9.0.2.0) and external (M60:D1). To avoid painting one could specify, for instance, a Marley factory applied polyester coated (PPC) satin matt finish to BS 6496, in one of 18 standard colours.

Drg. 10

<u>Asphalt and</u> | <u>Concrete Flat Roof 1</u>

Example 17

Construction

	7.600	5.400

add o'hgs.
²/100-32 | 136 | 136

7.736 | 5.536

<u>Conc. (21 N/mm²)</u>

7.74
5.54
0.15

Slabs
  reinfd.; ≤ 150 th.

	7.600	5.400

less
walls ²/255 | 510 | 510

7.090 | 4.890

<u>Fwk.</u>

7.09
4.89

Soffits of slabs
  slab ≤ 200 th.; hor.;
  ht. to soff. 1.500 - 3.000

7.600
5.400
²/ 13.000
26.000

less curs. ⁴/255 | 1.020

gth. on ₵ cav. wall | 24.980

24.98

<u>Ditto</u>
  over cav. of holl.
  walls; ≤ 150 wide;
  left in.

ext. gth. 26.000
add o'hg. of slab
⁴/68 | 272
₵ gth. of slab o'hg. 26.272

26.27

Soffits of proj. eaves
  ≤ 250 wide; hor.;
  ht. to soff. 1.500
  -3.000

This example uses the same plan on drawing 10 as the previous example. The overall dimensions of the concrete slab are calculated in waste by adding the overhang to the dimensions on the external faces of the enclosing walls.

In-situ concrete slabs are measured in m³, stating the thickness range and reinforced classifications as E10:5.1.0.1.

Formwork is classified as E20:8.1.1.2, according to the slab thickness and height to soffit. Where the thickness, of slab is >200, the formwork is given separately in 100 stages of slab thickness, while the height to soffit of slab is given in 1.500 stages.

It is necessary to seal the top of the cavity in the walls under the roof slab to prevent damage to the insulation and bridging of the cavity. SMM7 contains no specific provision for this situation and it is necessary to create an appropriate description which as far as possible is consistent with the rules contained in E20. The inclusion of locational information in the description will be helpful to the contractor in pricing. The formwork has to be left in as it cannot be recovered after the concrete is poured.

Again there is no provision in SMM7 for formwork to soffits of projecting eaves and the same principles as outlined above will also apply here.

Asphalt and Concrete Flat Roof 2

$$2/\frac{\begin{array}{r}7.736\\5.536\end{array}}{\begin{array}{r}13.272\\ \hline 26.544\end{array}}$$

Fwk. a.b.

| 26.54 | Edges of susp. slabs |

plain vert.; ht. ≤ 250

$$\frac{\begin{array}{cc}7.736 & 5.536\end{array}}{\text{less cover}\,^2/_{40}\;\;80\qquad\quad 80}$$
$$\overline{\begin{array}{cc}7.656 & 5.456\end{array}}$$

Reinft. to BS 4483 ref.

| 7.66 | A193 mesh; weighg. |
| 5.46 | 3.02 kg/m²; min. 200 laps |

Fabric

| | | Screed |
| | | depth |

gutter end   15

fall to rf. 1 in 60

$$60)\underline{5.536}\qquad\underline{\begin{array}{r}92\\107\end{array}}$$
max. depth

Fine conc. screed; 1 : 1½ : 3

ct.; fine agg., coarse agg.

10 max. size

Rfs.

| 7.74 | lev. & to falls only ≤ 15° |
| 5.54 | from hor.; 15 - 107 th.; |

av. 61 th.; in 1 ct. on conc.

&

Bit. rfg. felt, type 3B

to BS 747; lapd. 150 at

jts.

Plain areas

hor.; bonded in hot bit.

on conc.; in one layer.

Temp. suppt. wk.

To edges of screeds

| 2/ 5.54 | 15 - 107 high (av. 61 high) |

107 high

| 7.74 | |

&

15 high

The perimeter of the concrete slab is calculated to provide the length of formwork to the edges of the concrete slab, measured as E20:3.1.2.0. The height range is given and this makes the inclusion of a dimensioned description superfluous.

It is usual to allow about 40 concrete cover to all reinforcement to prevent corrosion. Fabric reinforcement is measured in m², giving the particulars listed in E30:4.1.0.0 and E30:S4 (minimum laps, which may vary between 100 and 300). Tying wire, cutting, bending and spacers and chairs which are at the contractor's discretion are deemed to be included (E30:C2).

The fall to the concrete roof has been taken as 1 in 60, as compared with the recommended fall of 1 in 40 used for the timber roof in example 16, as the concrete structure is much more rigid than the timber construction. It is however an absolute minimum to ensure the effective removal of surface water.

A fine concrete screed has been taken instead of a cement and sand screed because of its substantial thickness. It is measured in accordance with M10:6.1.1.0.

A single layer of type 3 bitumen felt provides an effective vapour barrier. It is classified as P10:1.1.10, with the appropriate heading, stating the particulars required by P10:S1&S2. The remainder of the description includes the method of fixing (P10:S3).

The edges of screeds require separate temporary support from that for the edge of the concrete slab. This is covered by M10:26.0.1.0 which although referring to 'risers and the like' can be regarded as applicable.

Asphalt and Concrete Flat Roof 3

			7.736	5.536	The area to be covered by insulation board is reduced by the 150 wide battens fixed around the perimeter of the roof.

less battens
2/150   300   300
              7.436   5.236

Insulation

'Claycork' cork insul. in
bds. 1.000 × 500 × 40 th.

Insulation is covered by P10:3.1.1.0. It was decided to specify a particular product to show how it would be described to meet SMM7 requirements. It is also necessary to ensure that the supplementary information particulars are included.

7.44
5.24

Plain areas
hor.; bonded in hot
bit. on bit. rfg. felt.

Asp. Rf. Coverg.

           7.736
less kerbs   300
             7.436

          5.536
less kerb   150
            5.386
add eaves   100
            5.486

The asphalt covering to the kerb on three sides of the roof and the drip at eaves will be measured separately from the roof covering. The overall dimensions are adjusted accordingly.

Mastic asp. to BS 988,
table 4 on blacksheathing
felt isolatg. membrane to
BS 747, type 4A (i) weighg.
17 kg/roll w. 50 lapd. jts.,
ld. loose

The mastic asphalt roof covering is measured in accordance with J21:3.4.1.0, stating the pitch. J21:S1–3 requires the kind, quality and size of material, including the underlay, to be stated, together with the thickness and number of coats and the nature of the base on which the asphalt is to be laid.

7.44
5.49

Rfg.
> 300 wide.; pitch 1 in 60;
2 cts.; 20 th.; to
corkbd. insul.

             kerb
           7.800
2/5.600   11.200
           19.000
less crnrs. 2/2/70   280
           18.720

18.72

Covergs. to kerbs
225–300 gth.; 2cts.; 20 th.
to tbr; reinf. w. met. mesh
rib lath ref. 269, 0.3 mm
th.; fxd. w. galvd. nls.

The coverings to kerbs are measured as a linear item stating the girth range. The length of kerb on its centre line is calculated in waste. The expanded metal lathing reinforcement is included in the description in accordance with J21:S1.

7.80

Drips
reinf. w. met. mesh
rib lath a. b.

Drips are measured in accordance with J21:15.0.0.0 at eaves.

Asphalt and Concrete Flat Roof 4

edge trim

Silver anodised alum.

19.00	Edge trims 63.5 mm wide & 76.2 mm face depth; primg. w. bit. primer; butt jts. w. int. jtg. sleeves; fxg. w. alum. scrs. to tbr.

The trim to the outside face of the kerb encloses the asphalt and provides support and a tight finish. It is covered by J21:22.0.0.0 and is deemed to include ends, angles and intersections (J21:C9).

```
               7.600
      2/68      136    7.736
               5.400
      2/68      136    5.536
                    2/ 13.272
                       26.544
      less 4/150           600
                       25.940
```

Sn. swd. a. b.

25.94	Individ. suppts. 150 × 40; nlg. to tbr. (edge of insul. bd.
7.80	Ditto. 30 × 70; nlg. to tbr. (gutter side, outside fascia

The various battens above the concrete slab are measured in a logical sequence to ensure that none are missed. They are all classified as individual supports as G20:13.0.1.4, giving a dimensioned overall cross section description. It is important to distinguish between different types of timber and to state whether they are sawn or wrought (G20:S1). Finally, the varying depths of fascia are calculated.

```
                     7.736
      2/5.536       11.072
                    18.808
```

18.81	Ditto. 150 × 75; splyd.; nlg. to tbr. (kerb 75 ⌐\ 150

depth of fascia

```
gutter side      240
high side        410
              2) 650
  av. depth      325
```

Asphalt and Concrete Flat Roof 5

		Fascia Bds.
		<u>Wrot. swd.</u>
	7.80	Fascia bds.
		32 × 410 ; x. tgd.
		&
		Ditto.
		32 × 240 ; x. tgd.
		Ditto.
2/	5.60	32 × 410 taperg. to
		32 × 240 on top edge;
		av. 325 dp.; x. tgd.

The fascia boards are measured in accordance with G20:15.2.2.0, giving a dimensioned overall cross section description. It is unnecessary to describe fascia boards as >300 or ≤300 as it will be apparent from the dimensions listed in the bill descriptions.

<u>Painting</u>
<u>total gth. of fascias</u>

high side	2/410	820
	2/32	64
		884
short sides	2/325	650
	2/32	64
		714
gutter side	2/240	480
	2/32	64
		544

	<u>Knot & prime on wd.</u>
7.80	Gen. surfs.
0.88	> 300 gth.; on site
2/ 5.60	prior to fxg.
0.71	
7.80	
0.54	

Knotting and priming is taken to all surfaces of the fascia boards, made up of twice the depth plus twice the thickness, and measured in accordance with M60:1.0.1.4.

<u>Expsd. gth. of fascias</u>

high side	410-50 =	360
add 32(bott)+30 =		62
(rear)		422
short sides	325-50	275
		62
		337
gutter side	240-50	190
		62
		252

To arrive at the exposed surfaces of fascia to be painted after erection it is necessary to adjust the dimensions in the manner shown in waste.

	<u>Prep. & ③ gloss pt. on</u>
7.80	<u>prev. primed wd.</u>
0.42	Gen. surfs.
2/ 5.60	> 300 gth. EXT.
0.34	
	Ditto.
7.80	≤ 300 gth.;
	isolated surfs.
	EXT.

The paintwork is measured in accordance with M60:1.0.1–2.0. Where the surface girth is >300 it is measured in m², but where it is ≤300 it becomes a linear item described as isolated surfaces, and both clearly marked as external work. The measurement of the rainwater goods and their painting has been omitted as the work is identical to that measured in example 16.

9 Internal Finishes

In the measurement of this class of work it is essential to adopt a logical sequence in the taking off work. Where the work is extensive it is advisable to prepare a schedule of finishes in the form illustrated in table 9.2 unless supplied by the architect. This will considerably simplify the taking off and reduce the liability to error.

This section normally includes ceiling, wall and floor finishes and the rules of measurement of most of these items of work are covered in a variety of work sections ranging from M10 to M60 of SMM7, but with plasterboard dry lining in K10, timber board flooring in K20 and skirtings and similar members in P20. As indicated in chapter 7, floor finishes may be taken off when measuring the floors or be left to be taken later with internal finishes. On balance, the best arrangement is probably to take finishes to suspended floors (boarded flooring) with the floors, since they form an integral part of the floor construction and the work is related to an associated trade. Finishes to solid floors which can be of infinite variety are best measured with internal finishes.

The wide range of relevant work sections is shown in table 9.1. This does cause problems for many students, and so the SMM7 pages have also been given for ease of identification. Example 18 encompasses a significant variety of finishes to help the student to become familiar with the use of the relevant *Standard Method* provisions.

The taker-off should work systematically through the building, room by room, often working from top to bottom, and recording the details of finishes in a schedule in the majority of cases. It will probably prove helpful to number the rooms for this purpose and to mark each room in some distinctive way once the details have been extracted. In the case of a very small building with only a limited variety of finishes the details can often be transferred direct to the dimensions paper, without the need for a schedule, but the same systematic approach should be adopted.

The order of measurement of internal finishes on each floor will normally be: (1) ceilings, (2) walls and (3) floors. The measurement of the main areas of wall finishes will be followed by linear items, such as cornices, picture rails, dado rails and skirtings, and working in this sequence (top to bottom).

Work to walls, ceilings and floors in staircase areas and to plant rooms must be given separately, because of the higher costs generated.

Table 9.1 *Schedule of SMM7 work sections covering internal finishes.*

Work section classification	SMM page number	Title
K10	87	Plasterboard dry lining, partitions and ceilings
*K11	73	Rigid sheet flooring and linings, such as plywood and chipboard
*K20	73	Timber board flooring and linings
*K21	73	Timber strip and board fine flooring and linings
M10	103	Screeds and toppings
M12	103	Bitumen, resin and rubber latex flooring
M13	103	Calcium sulfate screeds
M20	103	Plastered coatings
M30	108	Metal mesh lathing and reinforcement for plastered coatings
M31	109	Fibrous plaster
M40	111	Stone, concrete, quarry and ceramic tiling and mosaic
M41	113	Terrazzo tiling and *in-situ* terrazzo
M42	111	Wood block, composition block and parquet flooring
M50	113	Rubber, plastics, cork, lino and carpet tiling and sheeting
M51	113	Edge fixed carpeting
M52	115	Decorative papers and fabrics
M60	116	Painting and clear finishing
P20	123	Unframed isolated trims, skirtings and sundry items

*Denotes included with H classification items in SMM7.

CEILING FINISHES

The ceiling area is measured between wall surfaces with the area of each type of finish measured separately in m^2, followed by any associated labours, such as arrises to beams. For instance, with a plaster ceiling, covered in the *Standard Method* under work section M20, the various particulars listed in M20:S1–6, such as kind, quality, composition and mix of materials, method of application, nature of surface treatment and nature of base are to be given. In addition, the thickness and number of coats are to be stated. Work to ceilings and beams over 3.500 above floor level (measured to ceiling level in both cases), except in staircase areas, shall be so described, stating the height in further stages of 1.500, that is 3.500–5.000, 5.000–6.500, and so on (M20:M4 and General Rule 4.4).

Plasterboard ceilings are dealt with similarly, stating the thickness of the plasterboard. Skim coats of plaster to plasterboard ceilings are included in the same items.

When measuring plasterboard or plaster to sides and soffits of attached beams, this work is classified as to ceilings. However, if the width of each face is ≤300, as is often the case, then each is measured in metres ≤300 wide as M20:2.2.1.0, M20:M6 and M20:D5. Where the work to beams is different from the ceilings, then the work is described as to isolated beams (M20:3). The following example illustrates the treatment of a beam, 6.500 long, with the dimensions shown on figure 33.

3/	6.50	Pla. 13 th. in 2 cts. a.b. Ceilings ≤300 wide; to conc. (beam

followed by a deduction of ceiling plaster previously measured for the area occupied by the beam, namely:

	6.50 0.30	Ddt. Pla. 13 th. in 2 cts. a.b. Ceilings >300 wide; to conc. (beam

It will be noted that work in staircase areas is given separately to enable the estimator to price working off staircase flights and/or in restricted areas.

A typical set of specification clauses relating to plasterboard and skim coat is given in Appendix 4 at the end of the book. Use of specification clause numbers in billed descriptions reduces substantially the length of the description. For example, the billed description for a ceiling could read '9.5 mm plasterboard; 5 plaster in 2 skim coats; specification M20/280'.

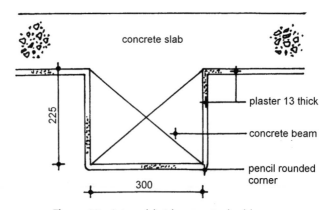

Figure 33 *Internal finishes to attached beam.*

WALL FINISHES

The measurement of wall finishes is taken from floor to ceiling, including the work behind wood skirtings and similar features, disregarding the grounds (M20:M2).

The girth of each room is usually built up in waste and the total girth of rooms of the same height and finish transferred to the dimensions column. The method of measurement varies with the type of finish, and reference needs to be made to the appropriate work section, such as M20 for renderings and plastered coatings, M40 for quarry and ceramic tiling and M41 for terrazzo tiling. Other relevant sections are identified in table 9.1 and the measurement of different finishes illustrated in example 18. The measurement of the main area of wall finish of each type is followed by its associated linear items.

Plaster to walls and isolated columns in widths ≤300 is measured in metres, and no deduction shall be made for voids ≤0.50 m² (M20:M2). Work to sides and soffits of openings and sides of attached columns shall be regarded as work to the abutting walls (M20:D5).

When measuring plaster to attached columns, a similar approach to that adopted for attached beams is taken. Hence if the column face on each return is ≤300, then a linear item will pick up the plaster to each face as M20:1.2.1.0; followed by the deduction of the wall plaster for the areas occupied by the attached columns, unless these areas had been adjusted previously when calculating the perimeter length of plaster to walls. The following example refers to four attached columns, each 2.800 long, of the cross section dimensions shown in figure 34.

$^4/_3/$	2.80	Pla. 13 th. in 2 cts. a.b. Walls ≤300 wide; to brickwork (col.
$^4/_4/$	2.80	Rdd. angles & intersecs.
4/	2.80 0.22	Ddt. Pla. 13 th. in 2 cts. a.b. Walls >300 wide; to brickwork (col.

Rounded angles and intersections to plaster in the range 10–100 radius are measured in metres (M20:16 and M20:M7), while those of smaller radius are deemed to be included in the plasterwork rates M20:C4. Angle beads

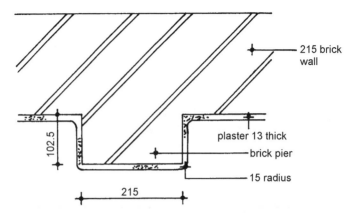

Figure 34 *Internal finishes to attached column/pier.*

·are often used to strengthen vulnerable external angles of plaster and these are measured in metres with a dimensioned description as M20:24.8.1.0 and illustrated in example 18. The need to include dimensions in the description is usually avoided by either quoting a British Standard or a manufacturer's reference. Other types of beads such as casing beads and stop beads may be required in other situations and in each instance the function must be stated.

Working plaster over and around obstructions, pipes and the like, and into recesses and shaped inserts is deemed to be included (M20:C1b).

Appendix 4 contains a typical plaster specification which will reduce the billed description to '15 plaster in two coats; specification M20/211'.

Where strips of expanded metal lathing are needed to provide a key for plaster over timber framing, a typical measured item description might take the following form (see M30:2.2.1.0):

Galv. plain e.m.l. to BS 1369: Pt. 1 ref. L2
w. butt jts.
Walls
    ≤300 wide; fxg. w. galv. clout nls. or galv. staples

Cornices, mouldings and coves are measured in metres (length in contact with base) stating the girth or giving a dimensioned description (M20:17–19.0.1–2.0). Ends, internal angles, external angles and intersections are each enumerated extra over the appropriate linear items, giving adequate details as M20: 23.1–4.1.0.

In addition to *in-situ* finishes covered by M20 and M41, wall finishes may also be of stone, reconstructured stone, concrete, clay, ceramic and terrazzo tiling and mosaic (M40&M41), or flexible and semi-flexible tile and sheet coverings in rubber, plastics, cork, lino and carpet (M50). In each case full particulars of the finishes are to be given, such as kind, quality and size of materials, nature of base, surface treatment, method of fixing, and treatment and layout of joints (M40/41/50:S1–8). Work >300 in width is measured in m², while narrower widths are measured in metres.

SKIRTINGS AND PICTURE RAILS

Timber skirtings, picture rails, dado rails, architraves and the like are measured in metres, giving a dimensioned overall cross section description as P20:1.1.0.1–4. The work is deemed to include ends, angles, mitres, intersections and the like, except on hardwood items >0.003 m² in sectional area (P20:C1). The description is to include the kind and quality of timber and whether sawn or wrought and method of fixing where not at the discretion of the contractor (P20:S1–9).

Inflexible tiled skirtings are measured in metres, stating the height or height and width as appropriate, and are deemed to include fair edges, rounded edges, ends, angles and ramps (M40:12.1–2 and M40:C8). Flexible skirtings are measured in a similar manner (M50:10.0.1–2 & M50:C7).

FLOOR FINISHES

With solid floors screeds are generally required, often consisting of cement and sand or other proprietary material of varying thicknesses to bring the floor finish to the required level. As floor finishes vary considerably in thickness from 2 for vinyl tiles up to 28 for terrazzo tiles, so the thickness of the screed will vary. A common mix is 1 : 3, but screeds thicker than 50 may be laid in two coats. A typical screed specification is given in Appendix 4, when the bill description might read '48 cement and sand screed, specification M10/110'. This item has a trowelled finish as specification clause M10/540; alternatively, a smooth floated finish may be specified as M10/530 to receive the thicker types of finish, although opinions vary as to the best surface finish to be adopted. It will be seen that the specification clause gives a minimum and maximum thickness for the screed to take account of deviations in the concrete base. In the item description a thickness mid-way between these extremes will normally be given.

Floor finishes are measured in m², irrespective of their width, except in the case of flexible finishes (M50 & M51), where widths >300 are measured in m² and widths ≤300 in metres. All types of finish are classified under three categories according to fall/slope (e.g. M10:5.1–3), although some minor variations in wording will be noticed between M50/51 and the other work sections. Where floors are laid in bays, sections M10/40/42 require the average size of bays to be stated. Screeds are measured in an identical manner to *in situ* floor finishings. The slope is the gradient to which the floor finish is laid, expressed in angular terminology or as the ratio of one vertical unit to a number of horizontal units, such as 1 in 40, as explained more fully in chapter 2. Dividing strips at door openings and the like, between different types of floor finish, are measured in metres giving a dimensioned description (M40:16.4.1.0), as illustrated in example 18.

Mat frames to mat spaces are often provided in domestic halls close to the front door and form enumerated items in accordance with N10.1 and SMM Appendix A. No deduction for floor finish will be made for the mat well as the area of the void is $\leq 0.50\,m^2$ (M40:M1). A typical measured item description follows for a galvanised steel frame, although it could also be formed of brass:

	1		Galv. m.s.
			Matwell
			    762 × 457 mat space; welded
			    fabrication; 40 × 40 × 6 r.s.
			    angle sectn. with mitred
			    angles; plain lugs; set
			    in screed

N10:1.1–2 requires either a component drawing or a dimensioned diagram, but in this case a dimensioned description seems more appropriate, although not strictly in accordance with SMM rules.

PAINTING AND DECORATIONS

The painting and decorating of ceilings, cornices and walls are classified as to general surfaces and measured in m^2, except for work on isolated surfaces ≤ 300 in girth which is given in metres, or work in isolated areas $\leq 0/50\,m^2$ which is enumerated (M60:1.0.1–3.0). Multi-coloured work is separately classified and is defined as the application of more than one colour on an individual surface, except on walls and piers or on ceilings and beams (M60:D2). Paintwork is deemed to include rubbing down with glass, emery or sand paper (M60:C1). Full particulars of painting, decorating and polishing shall be given in accordance with M60:S1–8.

Work in staircase areas and plant rooms is each to be kept separate because of the extra costs involved (M60:M1). Work to ceilings and beams over 3.500 above floor level (measured to ceiling level in both cases), except in staircase areas, shall be so described stating the height in further 1.500 stages (M60:M4).

The supply and hanging of decorative papers and fabrics require work to walls and columns to be given separately from that to ceilings and beams; with areas >0.50 m^2 measured in m^2 and those ≤0.50 m^2 enumerated. Where these items include raking and/or curved cutting and/or lining paper, these are included in the description (M52:1–2.1–2.0.1–2). Border strips are measured in metres, including cutting them to profile in the description where appropriate (M52:3.0.0.1). Material particulars are to include the kind and quality of materials, including the manufacturer and pattern (M52:S1), although these particulars may be included in a preamble clause or covered by a project specification reference. The *Code of Procedure for Measurement of Building Works* (*SMM7 Measurement Code*) states that the width of rolls and type of pattern would need to be given before wallpaper could be considered fully described. The type and size of pattern which requires to be matched on adjoining papers will determine the amount of waste. At the time of bill preparation, full details of decorative papers and fabrics are often not available. In such cases the measured items will be for hanging/fixing only and the supply and delivery to site of materials will be covered either by a prime cost sum for a nominated supplier or a provisional sum (M52:M1). The *Measurement Code* explains that it is still necessary to give the kind of materials and the length and width of the roll to enable labour costs to be assessed.

WORKED EXAMPLE

A worked example follows giving the dimensions of the ceiling, wall and floor finishes to an eleven roomed bungalow, covering a variety of finishes and including the use of a schedule of finishes (table 9.2).

Internal finishes   DRAWING 11

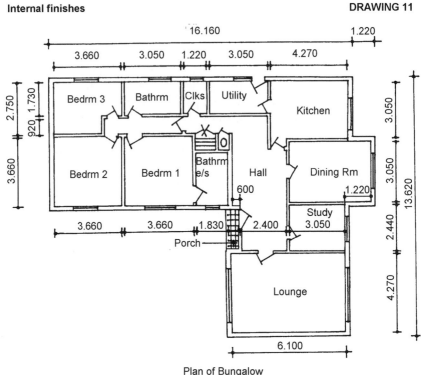

Plan of Bungalow
Scale 1:200

Notes

external cavity walls: 255 thick of faced brickwork
   and concrete insulating blocks.

internal walls: 100 thick concrete blocks.

internal finishes to ceilings, walls and floors
   as adjoining schedule (table 9.2).

height of rooms: 2.440 throughout.

adjustments to door and window openings
   to be taken with doors and windows.

walls partly tiled or boarded are to a height
   of 1.400 above floor level.

Table 9.2 *Outline schedule of internal finishes to bungalow.*

Room	Ceiling	Walls	Skirting	Floor finish
Hall	Plabd., skim ct. & emulsn.	Pla. & emulsn.	25 × 100 swd.	Vinyl tiles
Passage	Plabd., skim ct. & emuln.	Pla. & emulsn.	25 × 100 swd.	Vinyl tiles
Linen cupd.	Plabd., skim ct. & emulsn.	Pla. & emulsn.	25 × 100 swd.	Vinyl tiles
Lounge	Plabd., skim ct. & vinyl paper; cove	Pla. & textile coverg.; picture rail	25 × 100 maple.	Maple strip bdg. on swd. brrs.
Dining room	Plabd., skim ct. & vinyl paper; cove	Pla. & vinyl paper; picture rail	25 × 100 swd.	Edge fixed carpeting
Study	Beech plywood with decorative veneers	Mahogany match bdg. clear fin. & pla. with textile coverg. above	None	Edge fixed carpeting
Kitchen	Plabd., skim ct. & paint	Pla. & ceramic tiles	None	Flexible sheet linoleum
Utility	Plabd., skim ct. & paint	Pla. & ceramic tiles with paint above	None	Terrazzo tiles
Cloaks	Plabd., skim ct. & paint	Pla. & ceramic tiles with paint above	None	Ceramic tiles
Bathroom	Plabd., skim ct. & paint	Pla. & ceramic tiles	None	Rubber tiles
Bedroom 1	Plbd., skim ct. & emulsn.	Pla. & textile coverg.	25 × 100 swd.	Carpet tiles
Bedroom 2	Plabd., skim ct. & emulsn.	Pla. & emulsn.	25 × 100 swd.	Vinyl tiles
Bedroom 3	Plabd., skim ct. & emulsn.	Pla & emulsn.	25 × 100 swd.	Vinyl tiles
Bathroom e/s	Plabd., skim ct. & paint	Pla. & ceramic tiles with paint above	None	Cork tiles
Porch (external)	Swd. match bdg.; painted	—	—	Quarry tiles

Drg. 11

Example 18

<u>Internal Finishes 1</u>

<u>Ceilings</u>

<u>Plabd. of gyp. basebd.</u>
<u>to BS 1230, 9.5 mm th.;</u>
<u>fxg. to tbr. w. galv. nls.;</u>
<u>scrimmg. jts. & skim ct.</u>
<u>of Thistle bd. fin. pla. to</u>
<u>BS 1191, pt. 1, class B, 5th.</u>
<u>in 2 cts. fin. smth.</u>

Clgs.
          >300 wide

2.40		
2.23		(hall
3.00		
2.94		(hall
1.78		
1.74		(hall/passage
6.84		&
0.92		(passage
1.86		
0.80		(linen cupd.
2/	3.66	
	3.66	(bedrms. 1 & 2
	3.66	
	2.75	(bedrm. 3

<u>Emulsn. pt. to pla. in</u>
<u>2 cts.</u>
<u>Gen. surfs.</u>
          <u>> 300 gth.</u>

Ddt. last 2 items

| 1.10 | | |
| 1.02 | | (bedrm. 3 |

<u>Plabd. & skim a. b.</u>

Clgs.
          > 300 wide
                    bathrm. 3.050
                    clks.   1.220
          (kitchn.
                    utility 3.050
          (do.   7.320
                    &
4.27		
3.05		
1.00		
0.42		
7.32		(bathrm. clks. & utily.
1.73		
2.80		(bathrm. E/s
1.83		

<u>Gloss paintg. to pla.;</u>
<u>1 ct. primer, 2 u/cs,</u>
<u>1 ct. gloss fin.</u>
Gen. surfs.
          > 300 gth.

The work is deemed to be internal unless described as external (M20:D1). Adopt a logical sequence of taking off, such as ceilings, walls and floors. Figured dimensions are used wherever possible, although dimensions may have to be scaled in some cases, such as in the hall in this example. Plasterboard to ceilings is measured in m², where the width is >300, giving the thickness of plasterboard and the thickness and number of skim coats as M20:2.1.2.0, and the appropriate particulars listed in M20:S1–8. Alternatively, the type of plasterboard and associated particulars could be covered in a preamble clause or by cross reference to a project specification, as demonstrated in the text of this chapter.

Work to ceilings over 3.500 are so described in 1.500 stages, because of the higher costs involved (M20:M4), although it will not apply in this example as the room heights are 2.440. Work in staircase areas is given separately, irrespective of the height, because of the more difficult working conditions (M20:M3). Work to beams and columns is dealt with in M20:D5&6 and illustrated in figures 33 and 34 and in the text of this chapter.

Decorations to ceilings are measured in m² when the girth is >300 and classified as to general surfaces (M60:1.0.1.0) and the same provisions apply to high ceilings and those in staircase areas (M60:M1&M4). Note the method of combining the lengths of rooms of constant width in waste to reduce the number of entries in the dimensions.

Internal Finishes 2

	Plabd. & skim a.b.
	Clgs. > 300 wide
	d. rm.
	3.050
6.10	(lounge 1.220
4.27	
4.27	4.270
3.05	(d/rm. &

Note the use of abbreviated headings where a full description has been given previously. The length of the dining room is obtained by totalling the two dimensions figured on the drawing.

Fixg. only lining paper & washable vinyl paper, incl. sizing & applyg. adhesive; butt jts.
Clgs. & beams
areas > 0.50 m² to pla.

The paper is measured in accordance with M52:2.1.0.2 and giving in the heading the kind, quality, manufacturer and pattern of wallpaper where decided (M52:S1). Otherwise it is necessary to include a prime cost or provisional sum for the supply and delivery to the site of the papers (M52:M1), with the inclusion of a further item to enable the main contractor's profit to be added (A52:1.1.1.0 and A52:1.2.0.0).

Item

Prov. P.C. sum of £110 for lining paper & washable vinyl paper to be obtd. from a supplier nominated by the Arch.
&
Add for profit

Beech plywd. 4 th.; prefinished, decorative veneers; butt jtd.
Clgs.
> 300 wide; nlg. to tbr.

| 3.05 | |
| 2.44 | (study |

The plywood lining is measured in accordance with K13:3.1.1.0 as a rigid sheet fine lining, and having regard to the supplementary information in K13:S1–5&12.

Wrot. swd. matchbdg.; 16 th.; t. & g.; v-jtd. o.s. in 120 widths
Clgs.
> 300 wide; nlg. to tbr.

| 1.82 | (porch |
| 0.70 | & |

Gloss paintg. to wd.; 1 ct. primer, 2 u/cs, 1 ct. gloss fin.
Gen. surfs.
> 300 gth. EXT.

The matchboard lining is measured in accordance with K20:3.1.1.0 in a similar manner to the previous item. The painting is classified as to general surfaces as ceilings do not form a separate category in M60:1.0.1.0. It is important to classify the work as external as this is likely to be more expensive, being subject to the vagaries of the weather (M60:D1).

Internal Finishes 3

Coves

	lounge	6.100
		4.270
	2/	10.370
		20.740

	Ding. Rm.	4.270
		3.050
	2/	7.320
		14.640

Before proceeding to the measurement of wall finishes, the fibrous plaster coves at the junction of the walls and ceilings to the lounge and dining room should be taken. The coves are measured in accordance with M31:10.0.1.0 with an adequate description to identify them.

<u>Gyp. pla. preformed mldgs.</u>
<u>fxd. w. adhesive in accord.</u>
w. manufac's. instrs.

Gyproc coves
        100 gth.; to pla.

20.74	(lge.
14.64	(d. rm.

E.o. for

4	(lge. int. Ls
4	(d. rm.

Internal angles are enumerated as extra over the work in which they occur (M31:14.2.1.0). Then follows a deduction of the areas of paper already measured for the width occupied by the coves.

	20.740	14.640
less crnrs. 4/70	280	280
	20.460	14.360

Ddt.

20.46	(lge. Fixg. only washable
0.07	vinyl paper a. b.
14.36	(d. rm.
0.07	Clgs. & beams
	> 0.50 m² area;
	to pla.

Add

20.46	(lge. Emulsn. pt. to pla. in
	2 cts.

Gen. surfs.

14.36	(d. rm. isol.; ≤ 300 gth.

The coves are to be painted in emulsion paint and they are enclosed by paper on both sides. The painting will be measured as linear items as ≤300 girth (M60:1.0.2.0).

Internal Finishes 4

<u>Walls</u>
Pla. on blkwk.

	lens.
<u>Hall</u>	6.910
	3.000
2/	9.910
	19.820
<u>less</u> passage	0.920
	18.900
<u>Passage</u> 2/6.840	13.680
	0.920
	14.600
<u>Linen cupd.</u> 2/1.855	3.710
4/0.800	3.200
	6.910
<u>Study</u>	3.050
	2.440
2/	5.490
	10.980
<u>Kitchen</u>	4.270
	3.050
<u>add recess</u>	0.420
2/	7.740
	15.480
<u>Utility</u>	3.050
	1.730
2/	4.780
	9.560
<u>Clks.</u>	1.220
	1.730
2/	2.950
	5.900
<u>Bathrm.</u> As Utility	
<u>Bedrms. 1 & 2</u>	3.660
	3.660
2/	7.320
	14.640
<u>Bedrm. 3</u>	3.660
	2.750
2/	6.410
	12.820
<u>Bathrm. E/s</u>	2.800
	1.830
2/	4.630
	9.260
<u>Porch</u> 2/0.700	1.820
	1.400
	3.220

The lengths of walls to the lounge and dining room have already been measured in connection with the coves. The remainder will now be calculated adopting the same order of rooms as in the schedule of finishes. Note the different approach to the measurement of the lengths of walls to the passage and linen cupboard.

Follow a logical sequence in the calculation of the lengths of walls, normally comprising adding together the length and width on the wall faces and timesing the total by two to give the overall perimeter length. Also times the rooms with identical measurements such as the utility and bathroom and bedrooms 1 and 2. As the heights of all the rooms are the same, the totalled lengths can subsequently be multiplied by the common height to arrive at the area of plaster to the walls, thus simplifying the calculations. Note the method of recording the dimensions in waste.

Internal Finishes 5

	Pla. 13th. in 2 cts.; consistg. of Carlite premixed ltwt. browning pla. base ct. to BS 1191, pt. 2; 11 th.; fin. ct. of Carlite premixed ltwt. fin. pla. to BS 1191, pt. 2; 2 th.; trwld. smth.

Walls
> 300 wide; to blkwk.

18.90	(hall
14.60	(passge.
6.91	(lin. cupd.
20.74	(lounge
14.64	(ding. rm.
10.98	(study
15.48	(kitchen
2/ 9.56	(utility & bathrm.
5.90	(clks.
2/ 14.64	(bedrms. 1 & 2
12.82	(bedrm. 3
9.26	(bathrm. E/s

× 2.44 =   m²

Accessories
Galv. steel
Angle bds.
   to BS 6452 Pt.1 profile
   1b; fxg. w. galv. nls. or
   pla. dabs to blkwk.

3/ 2.44	(hall
2.44	(kitchen
2.44	(bedrm. 3

Emulsn. pt. to pla. in
2 cts.
Gen. surfs.
> 300 gth.

18.90	(hall
14.60	(passage
6.91	(lin. cupd.
14.64	(bedrm. 2
12.82	(bedrm. 3

                          2.440
              less
              sktg.   .100
                          2.340

× 2.34 =   m²

A full description of the wall plaster is needed in the absence of a suitable specification clause reference as illustrated in the text of this chapter and appendix 4. The rules for measurement are contained in M20:1.1.1.0. Locational notes are added for ease of identification and later reference if needed. Finally, the total length will be multiplied by the room height, in the manner shown, to give the area of wall plaster for billing.

Steel angle beads are taken to all external angles of plaster, as distinct from taking rounded angles to the plaster, to provide greater strength and less likelihood of damage at these vulnerable points (M20:24.8.1.0).

It will be noted that plaster has been measured to the whole surface area of the walls, even when they will be covered with another finish, such as ceramic tiles and timber board lining. This will be adjusted later and a cement and sand backing substituted for the areas covered by ceramic tiling, and no backing to the matchboarding to the study.

Internal | Finishes 6

		Ceramic tiling in crystal coloured tiles w. satin fin. suppld. by H&R Johnson Ltd.; 152 × 152 × 5.5 mm in size; w. cushion edges & spacer lugs; bedded in adhesive & ld. in accord. w. BS 5385, pt. 1 & manufac's. instrs.; & grtg. jts. on completn. w. Febtile Rainbow wh. ct.; strt. jts. both ways

Walls
>300 wide; to c/s backg

15.48	
2.44	(kitchen
9.56	
2.44	(bathrm.
9.56	
1.40	(utility
5.90	
1.40	(clks.
9.26	
1.40	(bathrm. E/s

&

Ddt.
Pla. in 2 cts. a. b.
Walls > 300 wide
&
Add
Ct. & sd. (1 : 4); one ct.; 12 th.; floated
Walls
>300 wide; to blkwk.

E.o. for
rdd. edge tiles

2.44	(kitchen
9.56	(utility
5.90	(clks.
9.26	(bathrm. E/s

ht.
less   2.440
cove   70
sktg. 100   170
                2.270

Hang'g. only lining paper & washble. vinyl paper, incl. sizing & applying adhesive; butt jts.

Walls & cols.
areas > 0.50 m²; to pla.

| 14.64 | (ding. rm. |
| 2.27 | |

| 14.64 | Fixg. only border strip w. adhesive to pla. |

A full description of the ceramic tiles is required to incorporate the kind and quality of materials, size, shape and thickness of units, nature of base, preparatory work, nature of finished surface, bedding and jointing (M40:S1–8). It is desirable to give the manufacturer's name so that an identifiable product can be used and this will assist the estimator in pricing the work. It is important to be precise, as in some cases price differences may occur within the same range of tiles with regard to alternative surface finishes and colours.

The measurement rules are contained in M40:1.1.0.0. The standard rules with regard to work in staircase areas apply (M40:M2) and curved work is so described with the radii stated, measured on the face of the tiling (M40:M4).

An adjustment is required on the previously measured plaster to provide a cement and sand backing behind the tiling.

Rounded edge tiles fall within the classification of special tiles in M40:15.1.1.0. These have been taken to the external angle in the kitchen and the top edge of the tiling in the utility, cloaks and en suite bathroom.

The wallpaper in the dining room is a 'hanging only' item and the cost of the paper is included in the PC item provided earlier in the example (see M52:1.1.0.2, A52:1.1.1.0 and A52:1.2.0.0).

Border strips are covered in M52:3.0.0.0. They are deemed to include mitres and intersections (M52:C2).

Internal Finishes 7

	ht. of wall paper

lounge		2.440
less cove	70	
sktg.	100	170
		2.270
Bedrm. 1		2.440
less sktg.		100
		2.340
Study		2.440
less match tdg.		1.400
		1.040

The height of the wallpaper varies from room to room and adjustments have to be made for the heights of coves, matchboarding and skirtings as appropriate.

Hang'g. only ling. paper & wide-width textile wall coverg. w. grade 3 adhesive; butt jts.

Walls & cols.
areas > 0.50 m²; to pla.

20.74	
2.27	(lounge
14.64	
2.34	(bedrm.1
10.98	
1.04	(study

Item

Prov. P.C. sum of £270 for textile wall coverg. to be obtd. from a supplier nominated by the Arch.
&
Add for profit

An expensive textile wall covering has been selected for the lounge, bedroom 1 and study as the most important rooms in the bungalow to give a feeling of uplift. The measurement rules are contained in M52:1.1.0.2. Where the manufacturer and pattern cannot be provided, work is measured as hanging/fixing only and the supply/delivery to the site of the fabrics is covered by a prime cost sum or provisional sum as adopted in this example (M52:M1). The prime cost sum can be ascertained by calculating the number of rolls required and multiplying this number by the cost per roll.

Paintg. to pla.; 1 ct. primer, 2 u/cs, 1 ct. gloss fin.

Gen. surfs.
> 300 gth.

9.56	
1.04	(utility
5.90	
1.04	(clks.
9.26	
1.04	(bathrm E/s

Paintwork to plastered walls remains to be taken for the areas above the ceramic tiling which have been measured previously to a height of 1.400. The painting is suitably described and classified as to general surfaces (M60:1.0.1.0).

<u>Internal Finishes 8</u>

<u>Wrot. Utile mahog.;</u>
<u>t. & g., v-jtd. o.s.</u>
<u>matchbdg. 16 th.;</u>
<u>in 120 widths</u>

Walls

> 300; nlg. to tbr.

10.98	
1.40	

(study

&

<u>Ddt</u>

<u>Pla. in 2 cts. a.b.</u>

Walls

> 300 wide

```
          1.400
add dado rl.   110
          1.510
```

<u>Clear finishg.; prep.;</u>
<u>2 cts. polyurethane</u>
<u>varnish on wdwk.</u>

Gen. surfs.

> 300 gth.

10.98	
1.51	

(study

<u>Diagram</u>
<u>P20/1</u>

matchbdg.
16 th.

10.98	

<u>Wrot. mahog.</u>
Dado rl.

63 × 38 rdd. &
grvd.; fxg. w. scrs.;
hds. of scr. sk. &
pelleted; as dimensioned
diag. P20/1

<u>Sn. swd. grade C16 to</u>
<u>BS 5268; vac. impregntd.</u>
<u>w. preservative</u>

3/

10.98	

Indiv. suppts.
25 × 38; plug'g. to
masonry

Full particulars are contained in the item description as required by K21:S1–3, 9&12. The measurement process follows the requirements of K21:1.1.1.0.

A clear finish is needed in order that the rich colour and grain of the mahogany remains visible. The finish is described as to general surfaces as M60:1.0.1.0.

The matchboarding is surmounted by a dado rail or capping to seal the space behind the matchboarding. The dado rail has been taken as a wrought hardwood linear item with a full description and a dimensioned diagram. It would seem that P20:2.1.0.0 is the most appropriate reference, although this item relates to unframed cover fillets, whereas the dado rail is framed to the matchboarding. In such cases it is best to give an explicit description as demonstrated here. The three rows of supporting battens at the back of the matchboarding are covered by G20:13.0.1.0. The ends, angles and mitres to the dado rail are deemed to be included as the cross sectional area does not exceed 0.003 m².

Internal Finishes 9

	Pict. rls.
	Wrot. swd.
	sktgs., pict. rls, archves.
	& the like
	25 × 56 mo.; plug'g.
	to masonry
20.74	(lnge.
14.64	(d. rm.

	sktgs.
	Wrot. swd.
	sktgs., etc. a. b.
	25 × 100 rdd.; plug'g.
	to masonry

18.90	(hall
14.60	(passage
6.91	(linen cupd.
14.64	(ding. rm.
2/14.64	(bedrms. 1 & 2
12.82	(bedrm. 3

20.74	Knot & prime on wd.
14.64	Gen. surfs.
18.90	isol. ≤ 300 gth.; on site
14.60	prior to fxg.
6.91	&
14.64	Prep. & ③ gloss pt. on
2/14.64	prev. primed wd.
12.82	Gen. surfs.
	isol.; ≤ 300 gth.

	Wrot. maple
	Sktgs., etc. a. b.
	25 × 100 rdd.;
	plug'g. to masonry
20.74	(lounge
	&
	Clear finishg.; prep.;
	2 cts. polyurethane
	varnish on wdwk.
	Gen. surfs.
	isol.; ≤ 300 gth.

Skirtings and picture rails are measured in metres giving a dimensioned overall cross section description as P20:1.1.0.0. The work is deemed to include ends, angles, mitres, intersections and the like, except on hardwood items >0.003 m² sectional area (25 × 100 mm = 0.0025 m²).

The lengths of walls which were calculated earlier can be used for the lengths of skirtings, as the difference is minimal. The method of fixing is described where not at the contractor's discretion (P20:S8). They may be fixed with wood grounds or be plugged to the masonry and the selected method needs to be stated.

The painting is classified as to general isolated surfaces ≤300 girth as M60:1.0.2.4. Priming the entire surface of 25 × 100 skirtings gives an overall girth of 250. It is important to separate the hardwood skirtings from softwood and to apply a clear finish, such as polyurethane varnish, so that the character of the hardwood is retained. The ceramic wall tiling is taken down to floor level and hence no skirtings are required.

Internal Finishes 10

Floors

screeds

Ct. & sd. (1 : 3); one
coat; floated

Flrs.
 34 th.; lev. or to falls
 only ≤ 15° from hor.;
 to conc.

4.27	
3.05	(ding. rm.
3.05	
2.44	(study
3.66	
3.66	(bedrm. 1
1.82	
0.70	(porch

Ditto.
 32 th.; ditto

1.22	
1.73	(clks.

Ct. & sd. (1 : 3); one
coat; trowelled

Flrs.
 42 th.; lev. or to falls
 only ≤ 15° from hor.;
 to conc.

3.05	(bathrm.
1.73	
2.80	(bathrm. E/s
1.83	

Ditto.
 44 th.; ditto.

2.40	
2.23	(hall
3.00	
2.94	(hall
1.78	
1.74	(hall/passage
6.84	
0.92	(passage
1.86	
0.80	(linen cupd.

Sand cement screeds to floors are dealt with in M10:5.1.1.0, with classifications as to falls and slopes. The thickness and number of coats shall be given and the method of surface treatment. The thickness of screed is related to the thickness of the finish and the deepest overall thickness.

In this example the deepest overall thickness above the concrete base is the strip boarding and bearers to the lounge, made up of 22 thick boarding and 25 deep bearers, giving a total depth of 47. It is essential that the different floor finishes have their upper surfaces at the same level to avoid the need for steps. From then onwards the thickness of the screed to each room is found by deducting the thickness of the finish plus a bedding allowance from 47. The thickness of the bedding will vary according to the type of finish.

A floated finish is normally used with inflexible quarry and ceramic tiles and for carpets, while a trowelled finish is associated with the thinner, flexible surfacing materials, such as vinyl tiles.

<u>Internal Finishes 11</u>

	<u>Ct. & sd.; trowelled a.b.</u>
4.27	
3.05	(kitchen Flrs.
1.00	44 th.; lev.
0.42	(kitchen
3.66	
3.66	(bedrm. 2
3.66	
2.75	(bedrm. 3
	<u>Ddt.</u> ditto.
1.10	
1.02	(bedrm. 3

The areas of screed are the same as those used for the ceilings of each room. The screeds have been taken complete as the first flooring component, to avoid having a much greater number of entries by following each floor finish item. It also enables screeds in rooms with an identical thickness and surface finish to be grouped together. The insertion of location notes is very important.

Flr. fins.

<u>Marley flex vinyl tiles, suppld. by Marley Floors; 300 x 300 x 2 mm; Tuscan patt.; fxd. w. Marley Embond adhesive; strt. butt jtd. both ways</u>

Each floor finish is now taken in turn, starting with the vinyl tiling which is being used in five separate areas and is a popular form of domestic floor finish to solid floors. Plastics tiling to floors is covered in M50:5.1.1.0. It is necessary to give the kind, quality and size of materials, nature of base, method of fixing and treatment of joints (M50:S1–8).

	Flrs.
	> 300 wide, lev. or
	to falls only ≤ 15° from
2.40	hor.; to screed
2.23	
3.00	(hall
2.94	
1.78	(hall
1.74	
6.84	(hall/passage
0.92	
1.86	(passage
0.80	
3.66	(linen cupd.
3.66	
3.66	(bedrm. 2
2.75	
	(bedrm. 3
	<u>Ddt.</u> ditto.
1.10	
1.02	(bedrm. 3

The dimensions of each room/area are methodically entered to arrive at the total floor area of this type of finish. No measurements have been taken for insulating layers and damp-proof membranes as these were adequately covered in examples 1 and 6.

Internal Finishes 12

Ceramic tiles to BS6431;
fully vitrified; red; 152
×152×12; jts. 3 wide;
symmet. layout; fxg. w.
thick bed adhesive 3
nom. th.; ptg. in c.m. (1:3).

Flrs.
    plain, lev. or to falls only
    ≤ 15° from hor.; to screed
1.22
1.73
(clks.

A full description of the ceramic tiles is needed to enable the estimator to price the work.
In this case a reference to the appropriate British Standard has been included instead of naming a product and the manufacturer, although the latter method is preferable whenever the information is available. Measurement rules are contained in M40:5.1.1.0, and the >300 wide classification does not apply to flooring measured under M40.

Terrazzo tiles to BS 4131;
300 × 300 × 28; agg. size
random; grd., grouted &
polished; stand. col. range;
jts. 3 wide; sym. layout;
bedd'g. in c.m. (1:3); 19
nom. th.; ptg. in nt. matchg.
ct.; pol. on completn. &
sealed in accord. w.
manufac's. instrs.

Flrs.
    plain, lev. or to falls only
    ≤ 15° from hor.; to conc.
3.05
1.73
(utility

Terrazzo tiles are covered by section M41, but measured in the same way as ceramic tiles in M40.
A full description is needed to give the estimator sufficient information for pricing purposes. Alternatively, this information can be provided in a project specification, when the item description can be reduced to quoting a specification reference or code, such as 'Terrazzo tiling, specification M41/112'. Terrazzo tiling is laid on a mortar bed without the use of a screed.

Clay quarry tiles, heather
brown, suppld. by Dennis
Ruabon Ltd.; 152 × 152
×16; jts. 6 wide; sym.
layout; fxg. w. thick
bed adhesive 3 nom. th.
& grtg. w. Febtile Rainbow
wh. ct., ld. in accord. w.
manufac's. instrs.

Flrs.
    plain, lev. or to falls
    only ≤ 15° from hor.;
1.82
0.70
(porch to screed

The quarry tiles are also measured in accordance with M40 and a specific product has been specified in this case. Although such products normally conform to relevant British Standards, it is unnecessary to quote these in the description once a specific product has been identified.

Internal Finishes 13

Wicanders cork flr. tiles;
ready sealed, med. density;
300 × 300 × 4.8 mm, fxd. w.
adhesive in accord. w. manufac's.
instrs.; strt. butt jtd., both ways

With the cork floor tiles it is also desirable to specify the actual product required. The measurement code in this case is M50:5.1.1.0 and the classification >300 wide appears in the description.

Flrs.
> 300 wide, lev. or to
falls only ≤ 15° from
hor.; to screed

2.80
1.83

(bathrm. E/s

Jaymart 'Sarina Zero'
studded synth. rubber tiles;
indoor qual.; 500 × 500 × 4;
fxd. w. adhesive in accord.
w. manufac's. instrs.; strt.
butt jtd. both ways

Rubber tiles are measured in accordance with M50 and a specific product is specified, in this case to meet the client's wishes. It is important that the tiles are laid in accordance with the manufacturer's instructions.

Flrs.
> 300 wide; lev. or to
falls only ≤ 15° from
hor.; to screed

3.05
1.73

(bathrm.

Carpet tiles; grp. 2 domestic
patternd.; 500 × 500; spot
bonded w. adhesive; strt.
butt jtd. both ways

Carpet tiles are covered by M50:5.1.1.0.

Flrs.
> 300 wide; lev. or to
falls only ≤ 15° from
hor.; to screed

No screed or floor finish has been measured to door openings and they are being left to take with adjustments for the door openings when measuring the doors. In practice it would be better to measure them at the same time as the doors, along with the dividing strips between different floor finishes. To have made the adjustments here would have lengthened considerably an already very long example, without any great benefit to the reader.

3.66
3.66

(bedrm. 1

Internal Finishes 14

Nairn 'Armourflex';
marble patt. sht. linoleum,
2.5 mm th.; nom. wt.
3.1 kg/m²; fxd. w.
adhesive; butt jtd.

Flrs.
>300 wide; lev. or to
falls only ≤ 15° from
hor.; to screed

4.27	
3.05	(kitchen
1.00	
0.42	(kitchen

Linoleum sheeting is also covered by M50:5.1.1.0, and a particular product is required and described.

Edge fxd. carpetg. 80%
wool, 20% nylon; cut &
looped pile, med. domestic
qual.; adhesive taped jts.
fxd. at perimeter w. met.
grippers; on Duralay crumb
rubber underly. ld. loose

Flrs.
>300 wide; lev. or to
falls only ≤ 15° from
hor.; to screed

4.27	
3.05	(ding. rm.
3.05	
2.44	(study

Edge fixed carpeting is measured in accordance with M51:5.1.1.3 and giving the relevant details prescribed in M51:S1–8, including the kind and quality of materials, nature and number of underlays, method of fixing and treatment of joints. Fixing and carpet grippers at the perimeter are deemed to be included (M51:C1d&M6).

Wrot. maple narr. strip
flrg.; 22 th., 75 wide
bds.; t. & g.; fxg. w.
secret nlg. to & incl.
50 x 25 impreg. sn. swd.
battens @ 400 c/s. plg'd.
to conc.; sanded fin.
& apply 2 cts. Bornseal
on completn.

Flrs.
>300 wide

| 6.10 | |
| 4.27 | (lounge |

Timber narrow strip flooring is measured in accordance with K21:2.1.1.0. The rules contain no reference to the supporting battens. Because these are inherently part of the flooring it is appropriate for them to be included as part of the description, rather than being measured separately.

Dividing strips between different floor finishes are classified as accessories (M10/20:24.7, M40:16.4 & M50/51:13.5), and measured in metres with a dimensioned description. A typical item in a door opening would be:

Accessories
0.75	Dividg. strip, plastic,
	6 × 16 flat sec.; set in
	c.m. (1.3)

10   Windows

ORDER OF MEASUREMENT

The normal order of measurement of windows is to take the windows first, followed by associated components, ironmongery, glazing and painting. This is followed by the adjustment of the window opening working in a logical sequence such as

(1) deduction of walling and finish on both faces;
(2) head of opening (arches and lintels);
(3) external reveals;
(4) internal reveals;
(5) external sill;
(6) internal sill.

By working systematically through the dimensions in this way, the risk of omission of any items is much reduced.

Some surveyors prefer to measure the adjustment of the openings in the first instance and then follow with the actual windows.

WINDOWS

Wood casements and sash windows together with their frames are enumerated and described, accompanied by a dimensioned diagram, as illustrated in example 19 (L10:1.0.1.0), giving the appropriate particulars listed in L10:S1–10. Where components are constructed using standard sections, these must be identified and described appropriately (L10:M1). Member sizes will normally be nominal sizes, but if finished sizes this must be stated (L10:D1). The need to give finished sizes usually arises when working from drawings which have their dimensions indicated in this manner. Metal and plastics windows are measured similarly to those in timber. The items include architraves, trims, sills, subframes and the like and finishes where they are part of the component, and also ironmongery and glazing where supplied with the component, and fixings and fastenings (L10: C2b–e&g). Details of such features must be included in descriptions in accordance with General Rules 2.11, otherwise the estimator may be unaware of their existence. The enumerated window item is followed by any additional items such as subsills, window boards, bedding and fixing frames, painting the backs of frames, and the like.

When measuring glazing, allowance must be made for the rebates in the enclosing members of the window and panes of irregular shape shall be classified and measured as the smallest rectangular area from which they can be obtained and so described (L40:1.1.1–2.2 and L40:M3). A full description of the glazing must be given in accordance with L40:S1–5, and standard plain glass (L40:D2) is measured in m^2, except for louvres which are enumerated, and giving the pane size classification as L40:1.1.1–2. Where glazing rebates are 20 or more deep, they are classified in 10 mm stages.

Ironmongery is enumerated in accordance with P21.1.0.0, and must be fully described where items are to be supplied and fixed. However, prime cost sums are often used to cover the supply of ironmongery, when separate items for fixing only will be required. A typical PC item is contained in example 19.

Painting is measured in m^2 to each side of windows and giving the pane size classification in the description as M60:2–3.1–4.1.0. Where panes of more than one size occur then the sizes are averaged (M60:M6). External painting must be kept separate and so described (M60:D1), in order that the estimator can allow in his price for the different working conditions. The work is deemed to include the edges of opening lights and portions uncovered by sliding sashes in double hung casements, additional painting to the surrounding frame caused by opening lights, cutting in next glass and work on glazing beads, butts and fastenings attached thereto (M60:C4). Work to associated linings and sills is measured as general surfaces (M60:M7).

Many windows installed in both new and old buildings are of the standard or stock pattern variety and so typical products have been included in example 19. In such cases General Rule 6.1 permits a precise and unique cross reference to a catalogue or standard specification to be given instead of the detailed description normally required by the rules. In many cases the standard product includes the provision of ironmongery, double glazing units and glazing beads and, in the case of PVC-U windows there is no need to paint the finished surface of the windows, culminating in a much reduced number of items to be measured.

There are also alternatives for the various components included in example 19, such as steel and precast concrete lintels over window openings, single and double glazing to windows and painting and staining of wood surfaces, to extend the coverage and usefulness of the example for the benefit of students.

ADJUSTMENT OF WINDOW OPENINGS

The adjustment of the superficial area of walling, including facework, and internal finishes for the space occupied by the window and frame is followed by the various items, mostly linear, that have to be taken around the opening. The worked examples that follow indicate the method to be followed with a variety of different forms of construction. This work needs to be taken off very carefully and systematically.

The adjustment of window openings is influenced considerably by the form of the construction and finishes, as illustrated in examples 19 and 20. Example 20 covers purpose made aluminium casements set in a stone surround, to illustrate the different problems posed in this situation and to give the student further guidance in the measurement of stonework. It seems desirable at this stage to define some of the principal labours on stone:

> *circular*: curved convex surface
> *circular sunk*: curved concave surface
> *sunk jointed*: sunk face adjoining brickwork or masonry
> *circular jointed*: ditto. with curved convex surface
> *circular sunk jointed*: ditto. with curved concave surface

When considering the stone dressings in example 20, it is assumed that the masonry terms faced one side or both sides relate to walls, whereas components such as sills, jambs and lintels are best described by the number of rubbed faces.

WINDOW SCHEDULES

In most cases, when measuring a complete building, it is desirable to compile a window schedule, if this has not already been prepared by the architect, on which all the essential details relating to the windows and finishings generally will be entered. Column headings could cover such matters as location of window, type of window, size of opening, ironmongery, glass size classification, type of glass, wall type and thickness, internal and external finishes, and possibly details at heads and jambs of openings and sills. This approach is illustrated in table 10.1. Architects' window schedules are normally prepared with the contractors' needs in mind and therefore are usually limited to information relating to the windows and their ancillary components. In these circumstances the quantity surveyor may wish to prepare a separate openings schedule. The benefits of this will be greatest where there are a large number of openings of varying size, types of construction and wall finish.

With the use of schedules all the windows with similar features can readily be seen and measured together. This simplifies the task of taking off and collation of similar items, and reduces the number of entries involved. Care must be taken to enter the correct timesing figures in each case.

WORKED EXAMPLES

Worked examples follow covering wood, plastics and metal casements, together with the adjustment of openings. The example of wood casement windows has been extended to cover eight windows in a small bungalow to provide a situation more akin to that occurring in practice and this is supported by a window schedule (table 10.1).

Wood casement windows   **DRAWING 12**

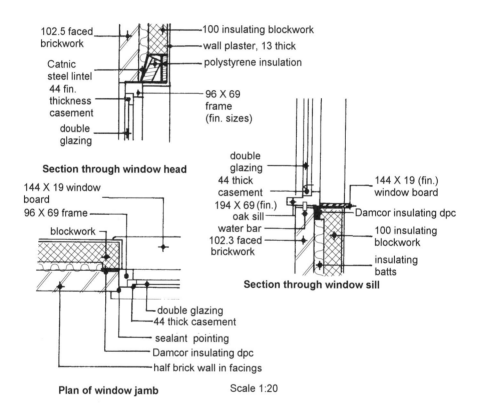

Window type A

Scale 1:50

Window type B

glazing
direct to
frame

1.350

1.770

1.200

102.5 faced
brickwork

Catnic
steel lintel

44 fin.
thickness
casement

double
glazing

100 insulating blockwork

wall plaster, 13 thick

polystyrene insulation

96 X 69
frame
(fin. sizes)

Section through window head

144 X 19 window
board

96 X 69 frame

blockwork

double glazing
44 thick casement

sealant pointing

Damcor insulating dpc

half brick wall in facings

double
glazing

44 thick
casement

194 X 69 (fin.)
oak sill

water bar

102.3 faced
brickwork

144 X 19 (fin.)
window board

Damcor insulating dpc

100 insulating
blockwork

insulating
batts

Section through window sill

Plan of window jamb   Scale 1:20

Table 10.1 Schedule of windows to bungalow.

Location	Window & opening size	Window type	Glazing	Finish to window	Ironmongery	Lintel details	External finish to wall	Internal finish to wall	External sill	Internal sill	Window head
Lounge	two 1.770 × 1.350	2 light 2 opg. wood casements	double glazing clear	gloss paint	Pr. 75 steel butts per casement (s and f) 2 nr. casement stays (PC) 1 nr. casement fastener (PC)	Catnic classic steel lintel ref. CN7F length 2.100	faced brickwk.	plaster & vinyl paper	hardwood sill 194×69 (fin.)	softwood window board 144 × 19 (fin.)	faced brickwork
Dining room	one 1.770 × 1.350	ditto.	ditto.	ditto.	ditto.	ditto.	ditto.	ditto.	ditto.	ditto.	ditto.
Kitchen	ditto.	ditto.	ditto.	ditto.	ditto.	ditto.	ditto.	ct-sd backing & ceramic tiles	ditto.	ceramic tiles	ditto.
Utility	one 1.200 × 1.350	ditto.	ditto.	ditto.	ditto.	ditto. ref CN.E length 1.500	ditto.	plaster & twice emulsion	ditto.	softwood window board 144 × 19	ditto.
Bathroom	ditto.	ditto.	double glazing obscured	ditto.	ditto.	ditto.	ditto.	ct-sd backing & ceramic tiles	ditto.	ceramic tiles	ditto.
Bedroom 1	ditto.	ditto.	double glazing clear	ditto.	ditto.	ditto.	ditto.	plaster & twice emulsion	ditto.	softwood window board 144 × 19	ditto.
Bedroom 2	ditto.	ditto.	ditto.	ditto.	ditto.	ditto.	ditto.	ditto.	ditto.	ditto.	ditto.

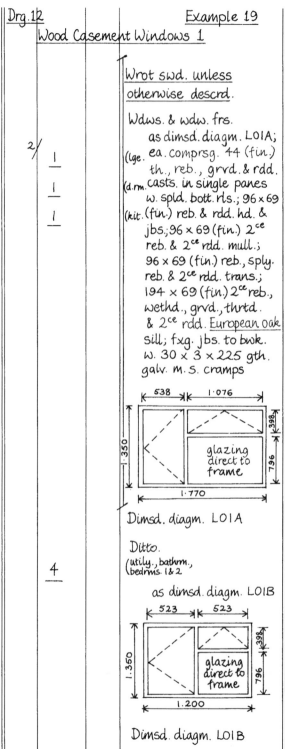

Drg.12

Example 19

Wood Casement Windows 1

Wrot swd. unless
otherwise descrd.

2/

Wdws. & wdw. frs.
as dimsd. diagm. LOIA;
(lge. ea. comprsg. 44 (fin.)
     th., reb., grvd. & rdd.
(d.rm. casts. in single panes
     w. spld. bott. rls.; 96 × 69
(kit. (fin.) reb. & rdd. hd. &
     jbs.; 96 × 69 (fin.) 2ᶜᵉ
     reb. & 2ᶜᵉ rdd. mull.;
     96 × 69 (fin.) reb., sply.
     reb. & 2ᶜᵉ rdd. trans.;
     194 × 69 (fin.) 2ᶜᵉ reb.,
     wethd., grvd., thrtd.
     & 2ᶜᵉ rdd. European oak
     sill; fxg. jbs. to bwk.
     w. 30 × 3 × 225 gth.
     galv. m.s. cramps

1

1

1

Dimsd. diagm. LOIA

Ditto.
(utily., bathm.,
(bedrms. 1 & 2

4

as dimsd. diagm. LOIB

Dimsd. diagm. LOIB

The eight windows to the bungalow listed in the accompanying schedule of windows (table 10.1) have been measured in accordance with L10:1.0.1.0, incorporating dimensioned diagrams. It is necessary to list the various members with their dimensions and the labours on them, as the small dimensioned diagrams could not possibly incorporate all the necessary details appertaining to the component parts.

All timber sizes are deemed to be nominal sizes, unless stated as finished sizes (see L10:D1). The latter option has been taken in this case because the drawing is dimensioned with finished sizes.

Where the ironmongery is supplied with the window, it can be included in the description of the item (L10:C2c). Similar provisions apply to sills, subframes, finishes, glazing, fixings and fastenings (L10:C2).

In this case the ironmongery is not supplied with the window and is measured separately in accordance with P21.

Wood Casement Windows 2

An alternative approach is to give a brief description and to refer to a more detailed drawing to be supplied with the bill of quantities at the tendering stage, as illustrated below

Wrot swd. w. European oak sills

4 | Wdws. & wdw. frs.
1.770 x 1.350;
type A as drg. 19;
fxg. jbs. to bwk.
w. 30 x 3 x 225 gth.
galv. m.s. cramps

Ditto

4 | 1.200 x 1.350
type B as drg. 19, do.

Another alternative would be to use stock pattern or standard windows to be obtained from the product range of a named joinery manufacturer or in accordance with a standard specification

Wrot. swd.

4 | Wdws. & wdw. frs.
std. unit; John
Carr code 313 cw.;
size 1.770 x 1.350;
incl. std. ironmgy.
& glazg. bds.; fxg.
jbs. to bwk. w. galv.
m.s. cramps

Some quantity surveyors may prefer the alternative approach as adopted here, as it eliminates the need for dimensioned diagrams and long descriptions. In practice large scale drawings would normally be produced, but limitation on the page size in this book prevents this. Nevertheless, most of the essential details have been illustrated. It is very important to recognise that copies of drawings must be supplied to the contractor with the bill of quantities and the descriptions must contain a cross reference to the drawing (SMM Measurement Code 4.4).

A high proportion of new buildings contains stock pattern windows, which can take a large variety of forms and often offer better value for money than purpose made windows. Hence an appropriate standard window has been described here in order that the student is made aware of a suitable approach. John Carr windows incorporate lockable fasteners and easy clean hinges, and can be pre-slotted for ventilators. The manufacturer's code or catalogue reference must be included in accordance with General Rules 6.1. Manufacturers frequently offer options with regard to type of timber, glazing, ironmongery and type of factory finish. Where this is the case the item descriptions must include the appropriate details.

<u>Wood Casement Windows 3</u>

Another alternative is the
use of PVC-U windows as
covered in the next item

Plastics

4

Wh. PVC-U cast. wdws. & wdw. frs
as dimsd. diagm. LO2A;
manfd. by Marco Glass &
Glzg.; using std. multi-chambd.
sectns. of high impact resistant
extrusn. & fully welded constrn.;
size 1.770 x 1.350, comprsg. 1 nr.
fxd. lt. & 2 nr. side hg. in 3 nr.
panes; incl. s.s. frictn. stay
hinges & high security shootbolt
lockg. mechanism; factory glzd.
w. d. glazg. units 28 th. to
Brit. kitemk BS5713-1979;
beaded extly.; fxg. to bwk. w.
cleats & scrs.

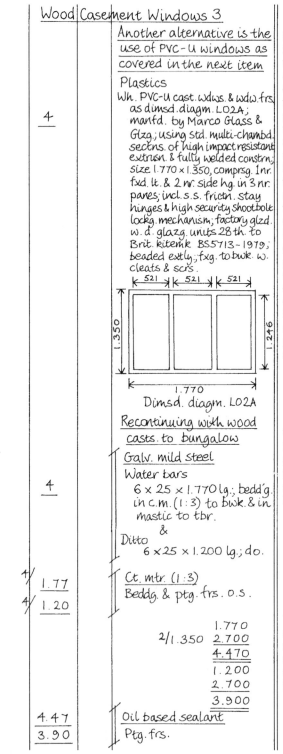

Dimsd. diagm. LO2A

Recontinuing with wood
casts. to bungalow

4

Galv. mild steel
Water bars
6 x 25 x 1.770 lg.; bedd'g.
in c.m. (1:3) to bwk. & in
mastic to tbr.
&
Ditto
6 x 25 x 1.200 lg.; do.

4/ 1.77
4/ 1.20

Ct. mtr. (1:3)
Beddg. & ptg. frs. o.s.

```
        1.770
2/1.350 2.700
        4.470
        1.200
        2.700
        3.900
```

4.47
3.90

Oil based sealant
Ptg. frs.

Plastics windows are being used increasingly as a substitute for wood, where permitted by the planning authority, as although more expensive they offer more resistance to vandals and eliminate the need for painting at regular intervals. PVC-U windows are measured in accordance with L10:1.0.1.0. L10:M1 requires the standard sections to be identified and this has been done in the accompanying description. The use of such sections does not mean that the window is a standard component as identified by General Rules 6.1, and therefore a dimensioned diagram must be provided or the description may contain a cross reference to a drawing supplied with the bill of quantities. Plastics windows include hinges, fasteners and double glazing units. This comprehensive item also picks up the fixing of the unit to the adjoining masonry.

Water bars are classed as ironmongery (see SMM7 Appendix A) and measured in accordance with P21:1.1.0.0. Grooves in brickwork to receive the water bar are deemed to be included (F10:C1c).

The bottom of the sill is bedded and pointed in mortar, while the outer edge of the jambs and head are pointed in a flexible sealant without any mortar bedding (see L10:8-10.0.0.0).

Wood Casement Windows 4

		1.770	1.200
	add ends 2/50	100	100
		1.870	1.300

Wrot swd.
Wdw. bds.

(lge. 19 × 144 (fin.);
(d/r reb.; rdd. one edge;
(bed plg'g. to masonry
rms.
& bath
rm.

3/ 1.87

3/ 1.30

The wood window boards follow the requirements of P20:4.1.0.0, with a dimensioned overall cross section description. All timber sizes are deemed to be nominal sizes unless described as finished sizes. The work is deemed to include ends, angles and the like except on hardwood items >0.003 m² sectional area (P20:C1).

Imgry.

Pressed steel

Prs. butt hinges,
75 to swd.

2/4/ 2

The timesing figures relate to the two different sizes of window with four windows of each size. There are two opening lights to each window with a pair of butt hinges to each light. The measurement follows P21:1.1.0.0, giving the type of unit and nature of base, and is deemed to include fixing with screws (P21:C1).

Include P.C. sum for
supply'g. only mats.

Imgry. £ _ _ _ _

&

Add for
    main contr's.
    profit   %

Item

The supply of ironmongery items covers the fasteners and stays to the opening lights, using the approach outlined in A52:1.1.1.0 and A52:1.2.0.0.

Fixg. only

Cast. fasteners
    to swd.

2/4/ 1

Separate fixing items follow the supply items, stating the nature of the base to which the ironmongery is to be fixed (P21:1.1.0.0), with a fastener and a stay to the side hung lights and a stay to the top hung lights.

Cast. stays
    to swd.

2/4/ 2

Glazg.
larger wdw.

	width	ht.
less stiles to opg.	538	1.350
lt. 2/34	68	2/52/34/86 172
	470 ×	1.178

It is necessary to deduct the width of opening light members and/or window frames, less rebates for glass, to arrive at the sizes of the sealed double glazing units.

<u>Wood Casement Windows 5</u>

		width	ht.
		1.076	398
less 2/34		68	68
		1.008 ×	330
		1.076 ×	796

smaller wdw.

		523	1.350
less 2/34		68	2/86 172
		455 ×	1.178
		455 ×	330
		523 ×	796

Special glass
Hermetically sealed double
glazg. units; comprisg. 2 × 4
mm clear float glass w. air
space cav. 6 wide.

To wd. w. scrd. bds. & non
setting compound
    size 470 × 1.178;
    30 – 40 glzg. rebs.
        &
Ditto.
    size 1.008 × 330; do.
        &
Ditto.
    size 1.076 × 796; do.

Ditto.
    size 455 × 1.178; do.
        &
Ditto.
    size 455 × 330; do.
        &
Ditto.
    size 523 × 796; do.

Special glass Hermetically
sealed double glazg. units;
compris'g. inner pane of
4 mm clear float glass and
outer pane of Pilkington
Cotswold texture glass w.
air space cav. 6 wide.

To wd. w. scrd. bds. & non
setting compound
(bath size 455 × 1.282;
 rm.   30 – 40 glzg. rebs.
        &
Ditto
    size 455 × 330; do.
        &
Ditto
    size 523 × 796; do.

4

3

1

The same procedure is adopted for each glazing unit for both window types in accordance with dimensioned diagrams LO1A and LO1B. Where the glazing is fixed direct to the frame in fixed lights, it would be wrong to deduct the net widths of opening light members as there are none.

As normal present day practice, the windows are glazed with hermetically sealed double glazing units. They are measured in accordance with L40:3.1.1.3. L40:D4g includes sealed double glazing units under the classification of special glass. The units are enumerated with a dimensioned description. Where glazing rebates are 20 mm wide or over, the width must be stated in 10 mm stages. The width referred to is the platform width or the face which the edge of the glass is in contact with. 14 mm double glazing units normally require 32 mm wide rebates.

The widespread use of the word ditto. should be noted as this permits a substantial reduction in the length of item descriptions.

The units to the bathroom contain Cotswold texture glass to a high obscuration classification and adequate privacy.

<u>Wood Casement Windows 6</u>

<u>As an alternative single glazg. could be used to reduce initial cost and a typical example follows</u>

An example of the measurement of single glazing is given as the approach is different from that for double glazing. It is measured in m², giving a pane size classification and stating the number of panes where ≤ 0.15 m² in area as required by L40:1.1.2.5. The kind, quality and thickness of glass, glazing compound, method of glazing and securing the glass and nature of frame or surround are to be stated (L40:S1–5).

<u>Std. pl. glass; 6 clear float to BS 952</u>

To wd. w. l. o. putty in panes, area 0.15 – 4.00 m²; 30 – 40 glzg. rebs.

4/	0.47
	1.18
4/	1.01
	0.33
4/	1.08
	0.80

```
1.770      1.200
1.350      1.350
2/3.120   2/2.550
6.240      5.100
```

<u>Paintg.</u>

<u>Paintg. wd.; 1 ct. primer on site prior to fxg.</u>

Gen. surfs.
   isol.; $\leq$ 300 gth.
(back of wdw. fr.

It is advisable to paint the back surfaces of wood frames with primer prior to fixing, to reduce the risk of moisture from adjoining construction being absorbed by the timber. This item is measured in accordance with M60:1.0.2.4.

4/	6.24
4/	5.10

<u>Pane sizes</u>
<u>larger window</u>

```
0.470 x 1.178 = 0.554
1.008 x 0.330 = 0.333
1.076 x 0.796 = 0.856
            3)1.743
0.50 - 1.00 m²   0.581
```

<u>smaller window</u>

```
0.455 x 1.178 = 0.536
0.455 x 0.330 = 0.150
0.523 x 0.796 = 0.416
            3)1.102
0.10 - 0.50 m²   0.367
```

In the case of glazed windows and screens the windows are to be classified according to the area of the glass panes which they contain. The range progresses from ≤ 0.10 m² to >1.00 m². The pane areas are those of individual panes (M60:D9), and where panes of more than one size occur, the sizes are averaged (M60:M6) as has been done in the calculations in waste of this example.

<u>Wood Casement Windows 7</u>

<u>Paintg. wd.; k.p.s.,</u>
<u>2 u/c, 1 f/c glass</u>

Glzd. wdws. & scrns.
> 300 gth.; in panes,
area 0.50 – 1.00 m²
&

4/ 1.77
   1.35

Ditto.
            EXT.

Glzd. wdws. & scrns. a.b.
in panes, area
0.10 – 0.50 m²
&

4/ 1.20
   1.35

Ditto.
            EXT.

3/ 1.87

Gen. surfs.
    isol. ≤ 300 gth.

3/ 1.30

(wdw. bds.

<u>An alternative treatment would be to apply</u>
<u>wood stains which are individually suitable</u>
<u>for either interior or exterior use, as</u>
<u>illustrated in the following example</u>

<u>Staing. wd.; prep., 2 cts.</u>
<u>Sadolin Harmoni w.</u>
<u>satin sheen fin.</u>

Glzd. wdws. & scrns.
> 300 gth.; in panes,
area 0.50 – 1.00 m²

4/ 1.77
   1.35

<u>Staing. wd.; prep.; 2 cts.</u>
<u>Sadolin New Base & 2 cts.</u>
<u>Sadolin Extra</u>

Glzd. wdws. & scrns.
> 300 gth.; in panes,
area 0.50 – 1.00 m²

4/ 1.77
   1.35

            EXT.

The painting of glazed windows is measured in accordance with M60:2.1–4.1.0, and clearly identifying any external work, because of the more variable conditions under which it is carried out. Work is deemed to be internal unless otherwise described (M60:D1). The area is measured to each side of windows (M60:M5) and is deemed to include edges of opening lights, additional painting to the surrounding frame caused by opening lights and work on glazing beads, butts and fastenings attached thereto (M60:C4).

The paint film includes knotting, priming (one coat) and stopping, to make good defects in the surface of the wood. Windows are often delivered to site ready primed. This would be included with the window items (L10:C2d). The subsequent on site painting items are then described as on previously primed surfaces. When measuring window boards the windows to the kitchen and bathroom are omitted as the internal sills are in ceramic tiles.

An alternative treatment is to use wood stains which are produced in a variety of colours, including several resembling hardwood finishes. Different treatments are usually advocated for interior and exterior use. It is necessary to refer to the manufacturers' product details when framing a description.

Wood Casement Windows 8

Adj of opgs.

Ddt. Fcg. bwk. in
g.m. (1:1:6) a.b.
Walls
    h.b.th.; facewk.o.s.

4/   1.77
      1.35

&

4/   1.20
      1.35

Ddt. Blkwk. in
Thermalite conc. blks.
in g.m.(1:1:6) a.b.
Walls
    100 th.
&
Ddt. Sundries
Formg. cav. in holl.
walls & insul. a.b.

Ddt. Pla. a.b.
Walls
     >300 wide; to blkwk.

3/   1.77
      1.35
      (lge.
      &dg.
      rm.

3/   1.20
      1.35
      (utility
      (& bedrms.

Ddt. Ct. & sd. (1:4) a.b.
Walls
     >300 wide; to blkwk.
(kitchn.

      1.77
      1.35
      1.20
      1.35

(bathrm.   &
Ddt. Ceramic tilg. a.b.
Walls
     >300 wide; to c/s backg.

Ddt. Hang'g only ling.paper
& washable vinyl paper a.b.
Walls & cols.
     areas > 0.50 m²; to pla.
(lge. & ding. rm.

3/   1.77
      1.35

Ddt. Emulsn. pt. to pla. a.b.
Gen. surfs.
(utility   >300 gth.
(& bedrms.

3/   1.20
      1.35

The sequence of items in the adjustment of the window openings follows a logical sequence, but having regard to the schedule of windows to the bungalow (table 10.1), as the internal finishes vary between rooms, with cement and sand backing to the kitchen and bathroom walls in lieu of plaster, as a base for the ceramic tiling. Similarly, the plaster receives emulsion paint in the utility and bedrooms and vinyl paper in the lounge and dining room.

The descriptions of the work are abbreviated substantially by the use of the letters 'a.b.' (as before), indicating that the items have been fully described previously and hence there is no need to repeat lengthy descriptions.

Standard Method classifications are given wherever possible to ensure uniformity and avoid possible confusion. It is helpful to other persons, who may refer to the dimensions at a later stage, to give a note of the locations of the work where appropriate.

<u>Wood Casement Windows 9</u>

<u>Proprietary items</u>

4	Catnic classic galv. stl. lintels, polyester ctd. 245 × 143 o/a. & 2.100 lg.; ref. CN7F w. cont polystyrene insul.; bldg. in.
	Ditto
4	245 × 143 o/a. & 1.500 lg.; ref. CN7E do.

<u>An alternative to steel lintels is precast concrete as illustrated in the next item</u>

<u>Pre. conc., designed mix C25 20 agg. to BS 5328</u>

4	lintels 230 × 215 × 2.100 lg.; boot sh.; reinfd. w. 4 nr. 12 ∅ m.s. bars & 6 ∅ stirrups at 150 ccs. to BS 4449; keyed for pla. on 2 nr. sides 365 gth.; fin. fair on 2 nr. sides 115 gth.; 2 nr. reb. ends; beddg. in g.m. (1 : 1 : 6)
	Ditto
4	230 × 215 × 1.500 lg. do.

<u>len.</u>
	2.100	1.500
add ends 2/75	150	150
	2.250	1.650

<u>width</u>
bwk.	100
slopg. sec.	160
blkwk.	100
	360

D.p.c.'s of single lyr. bit felt a.b.
Cavity trays
> 225 wide; stepped

4/	2.25
	0.36
4/	1.65
	0.36

The use of steel lintels to support masonry above window and door openings is becoming increasingly popular and extremely effective. The measurement requirements are contained in F30:16.1.1.0 and include a dimensioned description and the manufacturer's reference. Lengths are available in increments of 150 up to 3.000 and 300 increments for longer lintels. No deduction on the blockwork inner skin is required for the steel lintels as their height of 143 only displaces part of a 225 high block course, and deductions are only made for the full course displaced (F10:M3).

Precast concrete lintels are enumerated in accordance with F31:1.1.0.1, giving a dimensioned description and details of reinforcement. It is also necessary to include the kind, quality and mix of materials, bedding and fixing and surface finishes (F31:S1–5).

There must be adequate bearings at each end of lintels, preferably in multiples of brick and/or block lengths, but where standard products are being used this may not always be possible. Formwork is deemed to be included (F31:C1).

With the bungalow, no brick flat arches are needed as the top courses of brickwork are hidden behind the soffit boarding and fascia. Where required arches will be measured in metres giving the height on face, thickness, width of exposed soffit and shape of arch (F10:6.1.0.0). Adjustment of the faced brickwork will also be required.

The damp-proof course forming a cavity tray above the lintels is measured in accordance with F30:2.2.4.1. Unless required by the local authority or in exceptionally exposed situations, Catnic lintels themselves provide an adequate dpc.

		Wood Casement Windows 10

Working logically down the window leads to the reveals which are closed with a combination of blockwork and a proprietary product in the form of an insulating dpc which is necessary to meet Building Regulation requirements relating to the avoidance of cold bridging around openings. Closing cavities are measured in accordance with F10:12.1.1.0 with dpc's measured separately as F30:2.1.1.0, but as the dpc is also acting partially as a closure it is appropriate to measure a single item.

Six of the windows have plaster reveals measured in metres as ≤300 wide as M20:1.2.1.0. Note the extensive use of 'a.b.' to avoid repetition of descriptions which are given in full in example 18. The remaining two windows have cement and sand backing to receive ceramic tiles and the same measurement rules apply (M10:1.2.1.0).

The same finishing materials are now taken to the heads of the window openings, but had to be separated as the perforated bottoms of the Catnic lintels form the key for the plaster and backing, and this is included in the item description as nature of base (M10/M20:S5).

Calculate the combined length of head and reveals for use where common finishes apply.

Reveals

Blkwk. in g.m. a.b.

2/4/2/ 1.35

Clos'g. cav.
53 wide w. blkwk.
100 th. & Damcor insulating d.p.c. comprs'g. exp. polystyrene strip size 100 × 18 W. bonded polyethylene d.p.c. 165 wide; vert.

Pla. a.b.

6/2/ 1.35

Walls
≤ 300 wide; to blkwk.

Ct. & sd. (1:4) a.b.

2/2/ 1.35

Walls
≤ 300 wide; to blkwk.

Fins. to hd of opgs

Pla. a.b.

3/ 1.77
3/ 1.20

Walls
≤300 wide; to perf. met.

Ct. & sd. (1:4) a.b.

1.77
1.20

Walls
≤ 300 wide; to perf. met.

Combnd. len. (hd. & reveals)

hd.   1.770   1.200
reveals 2/1.350 2.700   2.700
                    4.470   3.900

Wood Casement Windows 11

3/ 4.47	Pla. a.b.
3/ 3.90	Rdd. angles & intersecs.
	Emuls. a.b.
3/ 3.90	Gen. surfs.
0.10	>300 gth.
	(utility
	(& bedrms.
	Hang'g. only vinyl paper a.b.
3/ 4.47	Walls & cols.
0.10	areas > 0.50 m²; to pla.
	(lge.&
	(dg. rm.
	Ceramic tilg. a.b.
	Walls
4.47	(kitch. plain width ≤ 300; to
3.90	(bathrm. c/s. backg.
	Polystyrene bd. insul. 25 th.
4/ 2.10	Pl. areas
0.14	vert.; fxg. to perf.
	met. w. adhesive
4/ 1.50	&
0.14	Ddt. Pla. a.b.
	Walls
	> 300 wide; to blkwk.
	&
	Add
	Ditto
	do.; to polystyrene insul.
	Sills
	Blkwk. in g.m. a.b.
4/ 1.77	Clos'g. cav.
	53 wide w. blkwk. 100 th.
4/ 1.20	& Damcor insulating
	d.p.c. comprs'g. exp.
	polystyrene strip size
	100 x 18 w. bonded
	polyethylene d.p.c.
	165 wide; hor.

Rounded angles exceeding 10 radius are taken to the external angles of the plaster around the six window openings as M20:16.0.0.0.

The emulsion paint to the plastered surfaces of the three window openings is taken as >300 girth as the work will be taken in with the same finish to the adjoining wall surfaces (M60:1.0.1.0).

A similar approach has been adopted for the vinyl paper (M52:1.1.0.2).

The ceramic tiling to the sides and soffits of the window openings is taken as linear items as M40:1.2.0.0. The tiling is classed as work to the abutting walls (M40:D5) and the width is the width of each face (M40:M5).

The eight lengths of polystyrene boards fixed to the inner face of the steel lintels are measured as P10:3.1.2.0.

The closing of the cavity below the sill is measured in the same way as for the reveals but it is now horizontal instead of vertical (F10:12.1.3.0).

Wood Casement Windows 12

Ceramic tilg. a.b.

Sills

1.77	(kit. 150 wide; to c/s backg.
1.20	(bathrm. &

Ct. & sd. (1:4) a.b.

Walls
≤ 300 wide; to blkwk.

1.770	1.200
1.350	1.350
2/ 3.120	2/ 2.550
6.240	5.100

Ceramic tilg. a.b.

E.o. for
   rdd. edge tiles

6.24	(kit.
5.10	(bathrm.

The ceramic tiled sill is a linear item, stating the width as M40:7.0.1.0. This work is deemed to include fair edges, and internal and external angles (M40:C4).

The cement and sand bed to the tiled sill is taken under margins (M10:11.0.1.0) as being the most appropriate item.

Metal casement windows

Portland stone keystone

float glass 6 thick

metal casement

Portland stone sill

brown facing bricks internally and red facing bricks externally

cavity

insulation

Section
Scale 1:20

300   300   900   225

255 cavity wall

150   225

50 X 25 copper dowels

H

G

J

F

E   D   C   B

A

Elevation

150   300   150   600   150   300   150

50

metal casement

100

Drg. 13		**Example 20**
		Metal Casement Windows 1

Wdws.

Include P.C. Sum for supplyg. only mats.

Item

3 nr anodised alum. casts. incl: 1 nr. w. semi-circ. hd. & opg. lt. £300
&
Add for main contr's. profit %

These windows are purpose made items and not stock pattern and hence their supply needs to be covered by a prime cost sum incorporating a description of the components in accordance with A52:1.1.1.0. This is followed by a main contractor's profit item with provision for a percentage addition (A52:1.2.0.0).

Fixg. only

Alum. wdws. & wdw frs. 325 × 925; as drg. 13; plug'g. & scrg. to stnewk.

2/ 1

Ditto w. semi-circ. hd. 625 × 1.225 o/a; do.

1

The supply items are followed by the fixing only items, giving the information prescribed in L10:1.0.1.0. The larger casement with its semi-circular head needs so describing because of the higher cost of fixing. The type and spacing of fixings will normally be covered in a preamble clause or the project specification, where available.

len.

4/300	1.200	
	600	
6/900	5.400	
½ × π D = ½ × 22/7 × 600	943	
extl.	8.143	

4/325	1.300	
	625	
6/925	5.550	
½ × 22/7 × 625	982	
intl.	8.457	

The external and internal lengths of sealant pointing to the window frames, including sills, are calculated in waste. They are measured in metres as L10:9.0.0.0.

8.14
8.46

Oil based sealant
Ptg. frs.

Glazg.

	width	ht.
side lts.	300	900
less frs. 2/10	20	20
	280	880

The glazing has been taken as single glazing to give the student experience in measuring this class of work.

Metal Casement Windows 2			

		width	ht.
	centre opg. lt.	600	900
	less frs. & cast.		
	membrs. 2/20	40	40
		560	860
	fanlight	600	300
	less frs. 2/10	20	20
		580	280

The glass dimensions are calculated in waste by deducting the width of frames and casement members where there is an opening light.

Std. plain glass; 6
clear float to BS 952

To met. w. clip on met.
bds. & extruded rubber
gaskets
(side in panes, area
(lts.   0.15 – 4.00 m²
(central
(opg.
(lt.)

2/ 0.28
    0.88
    ───
    0.56
    0.86
    ───

The definition of standard plain glass is given in L40:D2. The descriptions of the glass and glazing are given in accordance with L40:1.1.2.0.

Ditto
  in irreg. shaped
  panes, area 0.15 –
  4.00 m²

    0.58
    0.28
  (fault.

Panes of irregular shape are classified and measured according to the smallest rectangular area from which the pane can be obtained (L40:M3 and L40:1.1.2.2). Curved cutting to the glass around the semi-circular head is deemed to be included (L40:C1).

BS 4873 for aluminium windows requires one of the following finishes:
  • anodising
  • liquid organic coating
  • powder coating.
Such finishes do not normally require any further site painting. Steel windows may be finished either by powder coating or galvanising. In the latter case, site painting is required and is measured in accordance with M60:2.1–4.1.0.

		Metal Casement Windows 3

Metal Casement Windows 3

Adj. of opg.

Nat. stwk. dressgs. in P. st., Whitbed; rubd. on exposed faces; b. & j. in c.l.m. (1 : 2 : 9); flush smth. ptg. a.w.p.; cleang. on completn; to bk. faced walls

len. of sill
2/225	450	
2/300	600	
2/150	300	
	600	
	1.950	

1.95	Sills

1.95 — Sills
255 × 225; sk. wethd. 225 gth.; w. stool'gs. (4 nr.); ea. blk. >1.500 lg.; fcd. 3 sides
(A

2/2/ 0.23 — Isoltd. jamb. sts.
150 × 255 th.; reb. 55 gth; chmfd. 112 wide; fcd. 3 sides
(B & D

&

225 × 255 th.; ditto
(C & E

2/ 0.65 — Lintels (2 nr.)
255 × 225 high; reb. 55 gth.; chmfd. 112 wide; ea. sk. jtd., circ. sk., circ. reb. 55 gth. & circ. chmfd. 112 wide at one end; w. stool'gs. (2 nr.); fcd. 3 sides.
(F

The type of stone and method of jointing and pointing are given in a common heading, which encompasses all the individual stones that follow. Stone dressings, as the stones in this example, are where they are erected in walls of other materials (F21:D2). The extensive list of supplementary information in F21:S1–12 will preferably be covered in a project specification to which reference can be made in the bill of quantities. It is good practice to number or letter the stones, as inserted on drawing 13, for ease of reference. F21:D3 expects the use of dimensioned diagrams unless descriptions are sufficient for full clarity. In this case dimensioned diagrams have not been used and it is therefore important to ensure that all relevant information is given.

The sill is measured as F21:7.1.0.1–16, giving the function, size, labours and number of stoolings. Blocks >1.500 long or >0.50 m³ are so described.

The jamb stones are also taken as linear items as F21:11.3.1.0, giving a dimensioned description and stating whether isolated or attached. Isolated stones are those attached to another form of construction, as in this example, i.e. brick facework (F21:D13). Stones are deemed to be set on their natural beds unless otherwise described (F21:S9).

Lintels are measured in metres giving a dimensioned description as F21:6.1.0.0. It appears advisable to indicate that this linear item comprises two separate components. The various terms used in these descriptions are explained in the text of this chapter.

The lintels could alternatively be regarded as springers to a semi-circular arch when they would be categorised as special purpose stones (F21:33.1.1.0) and enumerated.

Metal Casement Windows 4

2		Voussoirs
		pl. cuboid 400 × 250 × 255;
		circ.; sk. jtd.; circ. sk.;
		circ. reb. 55 gth.; circ.
		chmfd. 112 wide; rad. nat.
		bed; fcd. 3 sides.
		(G
1		Keystones
		pl. cuboid 470 × 320 × 255;
		2ᶜᵉ sk. jtd.; circ. sk.;
		circ. reb. 55 gth.; circ.
		chmfd. 112 wide; rad.
		nat. bed; fcd. 3 sides.
		(H
2/ 0.90		Mullions (2 nr.)
		150 × 255 th.; 2ᶜᵉ reb.
		55 gth.; 2ᶜᵉ chmfd. 112
		wide; fcd. 4 sides; fxg.
		w. cop. dowels ea. end.

Centr'g.
Semi-circ. st. arches
600 span & 255 wide
on soff.

1		

Adjust of bwk.

	1.950
less 2/75	150
	1.800

Ddt. Fcg. bwk. in red

| 1.95 | | fcg. bks. in g.m. |
| 0.23 | | (sill (1:1.6) a.b. |

		Walls
2/ 1.80		(B&D h. b. th.; facewk. o.s.
0.23		&
2/ 1.95		Ddt. Ditto. in brown
0.23		(C&E fcg. bks.; do.
		Walls
2/ 0.39		(lintels h. b. th.; facewk. o.s.
0.23		&
½/		
2/ 22/		
7/ 0.54		(semi- Ddt. Sundries
0.54		(circ. (arch.
		Formg. cav. in h.w.'s,
0.44		(proj. incl. insul. a.b.
0.09		(to keyst.

Arches may be measured in accordance with F21:24.1.0.0 giving the quantity either in metres or enumerated. In the former case the mean girth or length is measured (F21:M8) and the number of arches is stated in the description (F21:M9). Alternatively where arches are enumerated, the span of the arch should be given although this is not specifically stated by the rules. In either instance the height on face, soffit width and shape of arch must be given. These rules are best applied where the stones comprising the arch are similar in size and shape. Where this is not the case, as in this example, each voussoir should be measured separately as special purpose stones in accordance with F21:33.1.1.0. Refer also to *SMM Measurement Code* which provides an explanation of special purpose stones and a list of applications. These include springers, voussoirs and keystones of arches. Descriptions are given as the smallest block from which each stone can be cut taking account of the plane of the natural bed (F21:M11), which radiates from the centre of the arch. The various labours on each stone must be given although the rules do not specifically state this.

Mullions are taken as linear items as F21:8.1.0.0 with a dimensioned description. Copper dowels and mortices are deemed to be included (F21:C1c & e), but the method of jointing together and fixing, and the type and positioning of dowels are required by F21:S7 and 10.

The centring which supports the arch during its construction is enumerated as F21:36.1.1.0.

Then follows the deduction of faced brickwork and cavity for the areas occupied by the windows and stone surround. Each dimension is taken to the nearest 10 mm, with 5 mm and over regarded as 10 mm in accordance with General Rules 3.2, even although this results in a small over-measurement of items deducted. There is faced brickwork to both inner and outer skins of the hollow wall. The void to the arch is measured up to the outer edge of the arch projected and the projection of the keystone above it added. Note the full use of locational notes to identify each set of dimensions. All fair cutting to the brickwork is deemed to be included in the brickwork rate (F10:C1b).

11 Doors

ORDER OF MEASUREMENT

The measurement of doors can be broadly subdivided into internal and external doors, and the dimensions of each of these two classes of door broken down into: (1) door and associated work and (2) adjustment of opening.

The door dimensions will include any ironmongery, glazing and painting required together with the frame or lining from which the door is hung. When adjusting the opening for an external door care must be taken to cover all the appropriate items by adopting a logical order of measurement such as deduction of wall, external and internal finishes, and measurement of external head or arch, internal head or lintel, external reveals, internal reveals and threshold. It is sometimes necessary to adjust the skirting and flooring in the door opening, where these have not been covered with finishes or floors, and to measure steps leading up to the door.

Where sidelights and/or fanlights are to be provided adjacent to the door, these will be measured at the same time as the door.

When taking off the dimensions of a number of doors, it is advisable to prepare a schedule where this has not already been done by the architect, detailing such information as location or number of each door, size and type of door, details of ironmongery, size of frame or lining, thickness and type of wall or partition, finishes to both sides of wall, details of head of opening, reveals and threshold. In this way the task of taking off the doors will be greatly simplified and similar items suitably grouped as the work proceeds, so avoiding duplication of items. The schedule will be marked as each item is extracted and entered on the dimensions paper. An internal door schedule is illustrated in table 11.1.

Most surveyors take off the doors first and then follow with the adjustment of the openings; the main justification for this procedure is that the size of the door determines the size of the frame or lining, which in its turn decides the size of the opening.

As with many classes of measured work, it is often helpful to commence with a take off list, as illustrated in example 21, to reduce the possibility of omitting items and give ample preliminary consideration to the sequence and nature of the component items.

DOORS

The principal rules governing the measurement of timber doors themselves are contained in work section L20. Doors are enumerated with a dimensioned diagram in accordance with L20:1.0.1.0, as illustrated in example 22. It should, however, be noted that in the *SMM7 Code of Procedure for the Measurement of Building Works* clause 4.4 it is stated that 'there may be occasions where it is more appropriate to issue the architect's or engineer's drawings with the bills of quantities rather than produce dimensioned diagrams. In such instances it will be necessary to identify the drawings in the bill description'. This option has been taken with regard to the flush door in example 21. Each leaf of a multi-leafed door is counted as one door (L20:M2). With glazed doors, the method of glazing, such as use of wood beads, and method of securing by brads or screws, and gasket where provided, are included in the description of the glass (L40:S3–4). The glazing beads themselves are included in the description of the door. The door item is deemed to include glazing which is supplied with the component (L20:C3e). Standard or stock pattern doors can be described by reference to a manufacturer's catalogue, as illustrated in examples 21 and 22, or the appropriate British Standard. Doors which are supplied with their frames or linings are measured as composite items in accordance with General Rule 9.1 (L20:M4). The precise details of stock pattern doors can be obtained from the manufacturers' catalogues and these contain a wealth of information on different door designs and their various components. Fitting and hanging of doors are deemed to be included with the items (L20:C1). It should be noted that all sizes of timber are nominal sizes unless stated as finished sizes (L20:D1).

The provision and fixing of units or sets of ironmongery are covered by enumerated and fully described items and the nature of the base must be stated (P21:1.1.0.0 and P21:S1–3). Where, as is often the case, the supply of the ironmongery is covered by a prime cost sum, which is likely to incorporate all ironmongery required for joinery throughout the whole building, separate fixing only items containing the necessary descriptive particulars are needed, as illustrated in examples 21 and 22. Ironmongery where supplied with the component is included with the description of the component (L20:C3c).

Painting to unglazed doors is classified as to general surfaces (M60:1.0.1.0), whereas painting of glazed doors is described separately in accordance with M60:4.1–4.1.3. The latter contains the appropriate pane area classification and will often be described as partially glazed as in example 22, where only the top panel of the door is open and glazed. Painting to panelled and matchboarded doors normally has its areas timesed by $1\frac{1}{9}$ to allow for the additional area in moulded surfaces and door edges, because the area measured must be the area covered, as indicated in M60:M2, and illustrated in example 22. The multiplier should be increased to $1\frac{2}{9}$ for a door with large bolection mouldings. With flush doors it is better to take the full area of door surfaces, including top and side edges, thus avoiding the need for the use of a multiplier, as illustrated in example 21.

DOOR FRAMES AND LINING SETS

Door frame and lining sets are grouped together and either measured in metres (L20:7.1–5.1.0) or enumerated (L20:7.6.1.0) at the discretion of the surveyor. If measured in metres the number of frame and lining sets must be stated. The extreme lengths of jambs, heads, sills, mullions and transoms which make up the frames and linings are each given separately in metres. The number of sills, mullions and transoms must also be stated. A dimensioned overall cross section description is required for the various frame and lining members. This means, for example, that the length of each differently sized jamb cross section must be measured separately. The frames and linings may include repeats of identical sets and because of possible cost savings, the number of repeats of each type of identical set must be given (L20:7.*.*.1). Conversely where the measured lengths of individual members include different cross section shapes, the number of shapes must be stated (L20.7.*.*.2).

The alternative to measuring frames and linings as composite sets is simpler, but is best applied where there is a significant number of identically sized and composed sets, otherwise a lot of separately enumerated items, each with a different dimensioned description may result. Standard frame and lining sets are used extensively and should always be measured as composite sets when supplied separately from the doors. The number of frame and lining sets need not be stated when measured as composite sets (L20:M5).

The method of fixing of door frames and lining sets, where not at the discretion of the contractor, is included in the item descriptions, in accordance with L20:S8. The method of fixing may take a variety of different forms, such as dowels, cramps, plugging, plugging and screwing, fixing bricks, blocks or grounds, and a selection of these is contained in the worked examples that follow.

Architraves are measured in a similar manner to skirtings and picture rails in metres, giving a dimensioned overall cross section description (P20:1.1.0.0). Ends, angles, mitres, intersections and the like are deemed to be included except on hardwood items $>0.003\,m^2$ sectional area (P20:C1). To put this into perspective, the sectional area of a typical present day architrave is $0.0018\,m^2$ and that of a 25×100 skirting is $0.0025\,m^2$. Hence even if in hardwood it would not be necessary to measure mitred angles and similar labours. However, a larger than normal hardwood skirting, say 25×150, with a sectional area of $0.00375\,m^2$ is over the prescribed limit and would entail enumerated labour items.

ADJUSTMENT OF DOOR OPENINGS

The golden rule in measuring this work is to adopt a logical sequence, working systematically through the adjustment from the main area of walling and finishings to the edges of the opening from top to bottom, and both inside and out in the case of an external door. This is illustrated in the two worked examples in this chapter.

Lintels to support brickwork or blockwork over door openings can be of precast reinforced concrete or steel and the latter is becoming increasingly popular. Both forms of construction are illustrated in examples 21 and 22. The precast concrete lintels are enumerated with a dimensioned description which includes surface finishes and reinforcement as F31:1.1.0.1 and F31:S5. Steel lintels are classified as proprietary items under F30:16.1.1.0 and enumerated, with a dimensioned description and the manufacturer's reference. It is necessary to refer to manufacturers' catalogues for precise details of steel lintels. The end bearings of both types of lintel are normally in the 100–150 range.

PATIO AND GARAGE DOORS

Patio door sets are commonly supplied in aluminium and PVC-U and examples follow of typical items in each material as proprietary items, measured in accordance with L20 and General Rule 6.1.

1			Alum; polyester powder ctd.
			Dr. sets
			std. unit; Crosby code 54691; 2.387 × 2.085; slidg. w. thermal break; hwd. sub-frs. code 54791; 2 pt. lockg. system; night ventilator; s.s. track; dble. glazg. units 20 th. toughened glass to BS 6206 class A; fxg. to bwk. w. g.m.s. clamps
1			Plastics
			PVC-U dr. sets
			'Tewkesbury 2100'; manfd. by John Carr; 2.073 × 2.095; integral sub-fr.; 4 pt. lock system; ventilator system; dble. glazg. units 22 th. to BS 6206 class A; plg'g + scr'g to bwk.

Garage doors are available in a variety of materials, including wood, galvanised steel, PVC-U and glass fibre. The measured example encompasses a galvanised steel door selected for durability, subject to regular painting.

	1		<u>Galv. stl.</u> Drs. garage 'up & over'; Apex Berry mark II; 4.267 × 2.134 powder ctd.; remote control; galv. stl. tracks, supptd. by adjust. track hangers; non-corrosion nylon rollers; scr'g to tbr.

WORKED EXAMPLES

Worked examples are now given which illustrate the method of taking off the dimensions for internal doors, using a door schedule, and an external door, including the adjustment of openings in both cases.

Internal doors   **DRAWING 14**

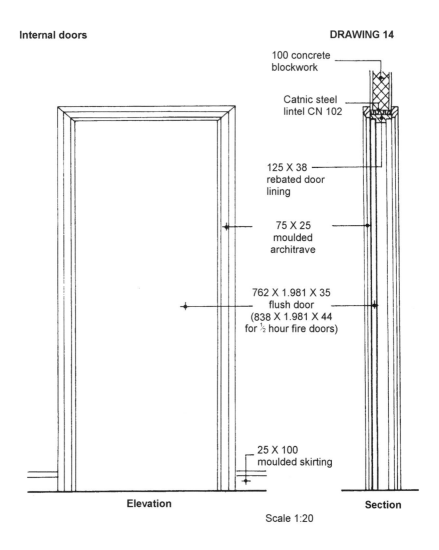

100 concrete blockwork

Catnic steel lintel CN 102

125 X 38 rebated door lining

75 X 25 moulded architrave

762 X 1.981 X 35 flush door
(838 X 1.981 X 44 for ½ hour fire doors)

25 X 100 moulded skirting

Elevation   **Section**

Scale 1:20

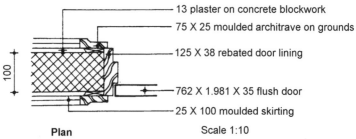

13 plaster on concrete blockwork

75 X 25 moulded architrave on grounds

125 X 38 rebated door lining

762 X 1.981 X 35 flush door

25 X 100 moulded skirting

100

Plan   Scale 1:10

Table 11.1 *Schedule of internal doors.*

Location	No. of doors	Door size	Door type	Door lining	Door finish	Wall details	Architraves	Lintel	Skirting	Wall finishes
Lounge	1	838 × 1.981 × 44	flush plywood-faced 2 lipped edges	125 × 38 twice rebated (type A)	gloss paint	100 concrete lightweight blockwork	75 × 25 moulded softwood	Catnic CN102	25 × 100 moulded softwood	Plaster (2 coats 13 thick) & twice emulsion
Dining room	1	ditto.	ditto.	ditto.	ditto.	ditto.	ditto.	ditto.	ditto.	ditto.
Kitchen	1	ditto.	ditto.	ditto.	ditto.	ditto.	ditto.	ditto.	ditto.	ditto.
Bathroom	1	762 × 1.981 × 35	ditto.	ditto. (type B)	ditto.	ditto.	ditto.	ditto.	ditto.	ditto.
Bedroom 1	1	ditto.	ditto.	ditto.	ditto.	ditto.	ditto.	ditto.	ditto.	ditto.
Bedroom 2	1	ditto.	ditto.	ditto.	ditto.	ditto.	ditto.	ditto.	ditto.	ditto.

Drg. 14   Example 21
    Internal Doors 1

The example starts with a take-off
list to illustrate the approach
        (1)   (2)
Construction   Adjustment of openings
doors   concrete blockwork
painting doors   wall plaster
ironmongery   decorations to plaster
door linings   lintels
architraves   skirtings and painting
painting of linings   floor finish in door
   and architraves   openings, if not taken with
                                          floors

                        Doors

                    ⌐ Flush
   3   Doors
   ─   838 × 1.981 × 44 th.;
                    (Lge. as drg. 14; wrot. swd.
                    (D/rm semi-solid core faced
                    (Kit. b.s. w. plywd. 5th, glued
                          to the core under
                          pressure; lipped & edged
                          in hwd. strips to both
                          vert. edges to BS 459
                          pt. 2A
                                    &
                      Doors
                          762 × 1.981 × 35 th.;
                              ditto
                    (Bathrm.
                    (& Bedrms. 1 & 2

                    Alternatively a stock patt.
                    dr. may be specified and
                    the follg. entry illustrates
                    a typical product
                    ⌐ Flush
   3   Doors
   ─   std. unit; John Carr
                          code 29SDLF; size
                          838 × 1.981 × 44 th.;
                          Sapele veneered.
                                    &
                      Doors
                          ditto size 762 ×
                          1.981 × 35 th.; do.

In most cases it is helpful to prepare a take off list prior to measuring the work as this concentrates the mind on the work ahead, ensures that the constructional details are examined closely and any queries raised with the architect, and it provides a valuable checklist as the work is measured.

The flush doors are enumerated and described in accordance with L20:1.0.1.0. However instead of including a dimensioned diagram, reference has been made to the architect's drawing as explained in *SMM7 Measurement Code* clause 4.4. The inclusion of the leading dimensions and constructional details in the item description would seem adequate. Fitting and hanging the doors is deemed to be included (L20:C1). The internal doors schedule (table 11.1) shows that half of the six doors are of different dimensions from the remainder and require a separate abbreviated item. This will also influence the dimensions of work that follows.

In the case of catalogued or standard doors, it is necessary to make a unique cross reference to the manufacturer's catalogue or to a standard specification in accordance with General Rule 6. This replaces the need to give a description containing the normally required details expressed in General Rules 2.6 and 2.12. The approximate weight requirement in L20:1.0.1.1 only applies to metal doors (L20:M3). To achieve variety, stock pattern internal doors can be obtained in a wide range of panel layouts with different hardwood veneers and types of core such that only a minority of doors are of the purpose made type.

Internal Doors 2

Paintg.

	838		762
v. edges ²/₄₄	88	²/₃₅	70
	926		832
	1.981		1.981
top edge	44		35
	2.025		2.016

Paintg. wd.; k.p.s., 2 u/c,
1 f/c gloss
Gen. surfs.
>300 gth.

3/	0.84
	1.98
3/	0.93
	2.03
3/	0.76
	1.98
3/	0.83
	2.02

The painting of door surfaces is classified as general surfaces under M60:1.0.1.0, whereas glazed doors are taken separately as M60:4.1–4.1.3 and an example of this will follow in example 22. The extra girth of edges must be included in the dimensions (M60:M2) and this can be done by increasing the width and height dimensions by the width of the edges in one surface area of the doors, as has been done in this example.
An alternative approach is to enter separate dimensions for the top and two vertical edges multiplied by the thickness of the door.
In the case of the larger doors, the addition will be as follows:
$838 + 2/1.981 (3.962) = 4.800 \times 0.04$.

Imgry.

6/	1

Pressed steel
Prs. butt hinges
75 to swd.
Include PC sum for
supply'g. only mats.
Imgry. £ _ _ _
&
Add for
main contrs. prof. _ _ %

It is necessary to state the nature of the hinges and the base to which they are applied (P21:1.1.0.0).

The prime cost sum for ironmongery will embrace all the ironmongery required in the bungalow, including that to windows, doors and fittings. The PC item follows the procedure outlined in A52:1.1.1.0, with provision for entering the main contractor's profit as a percentage of the PC sum as A52:1.2.0.0.

Item

6/	1

Fixg. only
Mtce. locks
to swd.

&

Sets lever furn.
alum.; to swd.

Fixing items for the mortice locks and sets of lever furniture, incorporating a general description and the nature of the base as P21:1.1.0.0.

Internal Doors 3

		Lings.
th. of ling.	38
less reb. for dr.	12
	26
	1.981
add tgd. jt.	12
	1.993

	762	838
add 2/26	52	52
	814	890

Wrot. swd.
Dr. lings. (6 nr. sets, incl.
repeats of identical sets,
3 nr. type A & 3 nr. type B)
   125 x 38 jambs; 2ce.
   reb.; plug'g. to blkwk.

   125 x 38 hds.; 2ce. reb.

6/2/ 1.99

3/ 0.81
3/ 0.89

An alternative approach
with enumerated composite
sets follows

	1.981
	26
	2.007

Dr. lings.
   composite sets 814 x
   2.007 o/a; 125 x 38 2ce.
   reb. x-sec.; plug'g to blkwk.
                  &
Dr. lings
   ditto 890 x 2.007 o/a; do.

3

Archves.

	75
	13
	62
	2.007
	62
	2.069
	890
	124
	1.014

Wrot. swd.
sktgs., pic. rls., archves. &
the like
   25 x 75 mo.; fix'g. to 45x
   12 sn. swd. grds. plugd.
   to blkwk.

2/6/2/ 2.07
2/3/ 0.94
2/3/ 1.01

The linings will have tongued joints where the jambs and head meet and it is therefore the overall lengths that are required. When working from the door dimensions, it is necessary to allow for the depth of rebate in the linings to accommodate the door as shown in the waste calculations.

The door lining sets are numbered and the component members measured in metres as jambs and heads, giving a dimensioned overall cross section description (L20:7.1–2.1.1). The six sets include 3 identical sets for each size of door which have been referred to as types A and B.

An alternative approach, as illustrated in the example, is to enumerate door lining sets, when it is no longer necessary to state the number of sets in the description (L20:M5). It involves two separate bill items because of the variations in door size.

The architraves need measuring to their outer edges to allow for the mitred top corners. The outer edge of the linings has been taken as the datum line to which the projections in both directions have been added. Architraves are measured as linear items, with a dimensioned overall cross section description within a grouped heading of members as P20:1.1.0.0. Fixing with grounds is not at the contractor's discretion and is given as P20:S8.

Internal Doors 4

Paintg.

Painting with one coat of primer has been taken to the backs of linings and architraves, to be applied on site prior to fixing. The architrave girth dimensions are taken on the centre lines of their faces. These members are usually factory primed and it might be considered adequate to specify touching up previously primed surfaces. However, the quality of factory priming does vary significantly and it is important to give adequate protection to the vulnerable back surfaces. Any factory applied priming is included with the component (L20:C3d&L20:S3).

6/2/	2.01	(ling. jbs.
3/	0.81	(ling. hds.
2/6/3/	0.89	(ling. hds.
2/6/2/	2.03	(archve. jbs.
2/3/	0.86	(archve. hds.
2/3/	0.94	(archve. hds.

Paintg. wd.; 1 ct. primer on site prior to fixg.
Gen. surfs.
 isol.; ≤ 300 gth.

mean gths. of archve. & ling.

bk. of lings.	814	890	2.007
add 2/15	30	30	15
	844	920	2.022

total lens.

2/2.022	4.044	4.044
	844	920
	4.888	4.964

The mean girths of architraves and linings are calculated in waste to give the operative lengths for painting purposes. The overall cross sectional girth of the combined architraves and linings is shown on the sketch and the figured dimensions enable the girth of the paintwork to be determined.

25 ⊢75⊣ 15 13 125

paintg. gth.
archve. 2/115 230
ling. 125+26+30=181
 411

The paintwork is measured in accordance with M60:1.0.1.0 as it is >300 girth. The painting to the architraves and linings can be carried out in a single operation and they do not therefore constitute isolated surfaces.

3/	4.89	
	0.41	
3/	4.96	
	0.41	

Paintg. wd.; k.p.s., 2 u/c & 1 f/c gloss
Gen. surfs.
 >300 gth.

Then follow with the adjustment to the opening, starting with the deduction of concrete blockwork. The area is taken to the outside surface of the lining which has been previously measured. No allowance is required for the depth of the Catnic lintel as it does not displace a full block course (F10:M3).

Adj of opg.

3/	0.81	
	2.01	
3/	0.89	
	2.01	

Ddt. Conc. blkwk. in g.m. (1:1:6) a.b.
Walls
 100 th.

		Internal Doors 5	The areas of plaster and emulsion paint to both sides of the doors to be deducted are calculated from the dimensions to the outside faces of the linings, as no deductions are made for grounds when measuring plaster (M20:M2). The plaster is measured in accordance with M20:1.1.1.0 and the emulsion paint as M60:1.0.1.0, covering the same area; although this is not strictly correct, it is not worth splitting hairs. Abbreviated descriptions are used for these items but giving adequate information to identify them.

Ddt. Pla. a.b.

Walls
> 300 wide to blkwk.

3/2/ 0.81
     2.01

&

Ddt. Emuls. a.b.

Gen. surfs.
> 300 gth.

3/2/ 0.89
     2.01

Proprietary Items

Catnic galv. stl. lintels
98 × 25 × 1.050 lg.; ref.
CN102; bldg. in.

An alternative would be to use a precast concrete lintel as follows:

6/ 1

Catnic galvanised steel lintels are being used increasingly. They are enumerated as F30:16.1.1.0, giving a dimensioned description and the manufacturer's reference.

Prec. conc. (1 : 1½ : 3/20 agg.)

Lintels
100 × 150 × 1.050 lg.;
rect.; keyed for pla. on 2nr. surfs.; reinfd. w. 1nr. 10 ∅ m.s. bar to BS 4449

6/ 1

As an alternative, precast concrete lintels have been included and are enumerated and described in accordance with F31:1.1.0.1. The concrete could be described by its strength rather than the mix as given, such as 21.00 N/mm².

Sktgs.

Ddt. Wrot. swd.

Sktgs., etc.
25 × 100 mo.; fxg. to sn. swd. grds. plugd. to blkwk.

3/2/ 0.94
3/2/ 1.01

&

Ddt. Paintg. wd.; k.p.s.

& ③

Gen. surfs.
isol.; ≤ 300 gth.

&

Ddt. Paintg. wd.; primg.

Gen. surfs.
isol.; ≤ 300 gth.; on site prior to fxg.

The skirtings finish against the outer edges of the architraves and the dimensions are extracted from the architrave dimensions calculated previously. They are dealt with as P20:1.1.0.0, giving a dimensioned overall cross section description, while painting the skirting is prescribed in M60:1.0.2.0 and the priming as M60:1.0.2.4. The timesing figures relate to three doors of each size and skirtings on both sides of each door. The floor finishes to the door openings are assumed to have been taken with the floors.

This page has been left intentionally blank for make-up reasons.

External door   **DRAWING 15**

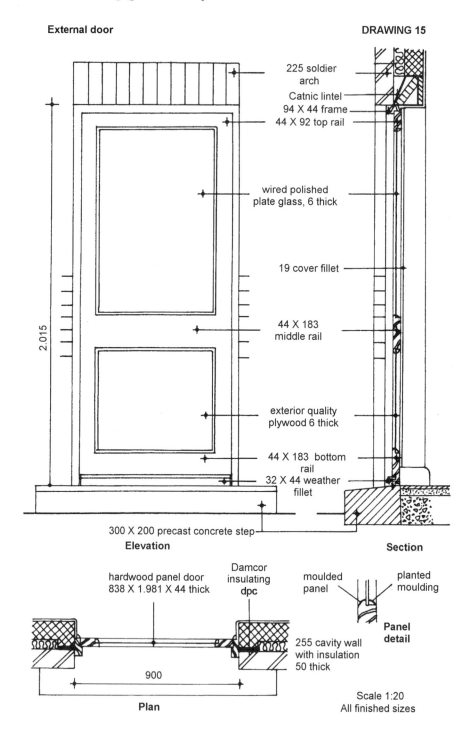

225 soldier arch

Catnic lintel

94 X 44 frame

44 X 92 top rail

wired polished plate glass, 6 thick

19 cover fillet

2.015

44 X 183 middle rail

exterior quality plywood 6 thick

44 X 183 bottom rail

32 X 44 weather fillet

300 X 200 precast concrete step

Elevation   **Section**

hardwood panel door 838 X 1.981 X 44 thick

Damcor insulating dpc

moulded panel

planted moulding

Panel detail

255 cavity wall with insulation 50 thick

900

Plan

Scale 1:20
All finished sizes

Drg. 15

Example 22

External Door 1

Panelled; wrot. Afromosia, selec. for clear fin.

Doors
  as dim. diagm. L20/1; mo., reb. for pan. & glzg.; 19 × 19 (fin.) plant mo. o.s. to plywd. pan. & glzg.; 32 × 44 (fin.) mo. & thrtd. weath. fill; reb. bott. rl.

1

The *Standard Method* and associated *Code of Measurement* permit two approaches to the measurement of purpose made doors. The most common approach is the first one adopted in this example, namely an enumerated item with a dimensioned diagram to support the general description as L20:1.0.1.0.

The diagram shows the door type with all the leading dimensions and the nature of the infill panels – in this case the upper panel is glazed and the lower one filled with plywood. The glazing will be measured later and so is not described in detail.

Dimensioned Diagram L20/1

OR

alternatively

Panelled; wrot. Afromosia, selec. for clear fin.

Doors
  838 × 1.981 as drg. 15

1

OR

suitable stock patt. product:

Panelled; wrot. Meranti

Doors
  std. unit; Crosby Code 37814; size 838 × 1.981 × 44 th. w. open pan. for glzg. & plywd. pan.; incl. std. imgry. & glzg. bds.

1

An alternative method is provided for in clause 4.4 on page 9 of the *SMM7 Measurement Code*, whereby on occasions it may be more appropriate to issue the architect's drawing(s) with the bill of quantities, rather than produce a dimensioned diagram. In such instances it is necessary to identify the drawing(s) in the bill description.

In many cases, purpose made doors have been superseded by stock pattern doors, as illustrated in this example. Such doors are described in accordance with General Rule 6.1. Furthermore, these doors can be supplied single or double glazed, and in a variety of glass types and door frames. All components which are supplied with the door are included with the item (L20:C3). Where doors are supplied with their frames or linings, they are measured as composite items in accordance with General Rule 9.1 (L20:M4), normally being described as door sets. Other types of standard doors include those made of PVC-U, steel and fibreglass.

External Door 2

Glazg.

	width	ht.
stile 92		990
less reb. 10	838	
2/82	164	rebs. 2/10 20
	674	1.010

Std. pln. glass; 6 Georgn.
wired pol. pl. to BS 952

0.67
1.01

To wd. w. l. o. putty &
bradded bds.
  in panes, area
  0.15 - 4.00 m²

OR

Special glass
Hermetically sealed double
glazg. units; comprisg.
2 x 4 mm obscure float
glass w. air space cav.
6 wide.

To wd. w. scrd. bds. &
non settg. compd.
  size 674 x 1.010;
  30 - 40 glzg. rebs.

1

Imgry.

Galv. m. s.
Water bars
  6 x 25 x 838 lg.; beddg.
  in c.m. (1 : 3) in
  prepd. grve. in conc.

1

Brass
Prs. d.s.w. butt hinges
  75 to hwd.

1½

Include PC sum for
supplyg. only mats.

Imgry. £ _ _ _
  &
Add for
  main contr's.
  profit _ _ %

Item

The glass area needs to be calculated from the dimensions supplied by allowing for the 10 deep rebate, to receive the glass. A full description of the glass must be given because of the wide range of glass types available. In this instance, Georgian wired polished plate glass, 6 thick, has been chosen on the grounds of safety and appearance. The police authority recommends laminated glass for front doors. The glass is measured as L40:1.1.2.0, and its area is 0.681 m² and thus falls into the 0.15–4.00 m² category.

An alternative is to fit a sealed double glazing unit, as illustrated in this example, measured in accordance with L40:3.1.1.3. It will be noted that L40:D4g includes sealed double glazing units under the classification of special glass.

The water bar is measured under P21:1.1.0.0, as an enumerated item with an adequate description of the bar (material and dimensions) and also the nature of the base (precast concrete step) in which it is to be bedded.

The butt hinges are a standard item of supply and fix as P21:1.1.0.0. 1½ pairs have been specified having regard to the size and weight of door.

The PC sum covers all the remaining items of ironmongery needed for this door and for other components, including windows (see A52:1.1.1.0), with the addition of main contractor's profit as a percentage of the prime cost sum as A52:1.2.0.0.

External Door 3

Fixg. only

2	Barrel bolts.
	100 brass; to hwd.
1	Door chain
	brass; to hwd.
	Mtce. locks
1	to hwd.
1	Sets lever furn.
	brass; to hwd.
1	Letter plates
	brass; to hwd.

All the remaining ironmongery items are 'fixing only', their supply and delivery to site being covered by the previous prime cost item. Fixing only such items is deemed to include unloading, storing, hoisting the goods and materials (where applicable) and returning packaging materials to the nominated supplier carriage paid and obtaining credits therefrom (A52:C4). The brief particulars given must be sufficient to enable the contractor to estimate precisely the amount and cost of the work.

The police authority recommends a door chain or similar device and bolts fitted a maximum distance of 150 from top and bottom of front doors. Letter plates to be a minimum of 400 from door locks, and locks to be an automatic deadlock rim lock or rim lock with separate mortice deadlock fitted below centre rail.

```
                    Frs.
                    hd.
                    900
    add horns 2/75   150
                   1.050
```

When calculating the length of the head to the door frame, an allowance of 75 is made for each horn which projects into the masonry when building in the frame.

Wrot. Afromosia

Dr. frs. (1 nr. set)

2/ 2.02	44 × 94 (fin.) jbs.; reb.; fxg. to blkwk. w. 30 × 3 × 250 gth. galv. m.s. cramps at 675 ccs. & 15 × 100 m.s. dowels
1.05	44 × 94 (fin.) hds.; reb.

Door frames are measured as linear items, but also giving the number of sets as L20:7.1–2.1.0, with a dimensioned overall cross section description. The jambs and head form separate items under this approach to measurement. The type of hardwood must be identified because of the variation in cost between species.

External Door 4

		<u>Alternative approach</u>

<u>Wrot. Afromosia</u>

1		Dr. frs.
<u> </u>		composite sets

900 × 2.015 o/a;
44 × 94 (fin.) reb.
x-sec.; fixg. jbs. to
blkwk. W. 30 × 30 ×
250 gth. galv. m.s. cramps
& 15 × 100 m.s. dowels

            900
2/2.015 4.030
        <u>4.930</u>

<u>Oil based sealant</u>

4.93		Ptg. frs.
<u> </u>		4.930

less pla. 2/2/13 52
~~sktg.~~ 2/100 200   <u>252</u>
             <u>4.678</u>

<u>Wrot. Afromosia</u>

4.68		Cover fillets
<u> </u>		19 (fin.) quad.

<u>Clear finishg. wd.;</u>
<u>prep. & 2 cts. polyurethane</u>
<u>varnish</u>

Glzd. drs.
  > 300 gth.; in panes
  0.50 – 1.00 m²; irreg.
  surfs.; partially glzd.

1⅑/	0.84	
	1.98	

               838
edges 2/44   <u>88</u>
              926
         1.981
top edge   <u>44</u>
         <u>2.025</u>

1⅑/	0.93	Ditto.
	2.03	
		EXT.

The alternative approach entails enumerating the door frame composite sets (L20:7.6.1.0), when it is no longer necessary to state the number of sets within the description (L20:M5). The first method of measuring frames is best suited where the same type of frame is used in conjunction with a number of different sized doors. The second method is particularly useful where there are a large number of doors of the same size.

Pointing the joint between the door frame and the brickwork on the external face to ensure a watertight yet flexible joint is a linear item (L20:9.0.0.0).

The length of the internal cover fillet is adjusted in waste for the thickness of plaster and height of the skirting. It masks the joint between the plaster and the frame and avoids the need for forming a rebate in the frame to receive the plaster. It is measured as a linear item with a dimensioned overall cross section description (P20:2.1.0.0).

The hardwood door and frame will be prepared for varnishing and then receive two coats of polyurethane varnish or lacquer, to protect the wood and show its natural grain. This work is measured on both sides of the door (measured flat) and includes edges (M60:M5). Because of the irregular surfaces to mouldings, it is timesed by 1⅑ and described as irregular surfaces and partially glazed as M60:4.3.1.3–4. External work is kept separate as it will probably give rise to higher costs (M60:D1).

<u>External</u> Door 5

		<u>Clear finishg. wd.; prep.</u> & 2 cts. polyurethane varnish

4.93		Gen. surfs.
		   isol. ≤ 300 gth.
		(fr.
		     &

		Ditto. EXT.

<u>Adj. of opg.</u>

		<u>Ddt. Fcg. bwk. in</u> <u>g.m. (1 : 1 : 6) a.b.</u>

0.90
2.02
		Walls
		   h.b. th.; facewk. o.s.
		    &

		<u>Ddt. Blkwk. in</u> <u>Thermalite conc. blks.</u> <u>in g.m. (1 : 1 : 6) a.b.</u>

		Walls
		   100 th.
		    &

		<u>Ddt. Sundries</u> Formg. cav. in hollow walls & insul. a.b. &

		<u>Ddt. Pla. a.b.</u>
		Walls
		   >300 wide; to blkwk.
		    &

		<u>Ddt. Emuls. a.b.</u> Gen. surfs. > 300 gth.

<u>Arch.</u>

		<u>Fcg. bwk. a.b.</u> Flat arches (1 nr.)
0.90		bk. on end; 225 hi. on face; 103 th.; 60 wide on exp. soff.; facewk. ind. ptg. o.s.

The preparation and varnishing of the door frame is classified as isolated general surfaces, ≤300 girth, as M60:1.0.2.0, with the external work separately classified. The length of frame to be painted is taken from the mastic pointing dimension, although it will be appreciated that it is a little on the generous side.

The first step in adjusting the door opening is to deduct the area of walling and internal finishes to the void so created. The area of the door opening is taken first as it applies to all five items, leaving subsequent adjustments to be made for the soldier arch and lintel.
The appropriate SMM7 clauses are as follows:
facing brickwork: F10:1.2.1.0;
blockwork inner skin: F10:1.1.1.0;
sundries (cavity to hollow wall): F30:1.1.1.1;
plaster: M20:1.1.1.0;
emulsion paint: M60:1.0.1.0.
All five items are bracketed together as they have a common set of dimensions.

The soldier arch is a linear item, stating the number of arches and giving the further details prescribed in F10:6.1.0.0. Arches are measured as the mean girth or length on face (F10:M6). The face width includes a mortar joint. Soldier arches are sometimes measured as ornamental bands (F10:13.1.3.1) on the grounds that they are not true arches because they are not self-supporting structures. If this approach is taken, items can either be 'full value' or extra over the work in which they occur.

External Door 6

		Lintel
		Proprietary items
1		Catnic classic galv. stl. lintels, polyester ctd. 245 × 143 o/a. & 1.200 lg.; ref. CN 73 w. cont. polystyrene insul.; bldg. in.

Steel lintels have become the most popular form of support for masonry over door and window openings. A dimensioned description and the manufacturer's reference are required (F30:16.1.1.0). The method of fixing is also needed (F30:S13). Lengths of Catnic steel lintels are available in increments of 150 up to 3.000 and 300 increments for longer lintels.

OR

Prec. conc. (1:1½:3/20 agg)

1	Lintels 150 × 215 × 1.200 lg.; rect.; keyed for pla. on 2 nr. adj. surfs.; reinfd. w. 1 nr. 12 ⌀ m.s. bar to BS 4449; bedd'g. in g.m. (1:1:6)

An alternative is to use a precast concrete lintel in place of the steel lintel to carry the inner skin of blockwork above the opening. The item contains a dimensioned description and reinforcement details as F31:1.1.0.1. Bedding and fixing details and surface finishes are also included as F31:S4–5.

```
                    900
add beargs 2/100   200
                  1.100
```

Galv. mild steel

1.10	Flat arch bar plain sec.; 32 × 6; bldg. into bwk.

A flat steel arch bar is needed to support the flat arch in the case of the concrete lintel approach. It would appear to be a F30 item but is not listed. The linear approach with a dimensioned description seems to be appropriate.

insul.

Polystyrene bd. insul. 25 th.

1.20	Pl. areas vert.; fxg. to perf. met. w. adhesive
0.14	(back of (Catnic Lintel

The length of polystyrene board insulation fixed to the inner face of the Catnic steel lintel is measured as P10:3.1.2.0.

Ddt. Fcg. bwk. in g.m. (1:1:6) a.b.

	Walls h.b.th.; facewk .o.s.
0.90	(bk. arch
0.23	

An adjustment is required for the brick arch, deducting facing brickwork for the area occupied by the flat brick arch, as it is not an extra over item (see F10:6.1.0.0). No deduction is required to the blockwork inner skin as the Catnic lintel does not displace a full block course (F10:M3).

External Door 7

<u>Reveals</u>

2/	2.02	<u>Blkwk. in g.m. a.b.</u> Clos'g. cav. 50 wide w. blkwk. 100 th. & Damcor insulating d.p.c. comprs'g. exp. polystyrene strip size 100 × 18 w. bonded polyethylene d.p.c. 165 wide; vert.

The closing cavity item to the two reveals follows the requirements of F10:12.1.1.0. A combination of blockwork and a proprietary insulating dpc is used to meet the Building Regulation requirements concerning cold bridging. This also obviates the need for a separate dpc item.

2/	2.02	<u>Pla. a.b.</u> Walls ≤ 300 wide; to blkwk.
	0.90	Ditto ≤ 300 wide; to perf. met (soff. of opg.

The plaster to the reveals is measured in metres as ≤300 wide as M20:1.2.1.0. The plaster is carried down behind the skirting. The work to the soffit is included in with the abutting walls (M20:D5), but has to be separated from the previous item as it is applied to a different base (M20:S5).

	1.20 0.14	<u>Ddt. Pla. a.b.</u> Walls > 300 wide; to blkwk. (back of (Catnic & lintel <u>Add</u> Ditto do.; to polystyrene insul.

The nature of the base to which plastered coatings are applied must be stated (M20:S5). Furthermore General Rule 8.2 indicates that each type of base must be identified separately. Consequently an adjustment is needed for plastering on the insulation at the back of the steel lintel instead of on blockwork as previously measured

```
              4.930
less sktgs. 2/100   200
              4.730
```

The dimension 4.930 represents the outer perimeter of the door frame, measured previously for mastic pointing.

	4.73	<u>Pla. a.b.</u> Rdd. angles & intersecs.

Rounded angles exceeding 10 radius are taken to the external angles of the plaster around the door opening as M20:16.0.0.0.

Emulsion paint to the reveals (head and soffit) are measured as M60:1.0.1.0. The classification >300 girth is selected because this work will be taken in with the same treatment to the adjoining wall surfaces. The length is the same as for the cover fillet.

The net length of skirting to be deducted for the opening is made up of the length in the opening less the returns to the two reveals.

The skirting is measured as P20:1.1.0.0, with a dimensioned overall cross section description. Mitres to the skirting are deemed to be included (P20:C1). The sizes are nominal as they are not described as finished sizes (P20:D1). Painting to the softwood skirting is measured as M60:1.0.2.0 and the priming prior to fixing on site as M60:1.0.2.4.

The nearest term in the *SMM7 Library of Standard Descriptions* to thresholds is steps, which are measured as F31:1.1.0.0, with a dimensioned description. The top surface is splayed and grooved to receive the water bar and the upper and front surfaces finished fair.

Finally, deduct facing brickwork and forming cavity in hollow wall below dpc due to displacement by threshold, concrete slab and flooring in door opening. Deductions on brickwork and blockwork for steps and the like are for the full courses displaced (F10:M3). Two courses of brickwork (150) are displaced in the outer skin, but a full course of blockwork is not displaced in the inner skin and consequently there is no deduction. A full deduction is required for the cavity. Any adjustment of the damp-proof course can be made when measuring the floor and associated damp-proof course.

External Door 8

Emuls. a.b.
Gen. surfs.
> 300 gth.

4.68	
0.09	

sktg.

	900
less revls. ²/₁₂₀	240
	660

Ddt. Wrot swd.
Sktgs., etc.
25 × 100 chfd.;
plg'g. to blkwk.

0.66	

&

Ddt. Paintg. wd.; k.p.s.
& ③
Gen. surfs.
isol. ≤ 300 gth.

&

Ddt. Paintg. wd.; primg.
Gen. surfs.
isol. ≤ 300 gth.; on
site prior to fxg.

Threshold

Prec. conc.(1 : 1½ : 3/20 agg.)
Steps
300 × 200 ×1.300 lg.;
rect., splyd. & grvd.;
fin. fair on 2 nr.
adj. surfs.

1	

Below d.p.c.

Ddt. Fcg. bwk. in c.m.(1:3)
a.b.
Walls
h.b.th.; facewk. o.s.

0.90	
0.15	

Ddt. Sundries
Formg. cav. in h.w.'s a.b.

0.90	
0.20	

12 Staircases and Fittings

TIMBER STAIRCASES

Most first examination measurement syllabuses include the measurement of relatively simple timber staircases. It is important to adopt a logical order of taking off in order to simplify the measuring process and reduce the possibility of omitting any items.

A good order to follow is

(1) staircase complete as a composite item;
(2) painting;
(3) associated items such as sloping ceilings and cupboards under the stairs.

Timber staircases are enumerated as composite items with the type stated and supported by a dimensioned description or component drawing (L30:1.1– 2.0.0). Typical approaches to both methods are given in example 23. Landings are included where they are part of the composite item.

It is sometimes necessary to calculate the amount of going or rise, by dividing the total length of a flight of stairs or the height between floors by the number of steps involved, where these are not shown on the relevant drawings. Part of a component detail of a tread and riser, illustrating the connecting joints is shown in figure 35.

Treads and risers are normally tongued and grooved and blocked together with their ends housed into strings. The bottom step may incorporate a semi-circular end. Wider staircases of 1.000 or more will normally include rough timber carriages under the stairs with brackets giving support to the treads. Landings, strings and cappings balusters or timber balustrades, together with attached handrails, where of the same materials as the staircase, newels or half-newels, and any spandril framing and panelling and soffit lining, where part of the component, are all included in a single enumerated composite item (L30:C1–2). Newels are normally morticed and draw-bored to receive strings and morticed for handrails in accordance with BS 585. Isolated balustrades which do not form an integral part of a staircase unit and associated handrails which are of a different material from the balustrade are measured separately in metres with a dimensioned description (L30:2–3.1.0.0). Isolated handrails are measured in metres in accordance with P20:7.1.0.0.

It is usually essential to make reference in the staircase description to the appropriate component drawing which should accompany the bill for

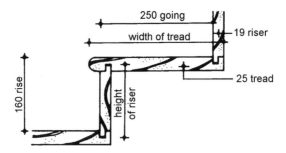

Figure 35 *Stair treads and risers.*

tendering purposes, and the drawing must contain all the information necessary for the manufacture and assembly of the staircase (General Rules 5.2). Detailed descriptive information is not required in the bill where a component drawing is supplied, since the purpose of the billed item will then be to identify the staircase with the relevant component drawing.

Where a component drawing is not provided, the work will be covered by an extensive dimensioned description as L30:1.1.0.0, but dimensioned descriptions are generally more appropriate for simpler structures such as loft ladders which are also covered by this section (L30:D1a).

Painting of staircase components will follow the enumerated composite item. Painting of skirtings, strings, cappings, handrails, newels, balusters, and margins of treads and risers will all be classified as general surfaces and will be measured in metres where $\leq$300 girth and in m^2 when >300 girth (M60:1.0.1–2.0).

Plasterwork and painting to walls and ceilings in staircase areas are kept separate because of the restricted working space and the additional difficulty of working on a stair flight (M20:M3 and M60:M1).

STANDARD TIMBER STAIRCASES

Long established and highly reputable joinery manufacturers supply a good range of stairs and stair parts. A typical measured item for a standard staircase supplied by John Carr is included in example 23. Furthermore, manufacturers usually offer 10 year guarantees for materials and workmanship and this may be extended to 30 years for softwood windows, doors and door frames, which are preservative treated against rot or fungal decay.

For example, Magnet provides standard staircases in whitewood, hardwood and clear timber (usually parana pine), all brush-sanded smooth and untreated ready for finishing on site with a wood stain or normal paint finish. The basic staircase is factory assembled with loose top nosing and the remaining components, where required, i.e. newel posts, spindles, handrails, base rails and cappings, etc., are available separately for assembly and fitting on site. The Magnet standard staircase components are: strings: ex 31 × 229; treads: ex 25; and risers: ex 9 plywood or medium density fibreboard. A 12 tread staircase has 2.700 going and 2.600 rise.

A Magnet balustrade staircase pack includes 28 × 28 square balusters, 44 × 65 handrail, 57 × 28 capping, 91 × 91 newel post and 120 × 120 × 34 newel cap. Magnet also supplies a good range of components in a variety of timbers, sizes and designs to meet the requirements of the majority of clients.

METAL STAIRCASES

Older large houses are increasingly being converted into flats (houses in multiple occupation: HMOs), to satisfy substantial demand from many sectors of the population. With these properties it is often necessary to provide external steel staircases as a secondary access forming a means of escape in case of fire. These staircases can be of two different types, namely a straight flight or flights and a spiral staircase. Measured items for these two types of standard staircase in taking off format, each making reference to detail drawings, follow and are in accordance with L30:1.2.0.0.

	1		Steel to BS 4360, grade 43A
			Strt. flt. staircases
			as drg. SS12; manfcd. by R. Glazzard; 900 wide btwn. strgs.; balustde. o.s. only; 5.300 going, 2.700 rise, w.14 nr. trds. & 1.800 lg. landg.; welded fabricatn.; 200 × 10 flat sec. strgs.; 250 × 6 cheqr. pl. trds.; 6 th. cheqr. pl. landgs. on bkt. suppts. to 100 × 75 × 10 'angle' sec. cantlvr.; 152 × 152 × 23 kg univ. col. suppts.; balustdes of 48.3 dia. ×4 CH sec. top & interm. rls. w. 42.4 dia. ×3.2 CH sec. vert. suppts. at 1.000 ccs.; trds. secrd. to strgs. w. 2 nr. 25 × 25 × 3.2 'angle' sec. bkts. fxd. to bwk. w. expsn. bolts at 600 ccs.; col. suppts. secrd. to fdn. pads w. 2 nr. h.d. bolts; galv. after fabrctn.

	1			Steel

Steel

Spiral staircases

    as drg. SES/1H, manfcd. by Crescent of
    Cambridge; Crescent 'H' range domestic type;
    2.604 rise w. 11 nr. trds.; trad. centre col. & Iroko
    trds.; core jsts. attachmt.; hsg. type infill patt.;
    riser bars; red oxide primer fin.; black PVC hdrl.
    cover

Both types of staircase must be supported by scale drawings showing plans
and elevations (not included). The amount of detailed particulars varies significantly between the two examples and illustrates how the form and content of
billed descriptions can vary for the two types of standard metal staircases in
providing the information which is needed for pricing.

It should be noted that the rules of measurement in work section L30 are
applicable to timber, metal and precast concrete staircases.

FITTINGS

Measurement of joinery fittings, such as cupboards, work-benches, counters,
dressers, bookcases, and the like, are taken as enumerated composite joinery
items. In the context of domestic work, such items are covered by work
sections N10 General fixtures, furnishings and equipment and N11 Domestic
kitchen fittings. These work sections share the same rules of measurement
and details of the relevant types of work are included in SMM7 Appendix A.
Alternatively, the supply of fittings by a nominated supplier could be the
subject of a prime cost sum as A52:1.1–2.1.0, including a description and
provision for the inclusion of the main contractor's profit. Fixing is measured
within the appropriate work section, whereas prime cost sums are usually
grouped together elsewhere in the Bill.

Where the exposed woodwork is to be stained or polished, the description of
the timber shall state that it is to be selected and protected for subsequent treatment, and this can often, with advantage, be included in the heading.

Stock pattern or standard cupboards and other fitments are enumerated
giving the appropriate supplier's catalogue reference as General Rules 6.1,
together with appropriate fixing details.

Isolated shelves, worktops and similar features are measured in metres,
giving a dimensioned overall cross section description as P20:3.1.0.0. All
labours will be included in the descriptions of these items, including any
stopped labours, the number of which must be given. The work is deemed to
include ends, angles, mitres, intersections and the like, except on hardwood
items $>0.003\,m^2$ in sectional area (P20:C1).

STANDARD JOINERY FITTINGS

Many joinery fittings including kitchen fitments used in domestic work are standard products, which can form enumerated items listing the relevant components by reference to the manufacturer's codes for identification purposes, measured in accordance with N11:1.1–2.0.0. The description needs to be supported by a component drawing or a dimensioned diagram (see figure 36) as shown in the following example.

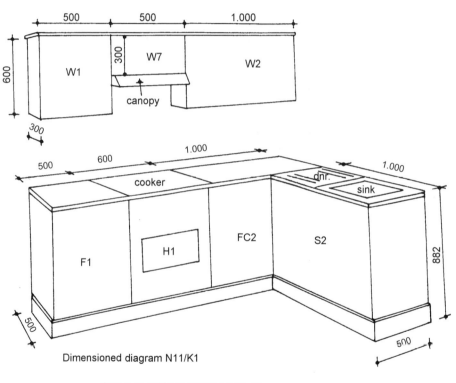

Dimensioned diagram N11/K1

Schedule of Marley kitchen fitments

Reference	Type of unit	Size of unit
F1	Single floor unit	500 X 500 X 882
H1	Single hob unit	600 X 500 X 882
FC2	Corner floor unit	1.000 X 500 X 882
S2	Sink unit	1.000 X 500 X 882
W1	Single wall unit	500 X 300 X 600
W7	Single top box unit	500 X 300 X 300
W2	Double wall unit	1.000 X 300 X 600

Figure 36 *Dimensioned diagram of kitchen fitments.*

	1		<u>Wrot oak</u>
			Kitchen fitments
			    suppld. by Magnet as dimensioned diagm.
			    N11/K1 made up of flr. units F1, H1, FC2 & S2,
			    and wall units W1, W7 & W2 & canopy unit, to
			    the dims. listed in the accompyg. schedule in
			    'Kings Oak' solid oak frd. drs., w. venrd, antique
			    centre pans. & scalloped dr. handles; matchg.
			    cont. plth. & cornice; p & s to blkwk.

Alternatively, the fitments could be covered by a prime cost supply item, with provision for addition of the main contractor's profit as a percentage of the PC sum, and a separate fixing only item, as illustrated in earlier examples.

WORKED EXAMPLE

A worked example follows covering the measurement of a simple timber staircase and shelving to a larder.

This page has been left intentionally blank for make-up reasons.

Timber staircase   **DRAWING 16**

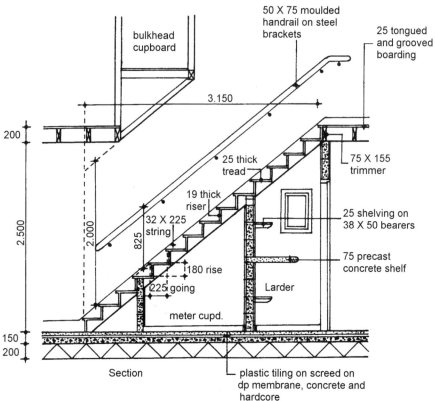

50 X 75 moulded handrail on steel brackets

bulkhead cupboard

25 tongued and grooved boarding

3.150

200

25 thick tread

75 X 155 trimmer

19 thick riser

32 X 225 string

2.500

2.000

825

25 shelving on 38 X 50 bearers

75 precast concrete shelf

180 rise

225 going

Larder

meter cupd.

150
200

Section

plastic tiling on screed on dp membrane, concrete and hardcore

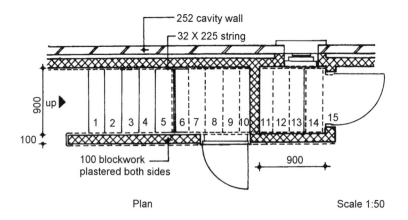

252 cavity wall

32 X 225 string

900

up ▶

100

| 1 | 2 | 3 | 4 | 5 | 6 | 7 | 8 | 9 | 10 | 11 | 12 | 13 | 14 | 15 |

100 blockwork plastered both sides

900

Plan   Scale 1:50

Drg. 16   Example 23

Timber Staircase 1

	Wrot swd.
	Staircases
1	strt. single flight
	as component drg.
	16.

One acceptable approach to measuring timber staircases is to give an enumerated composite item with a component drawing as L30:1.2.0.0, as shown in this example. Composite items are deemed to include newels, fixing items, nosings, cover moulds, trims, soffit lining and spandrel panels, where forming part of the component (L30:C1–2).

OR

	width
	900
less pla. 2/13	26
width over strgs.	874

Wrot. swd.

Staircases
   strt. single flight;
   3.150 going, 2.700
   rise; 874 wide;
   14 treads 25 th. &
   15 risers 19 th.;
   32 x 225 wall strgs.

Another alternative is to give a dimensioned description, as given in this example for the same staircase (L30:1.1.0.0). This gives all the essential dimensions and component parts. In a more complicated staircase, it would also include details of winders, balusters and handrails. In this example the bulstrade consists of a plastered block wall which will be measured under the brickwork and blockwork and finishes sections.

Wrot swd.

Std. staircases
   strt. flight; closed
   tread in parana pine
   to BS 585, pt. 1 &
   BS 5395, pt. 1;
   manufctd. by John
   Carr Code M; total
   going 2.700, rise 2.600;
   864 wide; 12 treads
   21 th. & 13 risers 9 th.;
   strgs. 27th.

A good selection of standard or stock pattern staircases is now used in many housing projects and a typical item is included to show the type of product on offer, which meets the requirements of many clients.

These descriptions can be extended to include balustrades as follows:

Wrot. hemlock 38 x 75
mo. hdrl., 41 x 41 turned
bals. at 150 ccs. hsd. ea.
end rakg.; o/a ht. 900;
82 x 82 turned newel
posts.

The staircase description has been extended to include common items of balustrading made up of handrail, balusters and newels. The constructional details of domestic timber staircases are illustrated in *Building Technology* by the present author.

Timber Staircase 2

		Wall strgs.	We will now continue with the

Wall strgs.

Paintg. wd.; k.p.s., 2u/cs & 1 f/c gloss

Gen. surfs.
   isol. ≤ 300 gth.;
   in staircase areas

2/ 4.40

Hdrl.
   4.000
     300
   4.300

We will now continue with the measurement of the staircase illustrated on drawing 16. The lengths of the wall strings are scaled from the drawing and the painting is measured as M60:1.0.2.0. The painting work is described as in staircase areas as M60:M1, as it is potentially more difficult and hence more costly to apply. The timber members are already included in the composite staircase item.

4.30

Wrot. African mahogany

Isol. hdrls.
   50 x 75, mo.;
   jtd. w. hdrl. scrs.;
   sel. & protectd. for
   subseqt. treatment;
   fxg. to stl. bkts.
&
Clear finish'g. wd.;
prep. & 2 cts. polyurethane varnish
Gen. surfs.
   isol. ≤ 300 gth.;
   in staircase areas

Isolated handrails, not forming an integral part of the staircase unit are thus described and measured under P20:7.1.0.0 (as indicated by L30:M2), giving a dimensioned overall cross section description. It is important to give adequate emphasis to this item as it is the only member on the staircase in hardwood.

Being in hardwood, the handrail is prepared and finished with two coats of polyurethane varnish and measured as M60:1.0.2.0, as the girth is around 250.

2

Wrot. hwd.
   E.o. isol. hdrls for
   ends

1

   angles

The ends and angle to the hardwood isolated handrail are enumerated as extra over items as P20:8.1–2.0.0, as the sectional area is $0.00375\,\text{m}^2$ and thus exceeds the prescribed $0.003\,\text{m}^2$. Such items are known as written short items, as they follow immediately after the handrail item in the bill of quantities.

Hdrl. bkts.
     Imgry.

Fixing only

6

Hdrl. bkts.
   s.s.; to hwd. & to
   blkwk.

Brackets are included in the term ironmongery in appendix A of SMM7 and are measured as P21:1.1.0.0 and taken as 'fixing only', assuming that their supply will be included in a PC item, encompassing all ironmongery needed for the project.

Timber Staircase 3

<u>Soffit</u>

⌐Plabd. of gyp. basebd.
to BS 1230, 9.5 mm th.;
fxg. to tbr. w. galv. nls.;
& skim ct. of Thistle bd.
fin. pla. to BS 1191, pt. 1,
class B, 5th. in 2 cts.
fin. smth.

Clgs.
  > 300 wide;
    in staircase areas
(meter cupd.

| 1.75 |
| 0.90 |

(larder   &
Emulsn. pt. to pla. in 2 cts.
Gen. surfs.
  > 300 gth.; in staircase
  areas

| 1.15 |
| 0.90 |

<u>Paintg. to margins</u>
o/a gth. of t. & r.

going	225
nosg.	20
riser	180
	425

⌐Painting. wd.; k.p.s., 2u/cs
& I f/c gloss
Gen. surfs.
(mgns. isol. ≤ 300 gth.;
(t.& r.s in staircase areas
(top
(riser

14/2/
2/ | 0.43 |
2/ | 0.18 |

<u>Shelvg. to larder</u>
⌐Wrot swd.
Isol. shelves
  225 × 25 on 38 × 50
  brrs. plugd. to blkwk.

2/ | 0.90 |

&
Paintg. wd. a.b.
Gen. surfs.
  isol. ≤ 300 gth.;
(shelf
(edges in staircase areas

Conc. shelvg.

	len.	width
add beargs.	900	700
2/100	200	
	1.100	

Prec. conc. (1 : 2 : 4/20 agg.)
Shelves
  1.100 × 700 × 75 th. w.
  smth. top & I nr. fair edge;
  bedd'g. in g.m. (1 : 1 : 6)

1
—

Plasterboard ceilings are measured in m² where the width exceeds 300, giving the thickness of plasterboard and thickness and number of skim coats as M20:2.1.2.0, and the appropriate particulars listed in M20:S1–8. Alternatively, this work could be cross referenced to a project specification, as illustrated in chapter 9, thereby reducing substantially the length of the description. Work in staircase areas is given separately as M20:M3, to enable the estimator to price working off staircase flights and/or in restricted spaces. The decorations are similarly described as M60:M1.

Painting may be needed to the margins of treads and risers to closed tread staircases, where carpets are to be fitted which do not cover the full length of the treads and risers. The work is measured as isolated general surfaces ≤300 girth in staircase areas.

Isolated shelves are measured in metres giving a dimensioned overall cross section description as P20:3.1.0.0, and including the method of fixing as P20:S8. The painting of shelf edges is taken as a linear item as M60:1.0.2.0 and including 'in staircase areas' as M60:M1.

The concrete shelf is enumerated with a dimensioned description as F31:1.1.0.0, and giving the relevant particulars listed in F31:S1–5. Moulds are deemed to be included with precast items (F31:C1).

Examples of metal stairs and kitchen fitments are given in the text of the chapter.

13   Water, Heating and Waste Services

ORDER OF MEASUREMENT

As with the measurement of most other types of building work, it is important to adopt a logical order of taking off. The following order is frequently followed in practice when measuring cold water supply and waste services:

(1) Connection to water authority's main and all work up to boundary of site, including reinstatement of public highway and provision of stopvalve near the site boundary.
(2) Underground service and rising service/main from site boundary up to cold water storage tank, including any stopvalves, holes through walls, ceilings and floors, insulation to pipes, and the like.
(3) Branches to rising service, such as supply to sink, including any associated work.
(4) Cold water storage tank or cistern and associated work, such as supporting bearers, overflow, cover and insulating lining.
(5) Down services with branches, including any stopvalves, holes through walls, ceilings and floors and insulation to pipes.
(6) Sanitary appliances such as sinks, washbasins, baths and water closets – supply and fixing, including supporting brackets, taps/bib cocks and the like.
(7) Traps and waste pipes to appliances.
(8) Discharge pipes (waste, soil and vent pipes) and associated work.
(9) Any other work connected with the installation, such as painting pipes, marking the position of pipes and testing the installation.

It will be noted that the *Common Arrangement of Work Sections for Building Works* (CAWS) replaced the term 'plumbing work' with disposal systems and piped supply systems, with mechanical and electrical (M&E) services covered separately.

DRAWINGS FOR WATER SUPPLY AND WASTE SERVICES INSTALLATIONS

The architect's drawings often give little information on the layout of the water service pipes in a building. The information supplied is frequently limited to the point of entry of the rising service into the building, the position of the cold water storage tank or cistern and the various sanitary appliances. In these circumstances the quantity surveyor must decide on a suitable layout of pipes and plot them on the various drawings before he can start the taking off work. Sometimes the pipework may be shown on a schematic isometric drawing as illustrated in drawing 18.

It is good policy to draw the various pipes in different colours, such as the rising service in red and down services in blue. The stopvalves required should also be indicated on the drawings, and it is good practice to provide one on each branch, so that the supply can be cut off when repairing or replacing appliances and their associated taps and ball valves.

CONNECTION TO WATER MAIN

The tapping and insertion of a ferrule to the water authority's main and the provision of the communication pipe from the main to the boundary of the site with a stopvalve provided at this point, including opening up and rein-statement of the highway and all watching and lighting, are usually covered by a provisional sum based on the cost of previous projects or quotations from the water authority and local authority who will normally carry out the work (A53:1.1–2.0.0).

PIPEWORK GENERALLY

SMM7 provides rules for the measurement of pipes and other services components under work group Y which are of general application to the various types of service installations covered by work groups R to W. However the measured work should be grouped in a bill of quantities according to the work section representing the type of installation concerned, such as S10 cold water, S11 hot water etc. Pipes are measured over all fittings and branches in metres, stating the type, nominal size, method of jointing and type, spacing and method of fixing supports, and distinguishing between straight and curved pipes (Y10:1.1.1.1 and Y10:M3). Pipes are deemed to include joints in their running length (Y10:C3). Furthermore, the provision of everything necessary for jointing is deemed to be included without the need for specific mention (Y10:C1). The type of background to which the pipe supports are fixed will be classified in one of the categories listed in General Rules 8.3.

Details of the kind and quality of materials used in the pipes, gauge and other relevant particulars listed in Y10:S1–6, may be included in preamble clauses or a project specification. Part of this information is likely to be included in a heading to the measured items when taking off, as illustrated in example 24, following the format used in the *SMM7 Library of Standard Descriptions*. This would seem to be the best approach for students to use when taking off quantities in the absence of a project specification.

Made bends, special joints and connections, and fittings such as Y-junctions, reducers, elbows, tees and crosses, are all enumerated as items extra over the pipes in which they occur (Y10:2.1–4). In the case of special joints, the type and method of jointing are to be stated and they comprise joints which differ from those generally occurring in the running length or are connections to pipes of a different profile or material, connections to existing pipes or to equipment, appliances or ends of flue pipes (Y10:D2). Comprehensive examples of pipe fittings and pipework ancillaries are given in the *SMM7 Measurement Code* (page 41), but reference to manufacturers' catalogues will also prove to be beneficial.

Pipe fittings ≤65 diameter are classified according to the number of ends, while those of larger diameter are described. The method of jointing is stated where different from the pipe in which the fitting occurs (Y10:2.3–4.2–6.2).

Stopvalves and cocks, draw-off taps, and drain and air cocks are classified as pipework ancillaries and are enumerated, stating the type, nominal size, method of jointing, type, number and method of fixing supports and type of pipe to which connected (Y11:8.1.1.0). Those located in ducts or trenches are each kept separate and so described.

WATER STORAGE TANKS OR CISTERNS

Water storage tanks or cisterns are classified as general pipeline equipment and enumerated giving appropriate particulars as Y21:1.1.0.0, or alternatively a cross reference may be made to the project specification, which will substantially reduce the length of the item description. A typical example of a storage cistern description is given in example 24. Timber supports to tanks are measured in metres with a dimensioned overall cross section description as G20:13.0.1.0. Insulation to tanks or cisterns is enumerated, giving the overall size of the tank, or loose insulation contained in casings can be measured in m^2, in accordance with Y50:1.4.1–2.0, and giving in both cases the particulars listed in Y50:S1–5. Any pipework located in the roof space should be insulated and this insulation is measured in metres giving the type of insulation and nominal size of the pipe (Y50:1.1.1.0). The insulation item is deemed to include smoothing the materials, working around supports, pipe flanges and fittings, excluding metal clad facing insulants (Y50:C1).

HOLES FOR PIPES

Cutting or forming holes through the structure for pipes and making good surfaces are enumerated, stating the nature and thickness of the structure and the shape of the hole, and classifying the pipes as to size in accordance with P31:20.2.1–3.2&4; i.e. pipes ≤55 nominal size, 55–110 and >110. The cutting of holes for pipes is best picked up when the various lengths of pipework are being taken off, rather than leaving all the holes to be taken off after the pipework has been measured complete. By contrast, painting of pipes may often, with advantage, be left to the end of the taking off. An alternative is to compile a schedule as illustrated in table 13.1.

SANITARY APPLIANCES

Sanitary appliances include low level WC suites, WC pans and cisterns, urinals and cisterns, sinks which are not supplied as part of a kitchen fitment, washbasins, bidets, baths, bath panels and trim, showers, vanity units, taps and waste fittings to appliances (*Common Arrangement* N13). These appliances are enumerated giving details of the type, size, pattern and capacity, as appropriate, and method of fixing, and may include a cross reference to the project specification (N13:4.1.1.1, 4&6). The supply of appliances is frequently covered by a prime cost sum, including relevant particulars and provision for main contractor's profit (A52:1.1–2.1.0) with separate enumerated fixing items, as illustrated in example 24. Another alternative is to give supply and fix items with full details including the supplier's catalogue reference numbers (General Rules 6.1).

BUILDER'S WORK IN CONNECTION
WITH SERVICES INSTALLATIONS

Builder's work in connection with a plumbing/services installation is identified under an appropriate heading (P31:M2). Unless identified in work sections P30 and P31, all other items of builder's work associated with a plumbing/services installations are given in accordance with the appropriate work sections (P31:M1). Builder's work includes holes for pipes, the rules for which were explained earlier. Where the hot water and heating installation is to be carried out by a nominated subcontractor, items will be provided to cover any specific items of special attendance required in accordance with A51:1.3.1–8.1–2, classified as either fixed or time related charges. General attendance by the main contractor on nominated subcontractors is measured as an item in accordance with A42:1.16/17.1–2.1. This covers a wide range of services and facilities, encompassing the use of main contractor's temporary roads, standing scaffolding, hoisting plant, temporary lighting and water supplies, clearing away rubbish, provision of space for subcontractor's own offices and storage of his plant and materials, and use of messrooms, sanitary accommodation and welfare facilities provided by the main contractor (A42:C3). In addition a separate item of General Attendance may be needed in meeting the requirements of the Joint Fire Code (A42:C4).

Unlike special attendances which are associated with a particular nominated subcontractor, general attendance is a single item which applies to all nominated subcontractors.

SCHEDULES OF WATER AND WASTE SERVICES

The measurement of this class of work can often be simplified by preparing a schedule of the type illustrated in table 13.1, encompassing the cold water services measured in example 24. This approach eliminates the repetition of items that often occurs in the traditional taking off system, as can be seen by comparing the entries in table 13.1 with the corresponding measured items in example 24, which follows the text of this chapter. It also simplifies the taking off and can help to eliminate possible errors and omissions.

HEATING AND HOT WATER INSTALLATIONS

These services were considered to be largely outside the scope of this basic text on building measurement and hence have been covered in outline only. Readers requiring more detailed information on the measurement of these classes of work are referred to *Measurement of Building Services* by G. P. Murray, in the Macmillan Building and Surveying Series.

Heating/hot water services are made up of the following three main components:

source	*distribution*	*outlets*
boilers and hot water cylinders	pipework	radiators

The pipework, pipework ancillaries, pipe insulation and cutting and forming holes for pipes will be measured in the same way as has been adopted for the cold water supply system in example 24. This leaves the boiler, hot water indirect cylinder and radiators to measure. The expansion tank is similar to the cold water cistern in example 24. Typical measured items, in take off format, for these categories of heating equipment are now provided.

Boilers can be fired by gas, oil or solid fuel, with gas being the most popular form of energy for practical reasons. The boiler can be free-standing or wall mounted and can be of the more efficient condensing type boiler. Typical descriptions for measured items of two wall mounted gas fired boilers follow (see T10 in Appendix B of SMM7). The *Measurement Code* (page 42) refers to boilers in the context of equipment with reference to Y20–25:1.1.1.1, 2&4. From a practical standpoint these rules are appropriate, but work sections Y20–25 do not apply to boilers.

	1		Plant and equipt.
			Boiler
			Glowworm, ref. 60 BF/SS for cent. htg. & hot water; gas fired; wall hg.; balance flued; automatically controlled by thermostat w. elec. control; max. output 60 000 Btu/h; sealed system; o/a dims. 360 × 700 × 297 dp.; w. expansn. vessel, pump, automatic air vent, pressure gauge & pipewk. in matchg. wh. case; fxg. to masonry

Typical equivalent condensing boilers are the Keston 60, ECOmax (Vaillant), the Trianco Heatmaker 60CB and Glowworm Energysaver 60P.

An alternative description might read as follows:

	1		Plant and equipt.
			Boiler
			packaged gas fired; rm. sealed; wall hg. as spec. clause H3/5; w. pumps; 4 nr. 28 male threaded conns. for flows & retns.; bal. flue assembly w. ext. grille to prefmd. hole in 255 ext. holl. wall; unit bolted to masonry (gas supply & conn. by others)

It will be noted that the first description contains specification details but gives little information on connections and fixing details, while the second description gives minimum particulars of the boiler, by referring to the appropriate specification clause and gives ample information about connections and fixing details.

A typical item for a hot water storage cylinder could read as follows (see S11 and Y23:1.1.1.4):

	1		Storage cylinders
			Hot water indirect cylinder
			136 litres cap.; cop. to BS 699 (grade 2) as spec. clause S11/5; w. 2 nr. 35 female thrdd. primary bosses; prectd. w. 32 th. insul.; supptd. on flr.

An alternative approach could be as follows:

	1		Storage cylinders
			Hot water indirect cylinder
			double feed; preinsul.; cop. to BS 1566 part 1, grade 3; ref. 8: 140 litres cap.; 4 nr. bosses & immersn. heater boss; placg. in posn.

The cylinder often contains a 2 or 3 kW immersion heater controlled by a thermostat and this is an enumerated item.

A typical radiator item could read as follows (see T32 in Appendix B of SMM7):

			Equipment
			Radiators
			enamelled m.s.; double panel; single convector; incl. 20 mm brass release air valve & plain plug; m.s. bkts. p & s to masonry
2			type 2/1) 600 × 600 hi.
4			type 2/2) 900 × 600 hi.
3			type 2/3) 1.200 × 600 hi.

The radiator items will be followed by the associated valve items:

			Ancillaries for equipt.
			Brass radiator valves
9			Lockshield valves to BS 2767; chrom. pltd.; angle patt.; jtg. to rads. & 15 conn. for cop. pipe.
			&
			Thermostatic control valves; chrom. pltd.; angle patt.; jtg. to rads. & 15 conn. for cop. pipe

WORKED EXAMPLE

A worked example follows covering the measurement of a cold water supply system, sanitary appliances, waste and discharge pipes, including a schedule of the cold water services.

This page has been left intentionally blank for make-up reasons.

Cold water and waste services

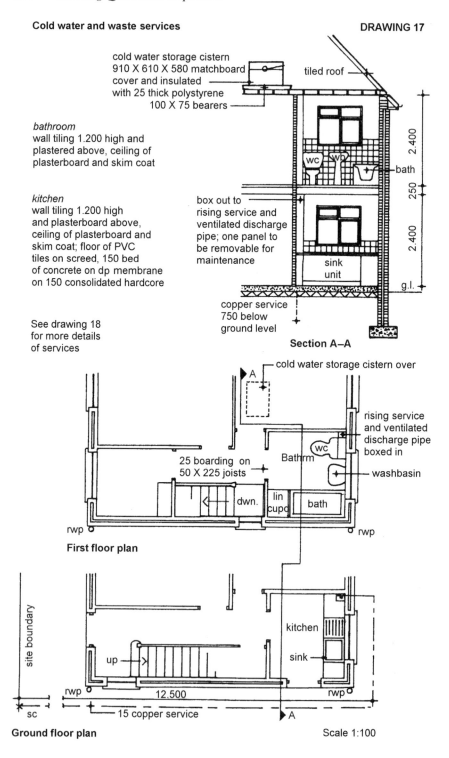

cold water storage cistern
910 X 610 X 580 matchboard
cover and insulated
with 25 thick polystyrene
100 X 75 bearers

tiled roof

bathroom
wall tiling 1.200 high and
plastered above, ceiling of
plasterboard and skim coat

wc wb

bath

2.400

250

kitchen
wall tiling 1.200 high
and plasterboard above,
ceiling of plasterboard and
skim coat; floor of PVC
tiles on screed, 150 bed
of concrete on dp membrane
on 150 consolidated hardcore

box out to
rising service and
ventilated discharge
pipe; one panel to
be removable for
maintenance

sink
unit

2.400

g.l.

See drawing 18
for more details
of services

copper service
750 below
ground level

Section A–A

cold water storage cistern over

A

rising service
and ventilated
discharge pipe
boxed in

wc

Bathrm

washbasin

25 boarding on
50 X 225 joists

lin
cupd

bath

dwn.

rwp

rwp

First floor plan

site boundary

kitchen

sink

up

12.500

rwp

rwp

sc

15 copper service

A

Ground floor plan

Scale 1:100

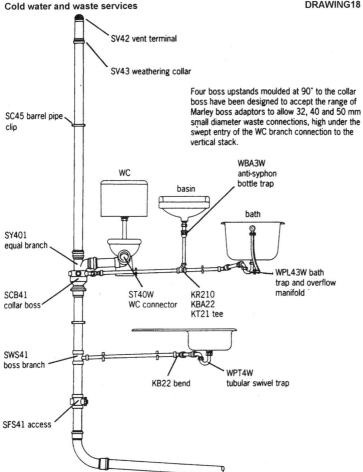

SV42 vent terminal

SV43 weathering collar

Four boss upstands moulded at 90° to the collar boss have been designed to accept the range of Marley boss adaptors to allow 32, 40 and 50 mm small diameter waste connections, high under the swept entry of the WC branch connection to the vertical stack.

SC45 barrel pipe clip

WBA3W anti-syphon bottle trap

WC

basin

bath

SY401 equal branch

SCB41 collar boss

ST40W WC connector

KR210 KBA22 KT21 tee

WPL43W bath trap and overflow manifold

SWS41 boss branch

KB22 bend

WPT4W tubular swivel trap

SFS41 access

Above ground waste drainage

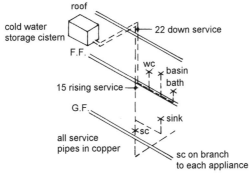

roof

cold water storage cistern

22 down service

F.F.

wc basin bath

15 rising service

G.F.

sink

all service pipes in copper

sc

sc on branch to each appliance

Cold water services

Table 13.1 *Schedule of cold water services to house.*

Cop. pipes in trs.

15	22
	16.05

Cop. pipes p&s to masonry

15	22
1.30	2.70
5.10	1.90
1.40	0.70
1.10	5.30
0.90	
9.80	

Cop. pipes scrd. to tbr.

15	22	28
3.40	3.00	2.20

E.O. cop. pipewk.

Made bends

15	22
2	3
1	2
3	1
2	6
8	

Spec. conns.

15	22
1	1
1	
1	
1	
6	

Fittings

2 ends	3 ends
15	15
1	1

22
1

28
1

Pipewk. ancillaries

S.V.s

15	22
1	1
1	
1	
1	
4	

Ball valves

15	22
1	1

Pillar taps

15	22
1	1

Excvtg. trs. for servs.

≤200 Ø

16.05

Pipe insul.

15	22
3.40	3.00

Cuttg. or formg. holes for pipes ≤55 Ø

255 wall	150 conc. flr.	rf. tilg.
1	1	1
22	15	22
	1	1

c.w. cistn.

1

Insul. to cistn.

1

Sn. swd. suppts.

75 × 100
2.00

Pipe sleeve cast into conc.

15	22
1	

Lead slate supply.

1

F.O.

1

282

Drgs. 17 and 18	Example 24

Cold Water and Waste Services 1

Cold Water Supply System

Conn. to main

Include the following defined Provisional Sums for work by Statutory Undertakings

Item

Conn. to water main in highway & bring'g. 15 cop. serv. to bdy. of site; incl. provsn. of s.v. & box by Water Authority: £200

Item

Perm. reinstatement of highway by Highway Authority following conn. to water main: £100

Cold water serv.

Cop. pipewk. to BS 2871, pt. 1, table X & EN 1057; comp. jts. to BS 864, pt. 2, type B

site bdy. to hse.
```
            12.500
             3.000
               550
            16.050
```

16.05

Pipes
15 nom. size; strt.; in trs.

&

Excvtg.
Trs. for servs.
≤ 200 nom. size;
500 - 750 av. depth

Cop. pipewk. a. b.

1

E. O. for spec. conn.; comp. jt. to xtg. fittg.
(WA's s.v.

2

E. O. for made bends

The use of the heading of Cold Water Supply System follows the terminology in *Common Arrangement*: S10. This avoids the need to describe the individual items that follow as cold water services.

Work to be carried out by the Water Authority and Highway Authority is covered by provisional sums as A53:1.1–2.0.0.

The measurement work then follows in a logical order from cold water services to sanitary appliances and foul drainage above ground, and finishing with associated items, such as pipe casings and sundries.

The copper pipe and compression joints are described in the main heading to avoid repetition in later items. The Standard EN1057 is a new European standard for copper and copper alloy seamless round copper tubes for gas and water in sanitary and heating applications. Another alternative is to name specific products such as Kuterlex and Kuterlex plus protected copper tube from Yorkshire Copper Tube and Tectite push-fit fittings from IMI Yorkshire Fittings Ltd.

The service pipe is measured in metres, stating the type, nominal size and method of jointing and including laying the pipe in trenches as Y10:1.1.1.3, while the excavation of pipe trenches is classified as P30:1.1.1–2.0. The excavation item is deemed to include backfilling and disposal of surplus excavated materials (P30:C1).

Special connections and made bends are enumerated as extra over the pipe in which they occur (Y10:2.1–2.1.0).

Cold Water and Waste Services 2

		thro. wall 300
		up to g.f. s.v. 1.000
		1.300

Cop. pipewk. a.b.

Pipes

1.30   15 nom. size; strt.; fxg. w. cop. clips p&s to masonry at 1.500 ccs.

1   E.O. for
     made bends

Rising serv.

Pipewk. ancillaries

Stop valves;

1   15 nom. size; brass h.p. comb. scrdn. & draw off tap to BS 1010; 2 nr. comp. jts. to cop. pipe

len.

g.f. sty. ht. from s.v.	2.300
ff jsts. & bdg.	250
ff sty. ht.	2.400
clg. jsts.	150
	5.100

Cop. pipewk. a.b.

Pipes

5.10   15 nom. size; strt.; clips p&s to masonry a.b.

across rf. sp.	2.800
up to inlet of cistn.	600
	3.400

Ditto.

3.40   do.; clips a.b. scrd. to tbr.
&

Foamed polyurethane preformed insul. 13th.; butt jts. in runn'g. len.; secured w. adhesive band

Pipelines
   15 nom. size

The measurement of the pipework is continued up to the ground floor stop valve on the rising main just above floor level.

The method of supporting the pipes and the background need to be stated as Y10:1.1.1.1. The item for the made bend is indented or written short below the 15 nominal diameter pipe item and it is then unnecessary to give the size of the bend.

Valves are enumerated and described as ancillaries in accordance with Y11:8.1.1.0. Cutting and jointing pipes to ancillaries are deemed to be included (Y11:C7).

The length of the 15 diameter rising service is calculated from the figured dimensions on the drawing (section A–A) and is recorded in waste for the benefit of those who may be working through the dimensions at a later date.

Similar copper pipework to that measured previously permits some measure of abbreviation, but sufficient information must be given to enable the work to be identified.

A further reduction in the length of descriptions is now possible as the one difference is the fixing to timber instead of masonry.

The insulation of the pipes in the roof space is measured as Y50:1.1.1.0, giving the type of insulation and the nominal size of pipe.

Co|ld Wat|er an|d Waste Services 3

		Gen. buildrs. wk.	Cutting and forming holes for pipes are enumerated as P31:20.2.1.2&4, giving the nature and thickness of the structure, pipe size classification and other relevant particulars, such as whether circular and making good. Items for holes are only required where the pipe passes through the structure of the building. A schedule would prove useful, especially on a larger project, wherein all the various pipe lengths grouped under their different sizes and backgrounds, fittings and cutting or forming holes can all be collated and the totals in each category transferred to dimension sheets (see table 13.1).
	1	Cuttg. or formg. holes for pipes
		≤ 55 nom. size; in holl. bk. wall 255 th.; circ.; m/gd.
		&
		ditto; in conc. flr. 150 th.; do.
		Steel sleeves to BS 1387, table 2
	1	For cop. pipe 15 nom. size; cast'g. into conc. 150 th.; m/gd.	An example of a pipe sleeve is given to show the approach to its measurement as P31:23.1.1.5. Alternatively, there could be two separate items – one for supplying the sleeve and the second for fixing by other trades.
		Cop. pipewk. a.b.
3/	1	E.O. 15 nom. size pipe for made bends	There are three made bends on the rising service in the roof space up to the inlet to the cold water storage cistern, measured as Y10:2.1.0.0.
		Pipewk. ancillaries
		Ball valves
	1	15 nom. size; h.p. piston type to BS1212, pt. 2; brass lever & PVC 112 ⌀ float to BS 1968; incl. strt. tk. conn. & strt. couplg. & comp. jt. to cop. pipe.	The ball valve to the cold water storage cistern is measured in accordance with Y11:8.1.1.0, giving the type and nominal size of the fitting, method of jointing and type of pipe to which it is connected. There are joints to the cistern and the copper supply pipe.

Cold Water and Waste Services 4

Br. from risg. serv. to sk.

Cop. pipewk. a. b.

Pipes
    15 nom. size, strt.; clips
    p & s to masonry a.b.

| 1.40 | |

E.O. for
    made bends

| 2 | |

E.O. for
    fittgs.; 3 nr. ends

| 1 | |

The pipework to the rising service, sometimes called the rising main, is completed by taking the branch pipe to the ground floor sink. This is normally the only connection from the rising service. The description of the pipe can be abbreviated as a similar pipe has been measured previously.

Pipewk. ancillaries

Pillar taps;
    chrom. pltd.; 15 nom.
    size; easy cln. patt.
    to BS 5412; indexed
    COLD; incl. strt. tap conn.
    & comp. jt. to cop. pipe

&

Stop valves
    15 nom. size; brass
    h.p. scrdn. type to
    BS 1010; 2 nr. comp.
    jts. to cop. pipe

| 1 | |

There are two pipework ancillaries consisting of a stop valve just below the sink and a cold tap over the sink, both measured in accordance with Y11:8.1.1.0. The tap forms a separate enumerated item as it is not usually provided with the sink. In practice a pair of taps would normally be provided (one for cold water and the other for hot).

Down serv.
In rf. sp.
  across rf. sp. 2.800
  up cistn.   200
          3.000

Cop. pipewk. a. b.

Pipes
    22 nom. size, strt.; clips
    scrd. to tbr. a. b.

| 3.00 | |

E.O. for
    made bends

| 3 | |

Now commence with the down service starting from the cold water storage cistern, taking the main run first, followed by the branches to the bath, washbasin and WC on the first floor.

An alternative to the copper pipes would be polyethylene pipes, when the description for internal pipes could be 'Polyethylene pipework, black, medium density to BS 6730, with brass and gunmetal compression joints to BS 864, Part 3; fixed with galvanised brackets to masonry'. Fittings could be Kuterlite 1700. Polyethylene pipes below ground will be blue pipes to BS 6572.

Cold Water and Waste Services 5

3.00	<u>Insul. a.b.</u> Pipelines 22 nom. size

All pipes in the roof space/void must be insulated, when the roof insulation is laid between or over the ceiling joists, to prevent freezing. The insulation is measured as Y50:1.1.1.0.

<u>dn. serv. & brs.</u>

clg. jsts. 150
ff sty. ht. 2.400
flr.   150
     2.700

The length of the main down service is calculated in waste and the copper pipework follows the same procedure with the item descriptions kept constant throughout. The branch pipe leading to the bath is scaled with the location indicated.

<u>Cop. pipewk. a.b.</u>
Pipes
   22 nom. size; strt.;
   clips p & s to masonry
   a.b.

2.70	(vert.
1.90	(hor. to (bath
0.70	(vert. to (bath
2	E.O. for made bends
1	<u>Pipewk. ancillaries</u> Stop valves 22 nom. size ; a.b.

The linear pipework is followed by enumerated items for the made bends, a stop valve, which is of similar pattern to the one measured previously, permitting the use of a much abbreviated description, and finally cutting or forming holes through the plasterboard and skim coat ceiling and boarded floor.

	<u>bath</u> <u>Cop. pipewk. a. b.</u> E.o. 22 nom. size pipe
1	for spec. conn. to pillar tap , incl. bent conn. & comp. jt.
1	<u>Pipewk. ancillaries</u> Stop valves 22 nom. size ; a.b.

The last two items pick up the remaining work around the bath, comprising a special connection to the pillar tap over the bath as Y10:2.2.1.0 and a stop valve on the vertical branch pipe to the bath as Y11:8.1.1.0.

Cold Water and Waste Services 6

			Washbasin

Cop. pipewk. a. b.

1.10   Pipes
     15 nom. size; strt.;
     clips p. & s. to masonry
     a. b.

1   E. o. 22 nom. size
     pipe for
       fttgs.; 3 nr. ends
(redcg.
(tee   &

   E. o. 15 nom. size
   pipe for
     made bends
       &
   Ditto
     spec. conn. to pillar
     tap, incl. bent conn.
     & comp. jt.

Pipewk. ancillaries

1   Stop valves
     15 nom. size ; a. b.

			WC

Cop. pipewk. a. b.

0.90   Pipes
     15 nom. size; strt.;
     clips p. & s. to masonry
     a. b.
     E. o. for
1   spec. conn. to cistn.
     a. b.

Pipewk. ancillaries

1   Stop valves
     15 nom. size ; a. b.

			c.w. storage cistn.

Gen. pipeline equip.

Galv. m.s.

1   Water cistn.; open top;
     910 x 610 x 580 o/a.;
     to BS417, pt. 2, grade A;
     ref. scm.; 227 litres cap;
     cover; perfs. for 2 nr.
     pipes; 2 cts. bit. pt. intly.

The water supply to the remaining sanitary appliances is now taken, starting with the washbasin and finishing with the WC. In each case a logical sequence is adopted starting with the pipework and then proceeding to measure extra over items such as fittings, made bends and special connections, and finally pipework ancillaries in the form of stop valves.

These items entail considerable repetition of earlier similar ones and the use of abbreviated descriptions and the letters 'a.b.' (as before) are used extensively. As stated earlier in this example, the use of schedules and grouping of like items, as shown in table 13.1, will assist in reducing the number of items entered on the dimension sheets.

The water storage cistern is enumerated, stating the type, size, pattern and capacity as Y21:1.1.0.0. The British Standard reference, number of perforations for pipes and any internal treatment are included in the description as prescribed in the *SMM7 Library of Standard Descriptions*.

Cold Water and Waste Services 7

	Glass fibre insul. encased in polythene to BS 5615; secured w. PVC bands	The insulation to the cold water storage cistern is enumerated, stating the type of insulations identifying the equipment and giving the overall dimensions of the cistern as Y50:1.4.2.0. The kind, quality and thickness of insulation and method of fixing are to be stated (Y50:S1–5).

1 —

C.w. cistn.
one-piece jacket
60 th.; 910 × 610 ×
580 o/a size.

Sn. swd. a.b.

2/ 1.00

Individ. suppts.
75 × 100
(tk. brrs.

The tank bearers are measured in metres giving a dimensioned overall cross section description as G20:13.0.1.0.

Overfl. to cistn.

Cop. pipewk. a.b.

Pipes

2.20

28 nom. size; strt.;
clips scrd. to tbr. a.b.

The pipework to the cistern overflow is measured in metres as Y10:1.1.1.1. An alternative to copper would be polythylene.

E.o. for
spec. conn. to cistn.,
incl. flgd. tk. conn.
& strt. conn. w.
comp. jt.

1 —

The connection of the overflow pipe to the cistern is an enumerated item as Y10:2.2.1.0, but with no requirement to give the nominal size of the connection, unless it differs from the pipe.

Gen. buildrs. wk.

Cuttg. or formg. holes
for pipes
≤ 55 nom. size;
thro. rf. tilg.
circ.; m/gd.

1 —

Forming holes for pipes in roof coverings are enumerated as P31:20.2.1.2&4.

&

Fixing only

Lead slate
300 × 300; w.
collar ard. 28
dia. pipe

The fixing of soakers and saddles (provided by other trades) is enumerated separately with a dimensioned description (H60:10.6.1.1).

&

Lead code 5 to BS 1178

Slate
300 × 300 w.
soldered collar ard.
28 dia. pipe in plain
tilg.; handed to others
for fixg.

A similar approach has been adopted for the lead slate. Metal slates, for fixing by the slater or tiler (roofer), are enumerated with a dimensioned description (H70:26.1.1.1).

Cold Water and Waste Services 8

San. appliances

Include PC Sum for supplyg. only components

Item | San. appliances £ _ _ _

&

Add for
 main contr's.
 prof. _ _%

These items follow the procedure for dealing with nominated suppliers as A52:1.1–2.1.0. The actual sum will be inserted in the bill of quantities and the main contractor will enter the percentage profit required on this sum.

1 | Assemblg. & fixg. only
Comb. drainer & bowl; 2.250 x 530 o/a; s.s. drainer 560 x 400 x rect. bowl 510 x 360 x 250 dp.; w. waste fittg., chain, stay & plug & fxg. to sk. unit

&

Bath, pressed stl. porcelain enam., rect., 1.700 lg. to BS 1189; w. 2 nr. chr. pltd. pillar taps, waste fittg., chain, stay & plug; met. cradles

&

Washbasin, ped. in vit. china 560 x 406; w. 2 nr. chr. pltd. pillar taps, waste fittg., chain, stay & plug & fixg. on pr. of wall hang'g. bkts. p. & s. to masonry & scrg. ped. to tbr.

&

WC suite, low lev. vit. china. ped., close couple, washdn.; dble. plastic ring seat; 9 litre glzd. flushg. cistn. on pr. of galv. m.s. concealed bkts.; enam. flushg. pipe; incl. scrg. ped. to swd., jtg. outlet to PVC-U disch. pipe; p. & s. cistn. bkts. to masonry & jt. flush pipe to cistn. & arm of ped.

(End of assemblg. & fixg.)

Fixing only these items is measured in the appropriate work sections (A52:M3), and is deemed to include such items as unloading, storing, hoisting and returning packaging materials (A52:C1). The measurement of sanitary appliances is covered by N13:4.1.1.1 and the appliances are as defined in Appendix A. An adequate description of each appliance needs to be given to enable the estimator to determine the cost of assembling and fixing each item. Hence the particulars generally include type of appliance, overall dimensions, materials, component parts and all jointing and fixing. Where the fixing background is to be given, this should follow the requirements of General Rules 8.3, and will usually be masonry or timber.

The stainless steel sink is fitted to the top of the sink unit and the latter forms a separate measured item. Bath panels and bearers are measured subsequently, as are also traps to fittings.

The WC pan, flushing cistern and flush pipe are combined in a single enumerated item.

It is advisable to indicate the end of the fixing items to avoid any confusion.

Cold Water and Waste Services 9

<u>Wk. assctd. w. san. appls</u>
Bath

<u>Vitrolite</u>
Panel to bath side; 1.700
× 500 × 3 th.; fxd. w. chr.
pltd. dome hdd. scrs. to swd.

1	

A panel is required to one side only of the bath and is best enumerated with a suitable description.

<u>Sn. swd. a. b.</u>
Framed suppts.
width > 300;
40 × 50 membrs. at
500 ccs. both ways

1.70	
0.50	

The bath panel bearers are measured in m² as the width of the supports >300, giving the size of the members and their spacing in both directions as G20:12.1.1.0, despite the fact that it could seem more logical to measure them as linear items.

WC overflow

<u>Cop. pipewk. a. b.</u>

Pipes
22 nom. size; strt.;
fixd. to masonry a. b.

0.60	

Note the use of subheadings to act as signposts throughout the dimensions. Instead of taking each overflow separately (cold water storage cistern and WC), they could be grouped together in the taking off to avoid the repetition of similar items. The pipework, which could be in polyethylene, is measured and described as Y10:1.1.1.1.

E.o. for
spec. conn. to
flush'g. cistn.; incl.
bent conn. & comp.
jt.

1	

&

fittg; 2 nr. ends;
hinged cop. flap
& fr. to sply. cut
end w. sold. jt.

The connection of the overflow pipe to the cistern with a bent connector, to obtain the desired direction, and the outlet fitting are both enumerated as extra over items, as Y10:2.2.1.0 and Y10.2.3.5.0. In practice, flaps of the kind described are little used nowadays, but it has been included to show the measurement approach.

<u>Gen. buildrs. wk.</u>

Cutt'g. or form'g. holes
for pipes
≤ 55 nom. size;
in holl. bk. wall
255 th.; circ. m/gd.

1	

An enumerated item follows to cover the overflow pipe passing through the hollow wall as P31:20.2.1.2&4.

Cold Water and Waste Services 10

			Waste pipes & traps
			Pipewk. ancillaries

Marley Universal plastics waste traps to BS 3943

Two piece traps; tub. swivel; code WPT4W; 40 nom. size; 75 seal; 'P' type outlet; 3 nr. comp. jts. & conn. to MU PVC pipe

| 1 |
(sk.   &

Two piece traps; Low inlet tub.; code WPL43W; 40 nom. size; 75 seal; 'P' type outlet; ov/fl. manifold code WB02W; flex. ov/fl. pipe code WOP1W; 5 nr. comp. jts. & conn. to MU PVC pipe

(bath   &

Anti-syphonage bottle traps; Code WBA3W; 32 nom. size; 75 seal; 'P' type outlet; 2 nr. comp. jts. & conn. to MUPVC pipe

(w. bsn.

Traps to waste appliances are classified as pipework ancillaries. As all three waste traps are of the same material, they can be grouped together under a common heading. The *SMM7 Library of Standard Descriptions* classifies traps into various types, such as two piece, anti-syphonage and bottle traps. Hence these terms have been used to head the item descriptions. The traps are enumerated giving the type, nominal size, type of pipe and method of fixing as R11:6.8.1.0. The Marley codes for traps are included in the descriptions to help identify them, and they are listed on drawing 18. Note that in the case of traps, it is necessary to give the depth of seal and type of outlet (P, Q & S).

MU PVC pipewk. to BS 5255 w. strt. couplg. solvent soc. weld jts.

Pipes
   40 nom. size; strt.; fxg. w. zinc electropltd.

| 1.40 |
| 1.85 |
(sk.   m.s. bkts. p & s to
(bath masonry at 1.500 ccs.

Ditto
   32 nom. size

| 0.70 |
(w. bsn.

E.o. 40 pipe for fitt'gs.; code KB22; 2 nr. ends.

| 3 |

fittg's; code KT21; 3 nr. ends.

| 1 |

E.o. 32 pipe for fittgs; code KB12; 2 nr. ends.

| 1 |

MUPVC is modified unplasticised PVC. The waste pipes and their jointing are described in the heading and then the pipes measured and described as R11:1.1.1.1, distinguishing between different nominal sizes, followed by fittings measured extra over pipes as R11:2.3.1–3.0. It is not necessary to describe the type of fittings when ≤65 and the term fittings includes bends (*SMM7 Measurement Code*, page 39), although made bends are so described.

<u>Cold Water and Waste Services 11</u>

		Foul drainage above grd.
		<u>Ventilated discharge pipe</u>
		len.
		above rf. 1.000
		in rf. space 2.600
		g.l. to rf. 5.200
		8.800

Marley PVC-U pipewk. to BS 4514 w. ring seal s & s jts. code SP403

8.80		Pipes 110 nom. size; strt.; fxg. w. PVC barrel pipe clips, code SC45, p. & s. to masonry at 1.500 ccs.
1		E.O. for fittgs.; pipe > 65 ∅; access pipe w. twist & lock access cap; code SFS41; 3 boss upstds; double solv. socs. & E.O. for ditto.; 110×40 boss br.; code SWS41; solv. socs. & E.O. for ditto.; 110 collar boss br.; code SCB41; 4 nr. boss upstds.; one open; solv. socs. & E.O. for ditto.; 110 plain equal br.; code SY401; rg. seal socs. & E.O. for ditto.; 110 weathg. collar; code SV43; solv. jt. to pipe
		<u>Pipewk. ancillaries</u>
1		110 vent terminal; Marley code SV42; PVC ctd. wire; solv. soc. to 110 PVC-U pipe

The length of the discharge pipe (the term used in BS 5572: Code of practice for sanitary pipework and subsequently in the Building Regulations, Approved Document H1 (1989) relating to the former combined soil and waste pipe) is built up in waste and stretches from ground level up to the top of the vent above roof level.

The measurement work commences with a description of the pipework and method of jointing as R11:S2. The pipes are measured in metres, stating whether straight or curved, the type, nominal size and jointing and fixing details (R11:1.1.1.1).

Fittings are measured extra over the pipes in which they occur and in the case of fittings to pipes >65 diameter, stating the type and whether it has an inspection door (R11:2.4.5.1). Each of the five fittings listed in the example is described in sufficient detail to identify it and to give its main features. Each description also includes the Marley code reference which will be of value to the estimator when pricing the bill. It also enables the student to identify all items against the components shown on drawing 18, and so obtain a better understanding of them and the ability to trace the items in the Marley catalogue.

The vent terminal, to prevent birds from nesting in the top of the discharge pipe, is classified as a pipework ancillary (R11:6.4.1.1).

Cold Water and Waste Services 12

Gen. buildrs. wk.

| | | Cuttg. or form'g. holes for pipes | The ancillary items to the pipework, such as cutting and forming holes for pipes, are now taken as P31:20.2.2.2&4. |

| 1 | | Cuttg. or form'g. holes for pipes |
| | | 55 – 110 nom. size; thro. rf. til'g. circ.; m/gd. |

The ancillary items to the pipework, such as cutting and forming holes for pipes, are now taken as P31:20.2.2.2&4.

Fixg. only
Alum. weathg. slate;

| 1 | | 450 × 450, w. rubber hd. to fit 110 ø pipe in plain tild. rf. |

The fixing only item by the roofing trade of the weathering slate is an enumerated item with a dimensioned description as H60:10.6.1.1 (saddles being the nearest equivalent listed item). The supply of the weathering slate forms a separate enumerated item as H70:26.1.1, with a dimensioned description, which includes the nature of the roofing and handed to others for fixing.

Alum. sheet

Marley weathg. slate;

| 1 | | 450 × 450 w. rubber hd. turned into soc. of 110 PVC-U pipe; handed to others for fixg. in pl. tild. rf. |

Br. to WC

PVC-U pipewk. a.b.

| 0.40 | | Pipes |
| | | 110 nom. size; a.b. |

Finally, the short branch pipe to the WC and the connector fitting are measured as R11:1.1.1.1 and R11:2.4.5.0.

E.O. for

| 1 | | fittings; pipe > 65 ø; WC connector code ST40W; w. solv. soc.; cap. & seal to WC spigot outgo |

Once again the Marley code reference for the fitting has been included in the description for ease of identification and to assist the estimator in pricing.

Alternative discharge pipe in cast iron

C.i. pipewk. to BS416
S & S jts. caulkd. in lead

Pipes

| 8.80 | | 102 nom. size; strt.; fxg. w.c.i. holderbats p. & s. to masonry |

An item has been included to cover the measurement of a cast iron discharge pipe instead of one in PVC-U. This item would be followed by the painting of the pipe and extra over fittings for a 100 diameter branch, two 40 diameter branches and an access pipe. As the 100 cast iron branch pipe is smaller than the 110 PVC-U connecting pipe from the WC, a plastic connector is set in the cast iron socket in mastic.

Cold Water and Waste Services 13	The pipe casings are now measured with allowance made for the shorter width of casing adjoining the WC on the first floor. The boarding is measured in metres as ≤300 wide as K20:1.2.1.0, with a dimensioned description.

Pipe casgs.
```
              1.600
              2.400
              4.000
```
Wrot. swd.

4.00   Bd. casgs.
     ≤ 300 wide; 220 × 16
     fxd. to swd.

2.40   ditto; 180 × 16; do.

The access panel is an extra over item as K20:12.1.1.0, with a dimensioned description. It is desirable to screw the access panel to the timber supports for ease of removal for inspection/ repairs. The number and lengths of the longitudinal and cross supports are calculated in waste.

1   E.o. for
     access panels;
     450 × 150; scrd. to swd.

```
                            suppts.
              2.400        1.600
less suppts.
    2/½/40      40            40
       500)2.360    500)1.560
            5+1          3+1
            ──          ──
            180          220
less vert. brrs.  65       65
            ──          ──
            115          155
```
Sn. swd.
Butt. jtd. suppts.
     ≤300 wide; 25 × 40; at
     200 ccs.; p & s to masonry

It is assumed that the supports will be butt jointed and not framed and will thus fall into the classification of butt jointed supports as G20:11.2.1.0. The description must contain a dimensioned description and include the spacing of the members. The method of fixing supports is to be stated, where not at the discretion of the contractor (G20:S2).

2/   4.00   (long
     (suppts.

10/ 0.16   Ditto.
4/   0.16   at 500 ccs.; a.b.
6/   0.12   (cross
     (suppts.

The painting is a general surfaces item as M60:1.0.1.0 and is described as >300 girth as the overall girth (both sides of the pipe casing) exceeds this dimension.

4.00   Paintg. wd.; k.p.s.
0.22   2 u/c, 1 f/c gloss
1.60   Gen. surfs.
0.22   > 300 gth.
2.40
0.18

An item is given for marking the position of holes as Y59:1.1.0.0.

Sundries

Item   Generally
   Markg. the posn. of holes,
   mortices and chases in the
   struct. for c.w. supply installtn.
     &
   Ditto for foul drainage
   above ground installtn.

Finally, an item is provided for testing and commissioning as Y51:4.1.0.0, stating the type of installation. Separate items for marking positions of holes and testing are required for each installation.

Installation generally
Item   Test'g. & commsng. as specified
   for c.w. supply installtn.
     &
   Ditto for foul drainage
   above ground installtn.

14   Electrical Services

GENERAL APPROACH TO MEASUREMENT

Where electrical circuits are to be measured in detail, such as circuits other than lighting and small power, the route of the conduit and/or cable should be plotted on the plan or a tracing overlay and the number of cables indicated. This sketch will then form a record of what is measured. An isometric sketch is often useful (as with pipework) to illustrate complex runs.

When plotting conduit and cables it is usual to draw runs at right angles to each other rather than running diagonally. This is usually necessary because of the nature of the structure through which the conduits and cables are passing, as for example following joists or beams. However, conduits and cables can sometimes be laid diagonally where running in floor screeds or in pitched roof voids. Once the route has been plotted and the specification fully understood, the measurement of the work is relatively straightforward, comprising enumerated items of equipment and final circuits and linear items of conduit, cable trunking, cable tray and cable on more complex systems, all measured in accordance with the rules prescribed in SMM7.

Within the constraints of this book, the relatively simple worked example can only provide an introduction to the subject. The measurement of electrical services, as with mechanical services, requires the traditionally trained quantity surveyor to develop his knowledge of the underlying technology. Readers requiring more detailed information on the measurement of this class of work are referred to *Measurement of Building Services* by G. P. Murray in the Macmillan Building and Surveying Series.

MEASUREMENT PROCEDURES

Conduit

Conduit (not in final circuits) is measured in metres, distinguishing between straight and curved, giving the radii, and stating the type, external size, method of fixing and background as Y60:1.1–2.1.1–5, and particulars of materials as appropriate (Y60:S1–6). The conduit is measured over all conduit fittings and branches (Y60:M2). The conduit is deemed to include all the labours and components described in Y60:C3. However, junction boxes and the like are enumerated as extra over the conduit in which they occur (Y60:2.1–6.1.1).

A typical measured item in take off format follows:

‖	18.50		PVC heavy gauge; push fit jts.; space bar saddles at 600 ccs.
			 Conduits
			    20 dia; strt; plug'g to masonry
			    surfs.

‖	3		E.o. for small circ. term. boxes

Cable Trunking and Tray
These are measured similarly to conduit, but in addition stating the method of jointing and spacing and method of fixing supports (Y60:5&8.1–2.1.1).

Cables
Cables (not in final circuits) are measured in metres, giving the type, size, number of cores, armouring and sheathing, method of support and background as Y61: 1.1.1–7.1–2, and particulars of relevant materials as Y61:S1–7. Y61:M3 prescribes the allowances to be made to cable lengths entering fittings and equipment.

Cable and Conduit in Final Circuits
These are enumerated on an enumerated points basis where they form part of a domestic or similar simple installation from distribution boards and the like (Y61:M7), as in the illustration shown in example 25. The *SMM7 Measurement Code* points out that this approach is appropriate for the majority of small power and lighting installations of a domestic or similar nature and also for the more simple installations in final circuits in other sections. Other types of final circuits are measured in detail.

Y61:P2 requires the following information to be given in connection with final circuits:

(a) a distribution sheet setting out the number and location of all fittings and accessories; and
(b) a location drawing showing the layout of the points.

Examples of these documents are illustrated in table 14.1 and on drawing 19 in this chapter. The *SMM7 Measurement Code* recommends that the distribution sheet should contain information relating to the location, number and type of lamps, the number of lighting, switch and socket points and the type of fittings, appliances and accessories, together with any information relevant to the circuit arrangement for each distribution board and the like.

Further clarification of the measurement of enumerated final circuit items contained in the *SMM7 Measurement Code* includes the following guidelines:

(1) It is not necessary to give the size of conduit as it will be at the discretion of the contractor.
(2) The 'distribution boards and the like' from which final circuits are measured include such control gear as control panels for boilers, fire alarms, or master clocks and similar items.
(3) The classification of points in the enumeration of final circuits relates to the terminations of the permanent wiring to switches and to outlet accessories and control gear for the connection of current using appliances or fittings.
(4) Flexible conduits, cables and the like between appliances or fittings and the associated terminal accessories or control gear on the permanent wiring of a final circuit should be included in the description of the relevant appliance or fitting.

Switchgear and Distribution Boards
These are enumerated, stating the type, size, rated capacity and method of fixing, usually with a cross reference to the specification, and giving details of fuses, supports provided with the equipment and method of fixing, and background as Y71: 1–2.1.1.1–3. The *SMM7 Measurement Code* states that fuse links and miniature circuit breakers (mcb's) supplied with the switchgear and distribution boards shall be included in the description of the control gear; whereas those supplied independently shall be measured separately.

Luminaires and Accessories for Electrical Services
These are covered in Y73 and Y74. Luminaires are enumerated stating the type, size and method of fixing, often with a cross reference to the specification, and any other relevant details listed in the fourth column (Y73:2.1.1.1–13). Pendants are also enumerated, distinguishing between those with a drop ≤1.000 and those >1.000. Lamps may be separately enumerated, stating the type, size and rated capacity (Y73:3.1.0.0), or alternatively they can be given in the description of the luminaires (Y73:M2). Accessories, which include lighting switches, socket outlets, thermostats, telephone cord outlet points and bell pushes, are enumerated stating the type, box, method of fixing, background and rated capacity, and may provide for plugs to be provided with socket outlets (Y74:5.1.1.1–2). The description of accessories should state the number of gangs comprised in the accessory.

Testing and Commissioning of Electrical Services
This is given as an item, stating the installation and any other relevant requirements listed in Y81:5.1.1–2.1–2. Provision of electricity and other supplies and of test certificates is deemed to be included (Y81:C1&C2).

Identification of Electrical Work

This is enumerated, where not provided with equipment or control gear, and is categorised as to plates, discs, labels, tapes and bands, arrows, symbols, letters and numbers, and charts, giving the type, size and method of fixing (Y82:4.1–6.1.1–2).

Sundry Items

These are covered in Y89. For example, marking positions of holes, mortices and chases in the structure is given as an item, stating the installation as Y89:2.1.0.1. Provision of electricity and other supplies required for the temporary operation of the installation is covered by Provisional Sums as listed in A54 (Y89:6.1.1.1–3&M3).

Preparing drawings can be included as an item, giving the details contained in Y89:7.1.1.1–2. Drawings include builder's work, installation drawings and records or 'as fitted' drawings (Y89:D1).

WORKED EXAMPLE

A worked example shows the method of measuring the electrical services to a small building, in accordance with the details shown on drawing 19.

Table 14.1 *Distribution sheet*

Location	Lighting					Power			Remarks
	Circt. nr.	Fittings		Switches one gang		Circt. nr.	13 A single SSO	13 A double SSO	
		nr.	type	one way	two way				
Entrance passage	1	1	100 W plain pendant		2	2	1		
Waiting room	12	1	150 W plain pendant	2		2	1	1	
Office	1	2	65 W fluor.		2	2	1	2	
Totals		4		2	4		3	3	

Electrical installation

DRAWING 19

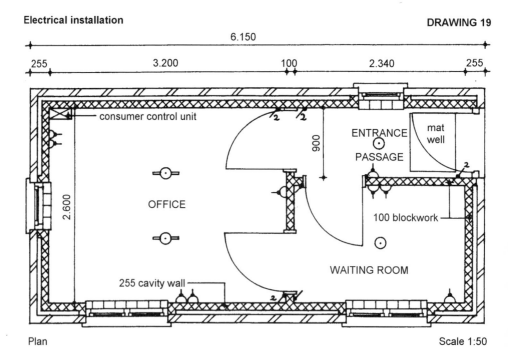

Plan   Scale 1:50

Drg. 19			Example 25	

Electrical Installation 1

Conn. to mains cable

Include the follg. defined Provisional Sum for works by Electricity Authority

| | Item | | Bring'g. mains cable to bldg. installg. meter and conn. to switchgear: £160 | The provisional sum for bringing the electric cable to the building, installing the meter and connecting to the client's switchgear, is in accordance with A53:1.2.0.0, covering work by statutory undertakings and General Rules 10. |

consumer unit

LV switchgear & disttn. bd.

| | 1 | | Distribution bd. 4 way surf. type met. clad. m.c.b. consumer control unit w. 100 amp main switch; manufac's. ref. KM 100; two 5 amp & two 30 amp m.c.b.s; plugd. & scrd. to masonry | The description of the distribution board is to include the type, size, rated capacity, fuses, method of fixing and background as Y71:2.1.1.1.–3. The reference to a manufacturer's product will ensure that all contractors tender on the same basis. The background classification is in accordance with General Rules 8.3b, whereby masonry includes concrete, brick, block and stone. M.c.b. is an abbreviation for miniature circuit breaker. |

General LV Power

Cables and conduit in final circuits

| | 1 | | Final circuits; cable & conduit in ring main circuit; concealed installation; consistg. of 2.5 mm² PVC insul. colour coded cable; drn. into black enam. heavy gauge m.s. conduit; in screeded flrs.; in circuit nr. 2, comprisg. six s.s.o's | Final circuits are measured on an enumerated points basis where they form part of a domestic or similar simple installation from a distribution board or the like (Y61:M7). The description is to include the cable and conduit installation, giving the size and type of cable and conduit and description of the final circuit (Y61:19.2.1.4). The circuit number is given for identification purposes although not specifically required by Y61:19. It is necessary to state whether the power circuit is a radial or ring main, whether attached to a surface or concealed and the number of sockets. The abbreviation s.s.o. stands for switched socket outlet. |

Accessories

Soc. outlet

| | 3 | | 13 amp; 3 pin shuttered flush; ivory plastic patt. single switched; stl. k.o. box; plugd. & scrd. to masonry | Accessories are enumerated and contain details of the type, box, method of fixing and background as listed in Y74:5.1.1.2. Accessories include switches and socket outlets (*SMM7 Measurement Code*: Y74:5). |
| | 3 | | ditto double switched; do. | The abbreviation k.o. refers to knock out box. |

Electrical Installation 2

General Lighting

Cables and conduit in final circuits

	Final circuits	
1	cable and conduit in lighting circuit; concealed installn.; consistg. of 1.00 mm² PVC insul. colour coded cable drn. into black enam. heavy gauge m.s. conduit; fxd. to tbr. w. clips in circuit nr. 1; comprisg. four lightg. pts. & two 1 way & four 2 way switch pts.	All final circuits measured on an enumerated points basis are shown on the distribution sheet (table 14.1) on drawing 19. The sheet sets out the number and location of all fittings and accessories, which include luminaires, lamps and lighting switches. The distribution sheet (table 14.1) is adequate for the purposes of Y61:P2 and must be inserted in the bill of quantities. The measurement of final circuits is prescribed in Y61:19.2.1–6.1–5. Colour coded cables must be so described (Y61:S7).

Luminaires and Lamps

	Luminaires	
2	pendant fittg.; comprisg. ivory plastic clg. rose & connector blk.; brass B.C. lampholder w. shade rg. & 3-core 0.75 mm² PVC insul. & protectd. flex. cable; drop ≤ 1.000, incl. 50 ∅ loop in conduit box, scrd. to tbr. (ent. passage & wditg. rm.	The type, size and method of fixing pendant luminaires are to be given, together with the length of drop (≤ or >1.000) and details of the ceiling rose, conduit box, connector block and lampholder as Y73:2.2.1.2,4,5,6,9&13. The locations of these fittings have also been included for ease of identification.
2	Luminaires fluorescent fittg.; 1.500 lg.; 65W single tube & diffuser, cat. ref. UK 654; incl. conn. blk.; 50 ∅ loop in conduit box & one suppt. bkt.; comprisg. 600 len. of 20 ∅ conduit & 2nr. 50 loop in boxes; one box scrd. to tbr. (office	Fluorescent fittings require two fixing points. One would be provided with the conduit installation and the other must be included in the description or, alternatively, measured separately. The bracket is made up in conduit and screwed to a timber ceiling joist. Timber bearers may be required if the bracket does not coincide with a ceiling joist.

Electrical Installation 3

lamps

1	lamps 100W B.C., G.L.S.; as specfd. (ent. pass.

Lamps are enumerated stating the type, size and rated capacity (Y73:3.1.0.0). It is common practice to take the lamps separately from the luminaires, although alternatively the lamps may be given in the description of the luminaires (Y73:M2). Many new lamps are now on the market aimed at giving greater efficiency and longer life, such as the Ecolamp™ 7, 9 and 11 equivalent to GLS 40, 50 and 60 W, by Solalight and Greenstock's triple life 26 000 hour fluorescent lamps.

1	Ditto. 150 W (waitg. rm.

2	Lamps fluorescent tube; 1.500 lg.; 65W; 'warm white' (office

Accessories

2	Lighting switches 1 gang, 1 way, 5 amp; single pole silent action; flush ivory plastic plate & stl. k.o. box; plugd. & scrd. to masonry (waitg. rm.

Lighting switches are classified as accessories in the *SMM7 Measurement Code*: Y74:5, and the descriptions are to conform to Y74:5.1.1.2. The descriptions of switches are to give the number of gangs in accordance with Y74:M3. The knock out box is included in the description. Alternatively, specific products can be specified, such as a 1 gang, 2 way stainless steel rocker switch, reference RIGTW/SS, supplied by Forbes and Lomax Ltd. As with all measurement items, work logically through the lighting switches, checking against the layout drawing and the distribution sheet. It is important to double check all quantities to ensure that they are correct.

2	1 gang, 2 way ditto (ent. pass.
2	(office

Item	Generally Markg. the posn. of holes, mors. & chases, in the structure for the general lighting & power (small scale) installtn.

The sundries item for marking the position of holes, etc. is required by Y89:2.1.0.0. In larger installations it is intended to apply to each installation separately, i.e. lighting and power. These have been combined in this example as it is a very small project and is classified under work section V90.

Electrical Installation 4

	Installn. generally	The testing and commissioning item is covered in Y81:5.1.1–2.1–2 and is deemed to include provision of electricity and other supplies and test certificates (Y81:C1&2).
Item	Testg. & commsng. as specified for general lighting & power (small scale) installn.	
	Identification	The identification label is not provided with the control gear and so is required to be measured under Y82:4.3.1.0, stating the type, size and method of fixing.
1	Label 125 × 100; white plastic; marked DIS BOARD 1/4; w. list of ea. circuit beneath; plug'd & scrd. to masonry.	
	(sited near (consumer unit	
	Sundries	The drawings item emanates from Y89:7.1.1.2, with one item only being required for the whole works. It is a common requirement for these drawings to be on linen and, if so, it must be stated in the description. The description is to include the number of copies and name of recipient.
Item	Preparg. drgs. 4 copies of 'as fitted' prints show.g. circuits & conduit runs & hand to maintenance engr.	
	Gen. buildrs. wk.	Cutting or forming holes, mortices, sinkings and chases for electrical installations is enumerated, distinguishing between concealed and exposed conduit/cable, and stating the type and giving the number of classified points and making good as P31:19.1–4.1–5.1–2, irrespective of the size, type and kind of points (P31:M10). Associated switch points are deemed to be included (P31:C3).
6	Cuttg. or formg. holes, mors., sinkgs. & chases for elec. installns. concealed m.s. conduit 20 ⌀; soc. outlet pts. & m/gd.	
	(circuit 2	
4	luminaire pts. & m/gd.	
	(circuit 1	

15   Drainage Work

ORDER OF TAKING OFF

A logical order of taking off and full annotation in waste is extremely important when measuring this class of work. Where separate drainage systems are to be provided for foul and surface water drainage, it may be considered desirable to measure the drains and associated work in each system separately, although it is usually quicker and simpler to take all the manholes together. However, in some situations it is simpler to take the whole of the work in the two systems together, as illustrated in example 26, showing that there is not a single universal approach and that it is better to take each case on its merits. The *Standard Method* does not require the separation of the two drainage systems.

A useful order of measurement is as follows:

(1) Lengths of main drain starting at the head of the drainage system and working downwards to the sewer or other point of disposal.
(2) Branch drains working in the same sequence.
(3) Connections, gullies and other accessories at the heads of the branch drains, linking up with the work previously measured under the drainage above ground installations. The building in of pipes at manholes/inspection chambers is usually taken with the manholes/inspection chambers.
(4) Manholes/inspection chambers measured in detail.
(5) Any other items, such as ventilating pipes, fresh air inlets, interceptors, connection to sewer, testing and commissioning drains, and the like.
(6) Any septic tank installations, cesspools or soakways, measured in detail.

It is good policy to check that all rainwater, soil and waste or discharge pipes and gullies are shown connected to the drainage system on the drawings, since some lengths of branch drains are occasionally omitted.

There is much to be gained on all but the smallest drainage systems by preparing drain schedules and inspection chamber/manhole schedules as illustrated in tables 15.1 and 15.2. These schedules offer greater flexibility, reduce the risk of omissions, enable the whole of the work to be visualised as a single entity, facilitate the summation of like items with consequently less taking off entries in the dimensions, and help in the identification and measurement of post contract work.

DRAINS

The measurement of drains may often with advantage be broken down into three principal items.

(1) Excavation of Pipe Trenches
This is measured in metres, stating the commencing level where >250 below existing ground level. The average depth range is given in stages of 250 (R12:1.1–2.1–2.1). For example, where a drain trench runs from 1.100 deep at one end to 2.700 deep at the other, the average depth is $\frac{1}{2} \times 3.800 = 1.900$, assuming that the ground has a uniform fall over the length of the trench, and this will be classified as ≤2.000 deep. Trenches to receive pipes ≤200 nominal size are grouped together and so described, whereas trenches for larger pipes are given separately for each nominal pipe size (R12:1.2.1–2).

When determining the average depth of trenches it is frequently necessary to make a decision as to whether the commencing level should be from existing ground level or a reduced level required in connection with pavings or landscaping requirements. The approach which is adopted is most appropriately based on whether the sequence of excavation chosen by the contractor is more likely to place the drainage excavation before or after any reduced level excavation. The effect of the latter will obviously be to reduce the depth of drainage excavation. In such circumstances it is necessary to include a preamble clause to the effect that the contractor will be expected to make allowance in his rates if it is proposed to use a different excavation sequence to that assumed in the measurement. Furthermore it should be stipulated that the original basis of measurement will be maintained in relation to any measurement of variations which may subsequently become necessary. Conversely where filling to make up levels is required it is reasonable to presume that excavation will commence at existing ground level.

The pipe trench excavation item is deemed to include earthwork support, consolidation of trench bottoms, trimming excavations, filling with and compaction of general filling materials, and disposal of surplus excavated materials, and so these items do not have to be repeated in the trench excavation items (R12:C1). Where trenches are excavated below groundwater level, next to roadways or existing buildings, or in unstable ground, this is to be incorporated in the item description. The method of determining the juxtaposition of the roadway or existing building to the pipe trench is illustrated in the *Code of Procedure for Measurement of Building Works* (*SMM7 Measurement Code*): figures 1 and 2 on page 20, prescribing that where the distance from the nearest side of the pipe trench to the foundation or edge of the roadway or footpath is less than the depth of excavation below the underside of the foundation to the building or the roadway or footpath, then it becomes next to existing building or roadway as appropriate. On this basis the top of the vertical dimension indicated in figure 1 should terminate level with the hard pavement and not the top of the kerb as shown. It will be noted that these two figures actually refer to the same terms used for earth-work support in D20:7.1–3.1–3.4&5. Items are taken as extra over excavating trenches, irrespective of depth, for breaking out existing rock, concrete, reinforced concrete, brickwork, blockwork or stonework, or coated macadam or asphalt, measured in m^3, except for existing hard pavings which are measured in m^2 stating the thickness (R12:2.1–2.1–5.0). The special requirements covering the excavation of pipe trenches next to existing live services or around existing live services crossing the trench should be noted (R12:2.6–7.1.0).

(2) Drain Pipes

These are described by the kind of pipe (such as clay, PVC-U and cast iron), quality of pipe (such as British Standard), nominal size and method of jointing (R12:8.1.1.0 and R12:S1&4). It is necessary to state that pipes are in trenches as distinct from in ducts, bracketed off walls or suspended from soffits, as these are each separately classified. Pipes are measured in metres over all fittings and branches (R12:M7). Iron pipe runs $\leq$3.000 in length are so described, stating the number, because of the relatively high cost involved. Vertical pipes, as in backdrops to manholes, as illustrated in figure 40, are also kept separate. Pipe fittings, such as bends, junctions and diminishing pipes, are each described and enumerated as extra over the pipes in which they occur (R12:9.1.1.0), with all cutting and jointing pipes to fittings deemed to be included (R12:C4). They are normally laid to gradients not flatter than 1 in 60, unless the peak flow is less than 1 litre/second.

(3) Pipe Protection

This may be provided to drains in three different forms which are described in SMM7 as beds; beds and haunchings; and beds and surrounds. These are

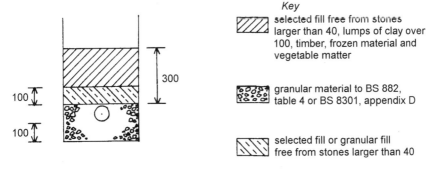

Key

selected fill free from stones larger than 40, lumps of clay over 100, timber, frozen material and vegetable matter

granular material to BS 882, table 4 or BS 8301, appendix D

selected fill or granular fill free from stones larger than 40

Figure 37 *Flexible drain pipe protection (Source: Building Regulations Approved Document H1 (1990)).*

each measured separately in metres, stating the width and thickness of beds and beds and haunchings, and the width, thickness of bed and thickness of surround in the case of beds and surrounds, and giving the nominal size of pipe in the latter two instances (R12:4–6.1.1.0). The protection may take the form of concrete with shallow pipes and granular fill with flexible jointed pipes to allow some movement without damage to pipes or joints. Both types of protection are measured in example 26. Any formwork to concrete protection is deemed to be included in the concrete rates (R12:C2). Figure 37 shows granular fill protection to drain pipes.

(4) Pipe Accessories

Pipe accessories such as gullies, traps, inspection shoes, fresh air inlets, non-return flaps and the like are each enumerated with a dimensioned description (R12:10.1.1.0). Dimensions stated for accessories are to include the nominal size of each inlet and outlet (R12:D8). These items are deemed to include jointing pipes thereto and bedding in concrete, without the need for specific mention (R12:C5). Another useful pipe accessory is sealed

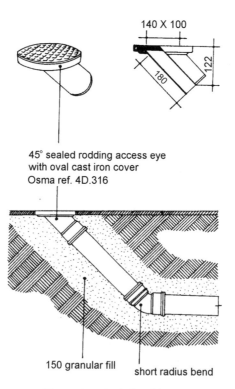

140 X 100

45° sealed rodding access eye
with oval cast iron cover
Osma ref. 4D.316

150 granular fill   short radius bend

Figure 38   *Sealed rodding eye.*

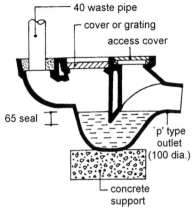

Figure 39 *Back inlet trapped gully.*

rodding eye as illustrated in figure 38, which provides a cheaper alternative to inspection chambers at the head of short surface water drain runs. The sealed rodding access eye and cast iron cover form an enumerated item. When connecting gullies to drains it is often necessary to insert one or two bends to obtain a satisfactory and smooth connection, and these are enumerated as extra over the pipes. A typical back inlet gully is illustrated in figure 39. Building in the ends of pipes at manholes/inspection chambers is enumerated, stating the nominal size of the pipe and including the provision of a brick arch where required, and the cutting of pipes is deemed to be included (R12: C6). It is not necessary to state the thickness of wall through which the pipes pass, as the building in item will follow that of the chamber walls.

MANHOLES/INSPECTION CHAMBERS

Manholes, as well as inspection chambers, septic tanks, cesspits and soakaways, are measured in detail under an appropriate heading, which will probably include the number of manholes, etc.

Where a considerable number of manholes/inspection chambers are encountered it is advisable to prepare a manhole or inspection chamber schedule, detailing plan sizes, depths, wall types and thicknesses, connections, channel particulars, number of step irons, and details of cover slabs, covers, backdrops and any other special features. The taking off work is then considerably simplified and all the manholes/inspection chambers are taken off together to avoid the repetition of similar items. A typical inspection chamber schedule is illustrated in table 15.2.

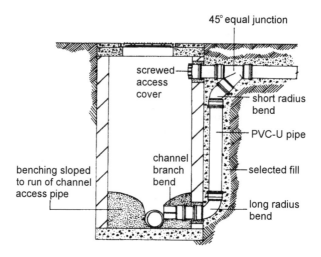

Figure 40 *Osma PVC-U backdrop to manhole.*

Figure 40 shows an external backdrop in PVC-U pipes to a brick manhole with a screwed access cover to the extension of the drain inside the manhole. Alternatively, the backdrop can be located inside the manhole which will need enlarging to accommodate it. When discharging into deep drains and sewers, the provision of a backdrop to the last manhole can reduce substantially the depth of the excavation to the previous drain run and consequently the cost.

Excavation for all types of manhole/inspection chamber is measured as excavating pits in m^3, stating the number, commencing level where >250 below existing ground level, and maximum depth classification in accordance with D20:2.4.1–4.1. Filling to excavations for manholes/inspection chambers,

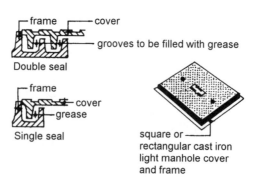

Figure 41 *Manhole covers.*

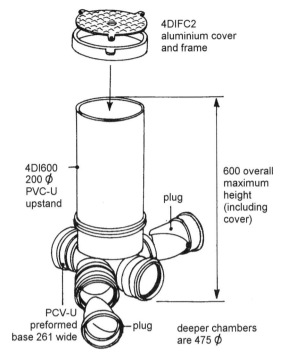

4DIFC2
aluminium cover
and frame

4DI600
200 ∅
PVC-U
upstand

plug

600 overall
maximum
height
(including
cover)

PCV-U
preformed
base 261 wide

plug

deeper chambers
are 475 ∅

Figure 42   *Terrain PVC-U shallow inspection chamber.*

disposal of surplus excavated material and earthwork support are all measured in accordance with the requirements prescribed in Section D20, as illustrated in example 26. Manhole/inspection chamber covers may be made of coated cast iron, galvanised steel or aluminium and may be single or double seal as illustrated in figure 41, and can be of light, medium or heavy duty.

Concrete bases to manholes/inspection chambers are measured as concrete beds as E10:4.1.0.5 and brick walls in accordance with the normal brickwork rules (F10:1.1–2.1.0). Ancillary work such as channels, benching, step irons and covers are each enumerated separately giving a dimensioned description as SMM R12:11–12.8–11.1.0, and illustrated in example 26.

Precast concrete rectangular and circular manhole/inspection chambers are measured generally in accordance with E50:1.1.0.0, with separate enumerated items for the various components such as base units, chamber rings, reducing or corbel slabs and cover slabs, as illustrated in example 26, and it is necessary to calculate the number and height of chamber rings required for each inspection chamber. Manholes of this type should be to BS 5911 Part 2 or 200, with Part 2 being for light loadings only. Where *in-situ* concrete surrounds and bases are used in conjunction with precast concrete manholes, the work should be measured in accordance with E10.

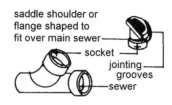

saddle shoulder or
flange shaped to
fit over main sewer
— socket
jointing
grooves
— sewer

Figure 43 *Sewer connection.*

PVC-U inspection chambers are now becoming quite common as they are quick to assemble and require less earthwork. They are classified under R12:12.14.1.0 as preformed systems and are enumerated as omnibus items/ units and the dimensioned description includes building in ends of pipes, channels, benching, step irons and cover. A PVC-U shallow inspection chamber is illustrated in figure 42.

Most manholes on domestic drainage systems are more correctly described as inspection chambers, as they are not large enough to permit the entry of a man for inspection and/or cleaning.

ASSOCIATED WORK

The work of connecting the drains to the public sewer is usually covered by a provisional sum in accordance with A53:1.1–2.0.0.

This item may include all the drainage work outside the boundary of the site in the public highway, with a separate provisional sum for permanent reinstatement of the highway. Sometimes the local authority is only prepared to make the actual connection, usually a saddle connection where an existing sewer is involved, as illustrated in figure 43. Where the contractor has to carry out the drainage work in the public highway, the excavation work will have to be separately classified with separate items for breaking up and temporary reinstatement of paved surfaces. General Rule 7.1c requires separate identification of work which is outside the curtilage of the site. R12:16.1.0.0 provides for the enumeration and giving of details of the work in connecting to the local authority's sewer, where it is executed by the contractor (R12:M10).

An item is taken for testing and commissioning the drainage system, stating the method of testing such as using water, smoke or air, and the number of stage tests (R12:17.1.2.0). The contractor is sometimes required to test the manholes/inspection chambers as well as the drains, and if this is the case it must be stated in the testing item. An alternative approach to that contained in example 26 could be testing the drainage installation after backfilling with a water test of not less than 1.500 head.

SEWAGE DISPOSAL PLANT

General Background

Sewage treatment and disposal units are measured in a similar manner to inspection chambers and manholes, as R12:13–15.1–6.0.0, measuring excavation, concrete, formwork, reinforcement, brickwork and rendered coatings in accordance with the rules for the appropriate work sections (R12:M8). While ancillary items such as building in ends of pipes, channels, benching, step irons, covers and intercepting traps are enumerated with a dimensioned description.

However, these structures are increasingly being provided as preformed sections using materials such as glass fibre reinforced plastics (GRP) to BS 4994. These are measured in accordance with the provisions of R12:13–15.14.1.1–6, and they are enumerated with a dimensioned description. These systems offer the advantages of speed and ease of assembly, which are particularly important in bad ground conditions. It is customary to give the capacity or size and the manufacturer's name and reference in the billed description.

Sewage Treatment Plant

An excellent biotec plant for installing below ground is available from Entec. The treatment process is in four stages: primary screening and settlement; carbonaceous biological treatment; nitrifying biological treatment; and final settlement. The GRP casing containing the plant is ideally sited in a pedestrian area at least 15 metres from the nearest house, seated on a concrete base with a surround of shingle on a dry site and concrete on a wet site.

The billed description could read as follows:

Preformed systems
Entec biotec sewage plant,
    comprising 100P GRP casing, weight of 2000 kg,
    7.600 × 2.800 × 2.500 deep; and motor rating of 1.50 kW;
    daily flow of 20 m³/day.

A much smaller alternative would be the 10P sewage treatment plant; weight 380 kg; 1.900 diameter ×2.500 deep.

Septic Tanks

These are ideal for serving small groups of residents from 4 to 38 in number in unsewered areas by means of a three stage process and emptied once a year. A billed description could read as follows:

Preformed systems
Entec GRP septic tank,
    comprising Agrément certificated tank; standard grade to BS 6297;
    6000 litres capacity; serving up to 22 persons; 2.390 diameter.

The effluent from the septic tank normally discharges into a looped system of rigid perforated or butted land drains laid in trenches surrounded with gravel, 20–50 grade, to a gradient of not more than 1 in 200, with polythene sheeting on top of the gravel to prevent clogging of the pipes. Typical land drains are concrete pipes with porous or non-porous inverts to BS 5911, Part 114, clayware field drains to BS 1196 or plastics field drains to BS 4962. Excavation is measured as R13:1.1.1–2.1–10 and pipes as R13:8.1.1.0, in a similar manner to house drains.

Cesspools

These are termed cesspits in SMM7 and are used to store the waste discharges from individual properties is unsewered areas. They should be sited a minimum of 15 metres from the nearest habitable dwelling and emptying vehicles should have access to within 35 metres of the tank. The tank should be vented independently or through the house drainage system. On a dry, well drained site the tank can be completely backfilled with pea shingle on a level concrete base. On a wet site, however, they are susceptible to buoyancy and, for this reason, should be backfilled with concrete. A typical billed description follows:

Preformed systems
Entec GRP cesspool
    comprising Agrément certificated factory sealed standard duty tank;
    24 000 litres capacity; 6.000 × 2.750 × 2.650 deep (inlet level
    to base of tank) to BS 6297; with access shaft and lifting eyes.

Entec also manufacture GRP petrol/oil separators (interceptors) and package pump systems. Another well known manufacturer of preformed GRP cesspools and septic tanks is Klargester, whose products are illustrated and described in *Building Technology*, fifth edition (1995) by the present author.

Soakaways

These can take various forms ranging from an excavated pit filled with hardcore and with a concrete cover slab at ground level, to precast concrete rings as described for inspection chambers in example 26, with the lower rings being perforated, and with a hardcore base and surround to the sides of the perforated rings. A precast concrete cover slab and cast iron access cover and frame are often provided to the latter type of soakaway. Soakaways can be used for the disposal of surface water from a domestic drainage system (R12) and for the disposal of groundwater in a land drainage system (R13). Soakaways are measured in the same way as inspection chambers and manholes.

WORKED EXAMPLE

A worked example follows covering the measurement of a house drainage system, encompassing separate arrangements for foul and surface water drainage and ten inspection chambers using three different constructional methods.

Table 15.1 *Drain schedule.*

Location	Type and size of pipe	Pipe fittings and accessories	Length between int. faces of I.C.s (pipes)	Length between ext. faces of I.C.s (excvtn.)
Surface water				
Rdp. to SIC 1	PVC-U 110	rdp. adaptor & bend	0.900	0.800
SIC 1–SIC 2	PVC-U 110	nil	6.600	6.500
Rdp.–SIC 2	PVC-U 110	rdp. adaptor & bend	0.900	0.800
SIC 2–SIC 3	PVC-U 110	gully junctn.	7.300	7.200
Gully branch	PVC-U 110	2 bends, 1 yard gully	1.100	1.000
Rdp.–SIC 3	PVC-U 110	rdp. adaptor & bend	1.150	1.100
SIC 3–SIC 5	PVC-U 110	junctn. for rdp.	7.400	7.300
Rdp. branch	PVC-U 110	rdp. adaptor & bend	1.300	1.200
SIC 4–SIC 5	PVC-U 110	junctns. for 2 rdps.	16.500	16.400
2 nr. rdp. branches	PVC-U 110	2 rdp. adapts. & bends	{ 2.900 2.900	2.800 2.800
Rdp.–SIC 4	PVC-U 110	rdp. adaptor & bend	3.400	3.300
SIC 5–SIC 6	PVC-U 110	nil	4.800	4.700
SIC 6–Sewer	PVC-U 110	saddle conn.	2.700	2.600

1 gully, 7 rdp. adaptors & bends, 2 bends, 1 saddle conn., 4 junctns.   59.850

Location	Type and size of pipe	Pipe fittings and accessories	Length between int. faces of I.C.s (pipes)	Length between ext. faces of I.C.s (excvtn.)
Foul water				
WC–FIC 1	Clay 100	1 bend, 1 WC conn.	0.500	0.400
FIC 1–FIC 3	Clay 100	1 gully junctn.	4.700	4.500
Gully branch	Clay 100	2 bends, 1 gully	0.400	0.350
3 brs.–FIC 2	Clay 100	4 bends, 1 gully, 2 conns. (SVP & WC)	4.000	3.700
FIC 2–FIC 3	Clay 100	1 gully junctn.	10.700	10.500
Gully branch	Clay 100	2 bends, 1 gully	0.500	0.450
FIC 3–FIC 4	Clay 100	nil	5.600	5.400
FIC 4–Sewer	Clay 100	saddle conn.	1.600	1.500

3 gullies, 9 bends, 1 saddle conn., 2 junctns.   28.000

Excavating pipe tenches

average depth ≤ 250	500	750	1.000	1.250
	0.800	0.800	7.300	4.700
	0.400	6.500	2.800	
	4.500	7.200	2.800	
	0.350	1.000	3.300	
	0.450	1.100	16.200	
	6.500	1.200		
		3.700		
		10.500		
		5.400		
		37.400		

Pipe protection		Depth of excvn. at upper end	Depth of excvn. at lower end	Av. depth of excvn.	Notes
concrete bed & surrd.	granular fill. bed & surrd.				
—	0.800	0.400	0.650	0.525	
—	6.500	0.650	0.600	0.625	
—	0.800	0.400	0.600	0.500	
—	7.200	0.600	0.750	0.675	
—	1.000	0.400	0.630	0.515	
—	1.100	0.400	0.660	0.530	
—	7.300	0.660	1.150	0.905	
—	1.200	0.400	1.000	0.700	
—	16.400	1.450	1.150	1.300	
—	2.800	0.400	1.450	0.925	
—	2.800	0.400	1.450	0.925	
—	3.300	0.400	1.400	0.900	
—	4.700	1.150	1.150	1.150	
—	2.600	1.150	2.900	2.025	under highway
	58.500				
0.400	—	0.400	0.400	0.400	
—	4.500	0.400	0.550	0.475	
0.350	—	0.400	0.300	0.350	
—	3.700	0.400	0.850	0.625	
—	10.500	0.850	0.550	0.700	
0.450	—	0.400	0.570	0.485	
—	5.400	0.550	0.800	0.675	
—	1.500	0.800	2.550	1.675	under highway
1.200	25.600				

Total length of pipe trench excavation = 85.300

1.500	1.750	2.000	2.250	Notes
16.400	1.500*		2.600*	*In public highway

Total length of pipe trench excavation = 85.300
(double check)

Table 15.2   *Inspection chamber schedule.*

I.C. ref.	Internal dims.	Constructn.	Cover	Approx. ground level	Cover level	Invert level
SIC 1	200 dia.	PVC-U Terrain preformed on 100 granular bed	Alum. Terrain cover & fr. 4DIFC2 204 dia.	90.100	90.150	89.600
SIC 2	200 dia.	ditto.	ditto.	89.700	89.750	89.250
SIC 3	200 dia.	ditto.	ditto.	89.500	89.550	88.900
SIC 4	675 dia.	Precast conc. chambr. rings 50 thick & cover slab	Coated c.i. cover & fr., 500 dia. opg. ref. MB2-50	90.300	90.350	89.000
SIC 5	675 dia.	ditto.	ditto.	89.500	89.550	88.500
SIC 6	675 dia.	ditto.	ditto.	88.400	88.450	87.400
FIC 1	600 × 450	Half brick walls on 150 conc. base	Coated c.i. cover & fr. 610 × 457 grade B, double seal, ref. MB2-60/45	89.500	89.550	89.250
FIC 2	600 × 450	ditto.	ditto.	90.200	90.250	89.500
FIC 3	600 × 450	ditto.	ditto.	89.500	89.550	89.100
FIC 4	600 × 450	ditto.	ditto.	88.400	88.450	87.750

I.C. ref.	Depth to invert from cover level	Depth of excavn.	Main channel	Branch channels	Step irons	Notes
SIC 1	0.550	0.650	110 PVC-U curved	nil	nil	2 plugd. branches, base of gran. fill 100 below invt.
SIC 2	0.500	0.600	110 PVC-U curved	1/110	nil	1 plugd. branch
SIC 3	0.650	0.750 $\frac{}{2.000}$ ($\leq$1.000)	110 PVC-U curved	1/110	nil	1 plugd. branch
SIC 4	1.350	1.450	110 PVC-U curved	nil	2	base of conc. 100 below invt.
SIC 5	1.050	1.150	110 PVC-U strt.	1/110	2	ditto.
SIC 6	1.050	1.150 $\frac{}{3.750}$ ($\leq$2.000)	110 PVC-U strt.	nil	2	ditto.
FIC 1	0.300	0.550	100 clay strt.	1/100	nil	conc. beddg. & base, total 250 thick
FIC 2	0.750	1.000	100 clay curved	2/100	nil	ditto.
FIC 3	0.450	0.700	100 clay strt.	1/100	nil	ditto.
FIC 4	0.700	0.950 $\frac{}{3.200}$ ($\leq$1.000)	100 clay strt.	nil	nil	ditto.

Drainage work

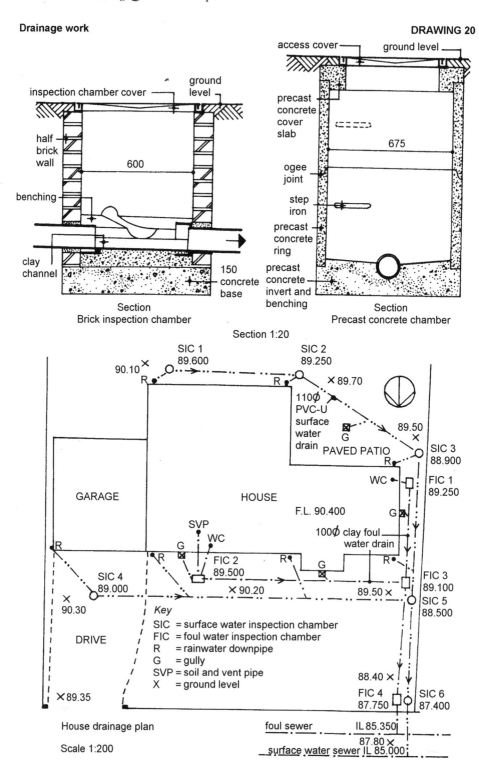

Section
Brick inspection chamber

Section
Precast concrete chamber

Section 1:20

Key
SIC = surface water inspection chamber
FIC = foul water inspection chamber
R   = rainwater downpipe
G   = gully
SVP = soil and vent pipe
X   = ground level

House drainage plan

Scale 1:200

<u>Drg</u>. 20		<u>Example 26</u>
	<u>Drainage Work 1</u>	
		Tr. excvn.
	<u>Excvtg. trs.</u>	
	For pipes	
6.50	≤ 200 nom. size; ≤ 500 av. depth	
37.40	ditto; ≤ 750 av. depth	
16.20	ditto; ≤ 1.000 av. depth	
4.70	ditto; ≤ 1.250 av. depth	
16.40	ditto; ≤ 1.500 av. depth	
1.50	ditto; ≤ 1.750 av. depth outside the curtilage of the site	
2.60	ditto; ≤ 2.250 av. depth; do.	
1.50 0.50	E.o. excvtg. trs. irrespec. of depth for breakg. out (final coated macadam pav'g. 100 th.; outside	
2.60 0.50	(s.w. curtilage of the site	
	<u>Disposal</u>	
Item	Surface water	
	<u>Granular mat. to BS882</u>	
	Beds & surrds.	
55.90	beds 400 × 100 th.; (s.w. surrds. 100 th.; to pipes 110 nom. size	
2.60	(s.w. ditto; do.; outside the curtilage of the site	
24.10	ditto; to pipes 100 (foul nom. size	
1.50	ditto; do.; outside the (foul curtilage of the site	

Excavating pipe trenches are measured as a linear item with the pipe sizes and average depth of pipe runs classified as R12:1.1.1–2.0. Excavating trenches is deemed to include earthwork support, consolidating and trimming bottoms and sides of trenches, filling with general filling materials and disposal of surplus excavated materials (R12:C1). Both foul and surface water drain trenches have been combined as there is no requirement to separate them when they fall into similar classifications. The use of a drain schedule, as illustrated in table 15.1, simplifies the measurement process significantly. For example, there are nine lengths of drain trench coming within the ≤750 range which, in the absence of a drain schedule would form nine separate dispersed items in the taking off with locational notes for each.

All work which is outside the curtilage of the site must be separately identified as General Rule 7.1c.

The breaking out of the coated macadam in the public footpath and road is measured as extra over the trench excavation in m^2 and giving its thickness (R12:2.2.5.0). In the absence of a bed width being given, the width is taken as the nominal size of the pipe + 300, with a minimum width of 500 in all cases (R12:M4).

Disposal of surface water item as R12:3.1.0.0.

The beds and surrounds of granular material are provided as protection to pipes with flexible joints to permit limited movement and are measured as R12:6.1.1.0. The 400 width of bed is calculated from the nominal size of pipe + 300 or thereabouts with a thickness of bed and surround of 100 (see figure 37).

Drainage Work 2

	<u>Conc. to BS 5328;</u> <u>ord. prescribed mix C15P</u> Beds & surrds.	The concrete beds and surrounds are provided to the shallow inspection chamber and gully branches to give additional protection to these vulnerable lengths. The same provisions would apply if the pipes were located within 1.000 horizontally and below the bottom of the house foundations (see R12:6.1.1.0). Note the alternative method of specifying the concrete.
	beds 400 × 100 th.; surrds. 100 th.; to pipes 100 nom. size	
1.20	(WC & (gully brs.	
	<u>Plastics pipewk. of</u> <u>PVC-U spigot & socket</u> <u>pipes to BS 4660; w.</u> <u>rg. seal push fit jts.</u>	All surface water drains are in PVC-U and the heading provides adequate particulars for pricing. A specific product such as Osma or Terrain pipes could be specified, where required. Instead of inserting 13 separate lengths, the combined length can be extracted from the drain schedule (table 15.1), and measured as R12:8.1.1.0, and is measured over all fittings and branches (R12:M7).
	Pipes	
57.15	110 nom. size in trs.	
2.70	(s.w. ditto; outside the curtilage of the site	
	E.O. for	
7	r.w. adaptors; 68 nom. size inlets; Osma ref. 4D.149	
	E.O. for	It is desirable to follow the pipework item with the associated extra over pipe fittings for bends and junctions. Each downpipe will be connected to the surface water drain with a 68 diameter adaptor and $87\frac{1}{2}$ long radius bend, as they do not need to be disconnected at ground level, as would be the case with a combined drainage system. Two bends are fitted to the patio gully – one for direction and the other for slope. See R12:9.1.1.0 for the measurement procedure. Pipe fittings are deemed to include cutting and jointing pipes to fittings (R12:C4). Note the use of Osma references to the required fittings which will assist the estimator in pricing.
7	lg. rad. rest bend $87\frac{1}{2}°$; Osma ref. 4D.581	
	E.O. for	
2	short rad. bend; (patio Osma ref. 4D.163 (gully	
	E.O. for	
4	equal junctn.; Osma ref. 4D.210	
	(Y-junctns.	

Drainage Work 3

	Plastics accessories	
	Gullies	
1	PVC-U trpd. yd. gully; 305 ⌀ × 600 dp.; 110 'P' type outlet; Osma ref. 4D.800; b&s w. gran. fill 100 th.; ductile iron gtg. & fr.; strt. bar hinged; grade B to BS EN 124; 303 × 325 o/a; Osma ref. 4D.810; galv. m.s. perf. catchmt. bucket; 225 ⌀ × 245 dp.; Osma ref. 4D. 815	

Clay pipewk. of 'Denseal' vitrified clay spigot & socket pipes w. polyester fairings & elastomeric 'O'rg. seals to BS EN 295-1; manufac'd. by Naylor Bros.

	Pipes
26.40	100 nom. size; in trs.
1.60	(foul ditto; outside the curtilage of the site
9	E.o. for bends
2	E.o. for junctns.; oblique 100 × 100

Clay accessories

Gullies

3	back inlet 68 ⌀; 225 × 225 top; 250 dp.; 100 'P' type outlet; coated c.i. cover & fr. 150 × 150; coated c.i. access cover 85 × 85; ref. Nat. 192 (Fig. 39); to BS EN 295, pt. 1

Pipe accessories are enumerated, stating the type with a dimensioned description (R12:10.1.1.0). Accessories include gullies, traps, inspection shoes, fresh air inlets and non-return valves (R12:D7). The dimensions shall include the nominal size of inlets and outlets (R12:D8) and accessories are deemed to include bedding in concrete (R12:C5). The granular surround to the largest gully goes beyond bedding a small gully in concrete and the fill has therefore been included in the description. An Osma sealed rodding access eye is shown in figure 38; the access fitting will be enumerated and described. If the drive sloped towards the garage, a box section drainage channel would be required in front of the garage door discharging into the surface water drain.

A full description of the clay pipes is given and a named product is specified (see R12:S1). A more general description would probably take the form of 'clay pipes to BS EN 295, jointed with flexible joints'. It is important to use pipes with flexible joints to permit some movement without the risk of fracture to pipes or joints.

Pipe fittings are enumerated as extra over the pipes as R12:9.1.1.0. Two bends have been provided for each gully, although Naylor's 275D universal gullies with clay hopper and base are capable of being rotated independently with an overlapping cast iron grate frame, which reduces the number of bends required, as the outlet can be directed towards the collection drain. The *SMM7 Library of Standard Descriptions* refers to junctions as branches; either term can be used.

The three back inlet trapped gullies collect the discharges from baths, sinks and washbasins and have provision for rodding the outlets. They are measured as R12:10.1.1.0 and are deemed to include bedding concrete (R12:C5). The reference Nat. 192 is that contained in the Clay Development Association's *Standard Illustrated Catalogue of Vitrified Clay Pipes and Fittings* (1992).

<u>Drainage Work 4</u>

<u>Inspection chbrs.</u>

<u>PVC-U</u>

chbr. dia.	200
<u>add</u> gran. fill	
2/100	200
o/a dia. of excvn.	400

<u>Precast conc.</u>

int. dia.	675
<u>add</u> thickness of	
rings 2/50	100
ext. dia.	775

<u>Brick</u>

int. dims.	600	450
<u>add</u> walls		
2/102.5	205	205
ext. dims.	805	655

<u>Excvtg.</u>

Pits (7 nr.)
  ≤ 1.000 max. depth

22/
7/ 1/4/ 0.40
  0.40
  2.00

(SIC
 1-3

&

(FIC
 1-4

  0.81
  0.66
  3.20

<u>Disposal</u>

Excvtd. mat.
  off site

<u>Excvtg.</u>

Pits (3 nr.)
  ≤ 2.000 max. depth

22/
7/ 1/4/ 0.78
  0.78
  3.75

&

(SIC
 4-6

<u>Disposal</u>

Excvtd. mat.
  off site

Excavating for inspection chambers is classified as pits and giving the number as D20:2.4.1–4.0, and is calculated using the depths listed in the inspection chamber schedule (table 15.2). The horizontal dimensions are worked up in waste and include an allowance for the 100 thick granular side fill around the PVC-U chamber. No filling has been taken at the sides of the precast concrete and brick chambers. It was not considered necessary to surround the precast concrete rings in concrete or granular fill as these chambers are shallow and constructed above groundwater level.

It is unnecessary to deal with the excavation to the different types of inspection chamber separately as its classification is the same in all cases. The dimensions of the volume of excavation to the circular chambers have been taken as $\pi D^2/4$ for the cross sectional area multiplied by the combined length for the group of inspection chambers, thus using the diameters entered in waste. Alternatively, πr^2 could be used in the calculations. These chambers conveniently fall into two of the prescribed maximum depth ranges (i.e. ≤1.000 and ≤2.000). As the volumes of excavation and disposal of excavated material are the same, these items can be grouped against the common dimensions. Disposal is covered in D20:8.3.1.0 and all excavated material is to be removed from the site.

It will be noticed that the diameter of the excavation for SIC 1–3 is no greater than the width of the adjacent drainage trenches. In such cases a permissible alternative would be to measure the trench excavation over the inspection chambers. If the bottom of the IC's is below the bottom of the trenches, then the extra depth only should be measured as pit excavation.

Drainage Work 5

E. W. S.

To faces of excvn.
   ≤ 1.000 max. depth;
   dist. btwn. opposg.
   faces ≤ 2.000;
   curved

22/7/ 0.40
      2.00   (SIC 1-3

Ditto.
   ≤ 2.000 max. depth;
   dist. btwn. opposg.
   faces ≤ 2.000;
   curved

22/7/ 0.78
      3.75   (SIC 4-6

Ditto.
   ≤ 1.000 max. depth;
   dist. btwn. opposg.
   faces ≤ 2.000

2/ 0.81
  3.20   (FIC 1-4

2/ 0.66
  3.20

Granular mat. to BS 882

Fillg. to excvns.
   > 2.50 av. th.; obt.
   off site; compactg.
   in max. 300 lyrs.

22/7/ 0.30
      0.10
      2.00   (SIC 1-3

Surf. treatmts.
Compactg. btms. of excvns.
      (SIC 1-3

3/22/7/1/4/ 0.40
        0.40
3/22/7/1/4/ 0.78
        0.78   (SIC 4-6

4/ 0.81
  0.66   (FIC 1-4

Earthwork support (e.w.s.) is measured in accordance with D20:7.1–3.1–3.1 in m^2. Note particularly the prescribed classifications of maximum depth and of the distance between opposing faces of the excavation. It is necessary to make reference to curved surfaces as the earthwork support will follow the surfaces of the curved excavation. Items can conveniently be grouped together under combined depths as for excavation and disposal of excavated material and so reduce the number of measured items. Locational notes are entered against the items for identification purposes. The foul inspection chambers are rectangular and the surface area is calculated as twice the length and width multiplied by the combined depth.

The granular fill to the sides of the PVC-U chambers, 100 wide, is measured on its centre line for the full combined depth of the chambers. It is measured in m^3, giving the type of filling and whether compacted in layers, as D20:9.2.3.0. No side filling is needed with the other types of inspection chamber.

Compacting the bottoms of excavations to the ten chambers is measured in m^2 as D20:13.2.3.0. The compacting is deemed to include levelling and grading to falls and slopes ≤158 from the horizontal (D20:C5).

<u>Drainage Work 6</u>

3 nr. PVC-U i.c.s

Granular mat. a.b.

Fillg. to make up levs.
≤ 250 av. th.; obt.
off site

3/22/
7/¼/ 0.40
       0.40
       0.10

> These chambers are illustrated in figure 42. The filling under the preformed bases of the PVC-U chambers is measured as filling to make up levels in m³ as D20:10.1.3.0.

<u>Preformed systems</u>

Terrain PVC-U shallow
inspectn. chbrs.
    comprisg. preformed
    base 261 wide; w. 4
    socs., 2 plugs; 110 ∅
    chans. & benchg.; 200 ∅
    upstd. ref. 4D1600; &
    alum. cover & fr. ref.
    4D1FC2; b.i. ends of
    2 nr. 110 ∅ pipes; depth
    to invt. 550
(SIC 1

1

Ditto
    ditto; b.i. ends of
    3 nr. 110 ∅ pipes;
    depth to invt. 500
(SIC 2

1

Ditto
    ditto; do.; depth
    to invt. 650
(SIC 3

1

> PVC-U inspection chambers of this type are termed 'preformed' and this becomes the main heading, as indicated in the SMM7 Library of Standard Descriptions. They are enumerated as R12:12.14.1.1–6, with details of building in ends of pipes, channels, benching, step irons and covers. No step irons are required for these shallow chambers and no concrete has been measured around the covers and frames, as this will be included in the concrete paving and patio adjoining the house. Each inspection chamber is different and this is largely accounted for by including the depth to invert in the description of each item. The varying depths are obtained by cutting the upstand to the required height using a fine toothed saw.

3 nr. precast. conc. i.c.s

Precast conc., designd.
mix C25, 20 agg. to BS 5328

Base units
    725 dia. × 200 hi.;
    reb. for chmbr. rgs.
    50 th. ard. perimeter;
    smth. falls to 110 ∅ h.r.
    PVC-U curved main
    chan.; b.i. ends of
    2 nr. 110 ∅ pipes
(SIC 4

1

> The precast concrete components are enumerated with a dimensioned description generally as E50:1.1.0.0. Most of the description for the base unit is common to all three chambers and they vary only in respect of branch channels, straight or curved main channel and number of pipes to be built in. Alternatively building in ends of pipes could be measured as a separate enumerated item.

Drainage Work 7

<u>Precast conc. a.b.</u>

Base units a.b.

1	110 h.r. PVC-U strt. main chan.; 1 nr. 110 ⌀ 3/4 sec. PVC-U br. chan. b.i. ends of 3 nr. 110 ⌀ pipes

(sic 5

> The enumerated items for the base units to the circular precast concrete chambers follow, using the basic description on the previous page with the variations listed for each chamber.

Ditto.

1	110 ⌀ h.r. PVC-U strt. main chan.; b.i. ends of 2 nr. 110 ⌀ pipes

(sic 6

> This entails the use of the terms 'ditto.' and a.b. (as before) to avoid the need to repeat descriptions that have already been provided.

<u>total ht. of chbr. rgs.</u>

	1.050	1.350
<u>less ddtns.</u> top & btm.	150	150
	900	1.200

<u>nr. of chbr. rgs.</u>

	600	300 hi.
sic 4	2	
sic 5	1	1
sic 6	1	1
	4	2

> The chamber rings follow on under the same precast concrete heading and it is necessary to determine the number and height of chamber rings required for the three chambers. These are set out in waste for ease of identification. The rings are then enumerated with a dimensioned description generally as E50:1.1.0.0.

4	Chamber rings to BS 5911, pt. 200; 600 hi.; 675 int. dia., 50 th.; b. & j. in c.m. (1 : 3) to ogee jts; 1 nr. galvd. step iron b.i.
2	ditto; 300 hi.; do .

3	Cover slabs to fit chbr. sec., 675 int. dia.; 125 th.; 500 dia. access opg. (sic 4 - 6

> The cover slabs are also covered by an enumerated item containing a dimensioned description, which includes the size of the access opening. It still follows on under the main heading of precast concrete.

	Drainage Work 8	

Covers

Cast iron

3   Access covers & frs.
coated; 500 dia. clr. opg.;
med. duty; ref. MB2-50
to BSEN 124; settg. in
posn.; beddg. fr. in c.m.
(1:3)

             4 nr. bk. FICs

In-situ conc.
(21 N/mm² - 20 agg.)

4/   Beds
0.81   ≤ 150 th.; poured on
0.66   or agst. earth
0.15   (FIC 1-4

     Benchings in btms.
4   600 x 450 x av. 230 dp.;
        rendered in c.m. (1:3);
        trwld. smth. to falls &
        crossfalls to chans.

             len. of bk. walls

                600
            2/   450
               1.050
               2.100

add crnrs. 4/102.5   410
mean gth. of walls   2.510

                 depth

comb. depth of
bk. i.c's from
cover to invt.   2.200
add beddg.   .100
less cov. & fr.   .030
        4/.070   .280
                2.480
less benchg. 4/.230   .920
ht. of facewk.   1.560

Engrg. bwk. in class B
clay eng. bks. to BS 3921
in stret. bd.; in Sulph.
resist'g. ct. & sd. mtr. (1:3)

2.51   Walls
0.92   h. b. th.

Engrg. bwk. in class B
clay eng. bks. to BS 3921
in stret. bd.; in Sulph.
resist'g. ct. & sd. mtr. (1:3);
flush ptd.

2.51   Walls
1.56   h. b. th.; facewk
        o.s.

The access covers and frames are enumerated with a dimensioned description as R12:12.11.1.0, and is further identified in this case by reference to the relevant European Standard, which has been adopted as a British Standard, and the appropriate cover reference.

The concrete bed is taken under a heading of *in-situ* concrete as E10 in SMM7, although the *SMM7 Library of Standard Descriptions* refers to it as 'plain concrete'. Reinforced concrete is so described and it does not seem necessary to further describe *in-situ* concrete as it is already adequately distinguished from precast concrete. The concrete bases are measured as beds under E10:4.1.0.5, and form a cubic item. Surface treatment of the underlying ground was taken earlier in the example.

Benchings are enumerated under the same concrete mix heading as R12:12.9.1.0.

The mean girth of the brick enclosing walls is calculated in waste in the normal way. The combined total depth to invert is taken from the inspection chamber schedule (table 15.2) and adjusted for the depth of cover and the concrete bedding for each chamber to arrive at the combined height of the brick walls. This is then subdivided to obtain the heights with and without facework.

A full description of the brickwork, including the bricks, bond, mortar and pointing is given in the heading, with the information probably extracted from the project specification. The brickwork is measured in accordance with F10:1.2.1.0.

Drainage Work 9

	B.i. ends of pipes
FIC	nr. of pipes
1	2
2	4
3	3
4	2
	11

11

Build in ends of pipes
100 nom. size

	chans.
FIC	nr. br. bds.
1	1
2	2
3	1
4	—
	4

Channel wk.
Clay chans. & bds. to BSEN
295 jtd. in c.m. (1:2)
H.r. secs
    100 nom. size ×
    600 effect. len.

3   (FIC 1,3,4
    100 nom. size ×
    900 effect. len.; curved

1   (FIC 2
Three quarter sec. br. bds.
    100 nom. size

4

covers
Cast iron

4   Access covers & frs.
    coated; 610 × 457; med.
    duty, double seal; ref.
    MB2-60/45 to BSEN 124;
    600 × 450 dr. opg.; settg.
    in posn.; beddg. fr. in
    c.m. (1:3); beddg. cover
    in grease & sd.

Include the follg. defined
Provsnl. Sums for wk. by
Staty. Undertakings

Item   Saddle conns. to 2nr. sewers
    by L.A. : £150

Item   Perm. reinstatement of
    highway by H.A.
    following conn. to sewers:
    £180

Instlns. Generally

Item   Include for testg. & commsg.
    both drainage instlns. as
    specfd.

The number of pipes to be built into the brick wall of inspection chambers can be obtained from table 15.2, and are listed in waste to arrive at the total number. The measured item is an enumerated one as R12:12.7.1.0 and is deemed to include cutting pipes (R12:C6). The description in the *SMM7 Library of Standard Descriptions* does not include the type and thickness of wall.

Channel work is also enumerated, stating the nominal size and effective length (R12:12.8.1.0). The *Standard Library* excludes the word 'straight' from the description, as if not described as curved then it must be straight. Half section or three quarter section branch bends are enumerated stating the nominal size but not the effective length as they are invariably stock items.

Covers are enumerated as R12:12.11.1.0. The size and type of cover with reference to the appropriate British Standard and size of clear opening are desirable inclusions. The remaining parts of the description follow the guidelines contained in the *SMM7 Library of Standard Descriptions*, which largely forms the basis of other Standard Libraries and the main computerised billing systems.

Work to be carried out by the Local Authority and the Highway Authority are covered by Provisional Sums as A53:1.1–2.0.0.

Finally, an item is provided for testing and commissioning both drainage installations as R12:17.1.0.0. Provision of water and other supplies and test certificates are deemed to be included (R12:C7–8). An item for the disposal of surface water on the surface of the site and excavations has been taken previously following the trench excavations.

16   External Works

This chapter is concerned with the measurement of roads, drives, paths, grassed areas, planting of trees, shrubs and hedges, and fencing and gates. Drainage work has already been covered in the previous chapter. Demolition work is considered to be outside the scope of this book and has accordingly been omitted, but is included in *Advanced Building Measurement*.

ROADS, DRIVES AND PATHS

The major components of roadworks are covered in work sections Q20–25, embracing the various types of pavings and their associated sub-bases; with kerbs, edgings and channels incorporated in work section Q10.

In general, the sub-base or foundation for the road or footpath and the surfacing are each measured separately with a full description. Filling material is measured in m³, classifying the average thickness as ≤ or >250 (Q20:10.1–2.1–3.1&3), while the approach to the measurement of the surfacing will vary according to the nature of the material. For example, *in-situ* concrete is measured in m³ as beds with a thickness classification as E10:4.1–3.0.1, while coated macadam and asphalt are measured in m², stating the thickness and number of coats (Q22:1.1.2.0), and paving slabs in m², stating the thickness (Q25:2.1.2.1). Compaction of filling or bottom of excavation is taken as a superficial item (Q20:13.2.2–3.0). Certain incidental items are measured separately, such as forming expansion joints including formwork in concrete roads (E40:2.1–2.1.0), and trowelling concrete in channels and around gullies (E41:3.0.0.2). The approaches to the measurement of coated macadam and *in-situ* concrete roads are illustrated in example 27, at the end of this chapter.

With concrete roads, the waterproof membrane, usually of polythene or building paper, and steel reinforcement are also measured separately (the waterproof membrane and fabric reinforcement in m² and bar reinforcement in tonnes). Formwork to the edges of concrete road slabs, except at expansion joints, is measured as edges of beds in metres and classified in the appropriate height category (E20:2.1.2–4.0).

When measuring footpaths, consisting of paving slabs, it is necessary to state the kind, quality, size, shape and thickness of slabs, nature of surface finish, method of bedding, treatment and layout of joints and nature of base (Q25:S1–7). This type of information is often incorporated in preamble clauses or may be the subject of cross references to the project specification where one has been prepared adopting the coordinated project information approach, using the same component references in drawings, specification and billed descriptions. This permits a substantial reduction in the length of billed descriptions, as shown in the examples in chapter 17.

The terminology of laying coverings to roads and pavings to falls and crossfalls and to slopes ≤158 was explained in chapter 2, and refers to falls in both directions (along the length and across the surface).

Precast concrete kerbs, channels and path edging are each measured separately in metres with a dimensioned description and often giving a catalogue or other reference number, and including the foundation and haunching (Q10:2–4.1.0.2). Curved members are so described stating the radii.

Any surface water drainage work to roads would be measured in the manner outlined in chapter 15, with road gullies fully described and enumerated as pipe accessories (R12:10.1.1.0).

GRASSED AREAS

Filling to make up levels with topsoil is measured in m^3 and stating the average thicknss as to whether ≤ or >250 and other relevant particulars as D20:10.1–2.1–3.3. Cultivating, surface applications, seeding and turfing are each measured separately in m^2, stating the depth of cultivating, the type and rate of surface applications and rate of seeding. Full descriptions accompany these items, paying particular attention to such matters as kind, quality, composition and mix of materials, method of application, method of cultivating and degree of tilth, timing of operations and method of securing turves (Q30:S1–5). The type of surface applications must be stated and they include herbicides, selective weedkillers, peat, manure, compost, mulch, fertiliser, soil ameliorants and sand (Q30:D1). Cultivating is deemed to include the removal of stones and seeding includes raking or harrowing in and rolling (Q30:C1&3). Useful examples of cultivating, fertilising, seeding and turfing are given in example 27.

TREES, SHRUBS AND HEDGES

Trees are enumerated, giving the botanical name, BS size designation and root system or, alternatively, the girth, height, clear stem and root system, and the description can include supports and ties, and refilling with special materials and watering where required (Q31:3.1.1–2.4–6). Shrubs are also enumerated stating the botanical name, height and root system (Q31:5.1.1.5–6). Where trees are to be provided with tree guards, the guards are enumerated with a dimensioned description under a heading of 'protection' (Q31: 10.1.1.0).

Hedge plants are usually measured in metres, giving the botanical name, height, spacing, number of rows and layout (Q31:6.1.2.0), as illustrated in example 27.

FENCING AND GATES

Fencing is measured in metres, stating the type and height of fencing (measured from the surface of the ground or other stated base to the top of the infilling or the top wire or rail where there is no infilling), and the spacing, height and depth of supports (Q40:1.1.1.0). The description of the fencing shall include the kind and quality of materials, construction, surface treatments applied before delivery to site and size and nature of backfilling (Q40:S1–4).

Fencing posts and struts occurring at regular intervals, classified as supports, are included in the description of the fencing, without the need for separate measurement (Q40:D1). Special supports such as end posts, angle posts, integral gate posts and straining posts, with supporting struts or back stays, are enumerated separately as special supports as extra over the fencing in which they occur (Q40:2.1–4.1.1–2). The excavation of holes for supports, special supports and independent gate posts, together with backfilling, disposal of surplus materials and earthwork support are deemed to be included (Q40:C1), while concrete surrounds to posts and stays are included in the descriptions of special supports and independent gate post descriptions, as illustrated in example 27. Gates are enumerated, stating the type, height and width, and are deemed to include gate stops, gate catches, independent gate stays and their associated works (Q40:5.1.1.0 and Q40:C5).

It should be noted that fencing to ground sloping >158 from the horizontal, fencing set out to a curve but straight between posts, curved fencing radius >100 m, curved fencing radius ≤100 m stating the radius, and lengths ≤3 m are each measured separately.

The measurement of brick and stone boundary walls has been covered in chapters 5 and 6.

WORKED EXAMPLE

A worked example follows covering the measurement of a road, channels, footpath, edging, fencing, gates, trees, shrubs, hedges and a grassed area.

This page has been left intentionally blank for make-up reasons.

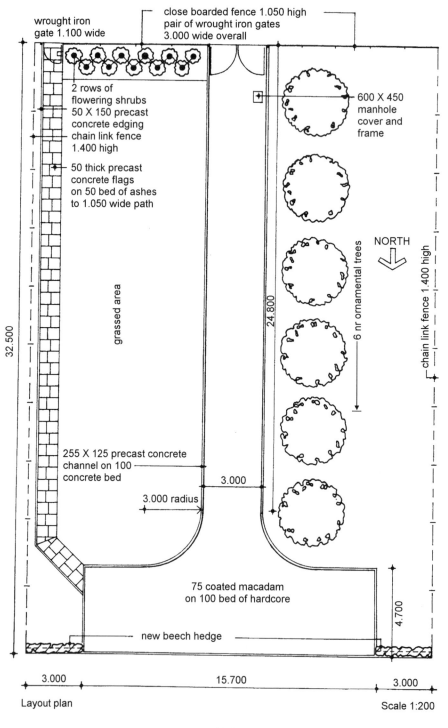

wrought iron
gate 1.100 wide

close boarded fence 1.050 high
pair of wrought iron gates
3.000 wide overall

2 rows of
flowering shrubs
50 X 150 precast
concrete edging
chain link fence
1.400 high

50 thick precast
concrete flags
on 50 bed of ashes
to 1.050 wide path

600 X 450
manhole
cover and
frame

NORTH

6 nr ornamental trees

chain link fence 1.400 high

grassed area

32.500

24.800

255 X 125 precast concrete
channel on 100
concrete bed

3.000

3.000 radius

75 coated macadam
on 100 bed of hardcore

4.700

new beech hedge

3.000

15.700

3.000

Layout plan

Scale 1:200

334

Drg. 21		Example 27

External Works 1

Access road
len.
24.800
add bellmth.   3.000
width 27.800
3.000
add chans.
2/255   510   bellmth.
3.510   3.000
less chan.   255
2.745

Note the use of subheadings to act as signposts throughout the dimensions. The areas of access road are built up in waste using figured dimensions from the drawing as far as possible. In arriving at the width of the road excavation, it is necessary to add the widths of the channels as the depth of excavation is the same for both. In like manner, the radius of the bellmouth has to be adjusted for one channel.

Excvtg.
Topsoil for presvn.
225 av. depth
&
Surf. trtmts.
Compactg. btms. of excvns
(bellmth. &
Excvtd. matl.
Fillg. to make up levs.;
≤250 av. th.;
arisg. from excvns.;
topsoil
Cub. × 0.23 = _____ m³

27.80
3.51
15.70
4.70
2/3/14/ 2.75
2.75

Excavating topsoil which is to be preserved is measured in m², giving the average depth (D20:2.1.1.0). The site is well endowed with topsoil which is taken for the full 225 in depth, thus avoiding the need to take part of the excavation to reduce levels in m³ as D20:2.2.1.0. Compacting bottoms of excavations is measured in m² as Q20:13.2.3.0. The additional area at each side of the bellmouth = 3/14 × radius².

The spreading of the excavated topsoil is assumed to be carried out direct from the access road without the need for on site spoil heaps, and this is measured in m³ in accordance with D20:10.1.1.3. Note the method used for converting superficial measurements into cubic ones, to avoid having to enter another set of dimensions.

hammer head
to access road
len.
15.700
less chans. 2/255   510
15.190
width
4.700
less chans. 2/255   510
4.190

It is necessary to deduct the width of channels where appropriate to arrive at the areas of hardcore required under the hammer head to the access road.

Exteŕnal Works 2

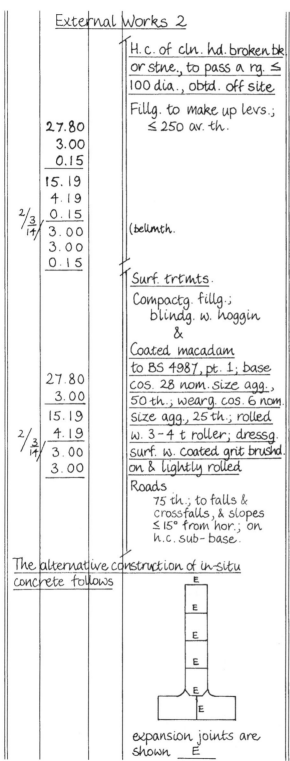

	H.c. of cln. hd. broken bk. or stne., to pass a rg. ≤ 100 dia., obtd. off site
	Fillg. to make up levs.; ≤ 250 av. th.
27.80	
3.00	
0.15	
15.19	
4.19	
2/3/14/ 0.15	
3.00	(bellmth.
3.00	
0.15	
	Surf. trtmts.
	Compactg. fillg.; blindg. w. hoggin &
	Coated macadam to BS 4987, pt. 1; base cos. 28 nom. size agg., 50 th.; wearg. cos. 6 nom. size agg., 25 th.; rolled w. 3–4 t roller; dressg. surf. w. coated grit brushd. on & lightly rolled
27.80	
3.00	
15.19	
2/3/14/ 4.19	
3.00	
3.00	
	Roads 75 th.; to falls & crossfalls, & slopes ≤ 15° from hor.; on h.c. sub-base.

The alternative construction of in-situ concrete follows

expansion joints are shown   E

This item commences with a description of the hardcore base under the access road. In practice, reference is often made to a project specification, which will contain much more comprehensive details. Excellent examples of typical clauses and requirements can be found in Laxton's *General Specification for Major Building and Civil Engineering Projects, Volume 1.*

The hardcore filling in making up levels as road sub-base is measured in m^3, in accordance with Q20:10.1.3.0. The compacting of the sub-base follows the procedure contained in Q20:13.2.2.1. The hoggin will consist of a combination of naturally occurring fine gravel, sand and clay of approved quality to fill the interstices in the hardcore.

The coated macadam item commences with a detailed description of the material and the thickness and number of coats, including the sizes of aggregates, weight of roller and any surface dressing required, as Q22.1.1.2.0. Supplementary information can include such items as surface finish and laying (Q22:S2–3). Working around the manhole cover is deemed to be included (Q22:C1b), and no deduction is required as the area is ≤0.50 m² (Q22:M1).

The concrete road is likely to be 150 thick on a 75 granular base, separated by a waterproof membrane and reinforced with steel fabric reinforcement. The accompanying sketch (NTS) shows the location of expansion joints. It is assumed that the precast channels on an *in-situ* concrete base will be laid first and the concrete tamped off them.

External Works 3

		Reinfd. in situ conc. to BS 5328; designed mix C20; 20 agg., min. ct. content 240 kg/m³; vibrated
		Beds ≤ 150 th.; slopg. ≤ 15°
27.80		Cub. × 0.15 = _____ m³
3.00		&
15.19		Reinft. to BS 4483
4.19		Ref. C283 mesh; whg. 2.16 kg/m²; min 100 side laps & 400 end laps
2/ 3/14	3.00 3.00	Fabric
		& Worked fins.
		Tampg. in situ conc. by mech. means slopg.
		& Sundries
		Polythene sht. u/ly., 500 g.; layg. on gran. base to rec. conc.

	expsn. jts	
trans. 4/3.000	12.000	
bellmth. 2/1.500	3.000	
	2.000	
	4.190	
	21.190	

Designed jts. w. 10 th. f. bd. & 10 × 25 Sealant of hot bit.

21.19		Formed ≤ 150 dp.; hor.

In-situ concrete roads are covered in section Q21 of the *Standard Method*, but the relevant measurement codes are contained in E10, 20, 30, 40, 41 and 42. The concrete road slab is therefore classified as a concrete bed ≤150 thick, reinforced and sloping ≤158 as E10:4.1.0.1&3. The road slab will not be cambered and will fall from one side to the other, and be tamped off the channels and formwork adjoining expansion joints. The concrete has been specified with a designed mix which is common with road projects.

The fabric reinforcement, of which there is a wide range available, is measured in m², stating the mesh reference and weight/m² (E30:4.1.0.0). Fabric reinforcement is deemed to include laps, tying wire, cutting and bending, and spacers which are at the contractor's discretion (E30:C2).

Worked finishes to the concrete surface are taken as E41:1.0.0.1, assuming mechanical tamping off the channels, etc.

The polythene underlay, to prevent the newly laid concrete drying out too quickly through the granular material beneath, finds no place in E10 and so the heading of 'sundries' has been adopted. Alternatively, a heading of polythene could have been used.

The length of expansion joints is determined using the diagram on the previous page and applying it to drawing 21. The appropriate measurement code is E40:2.1.1.0, and includes a description of the jointing material and sealant. Formwork is deemed to be included (E40:C1).

338   *Building Quantities Explained*

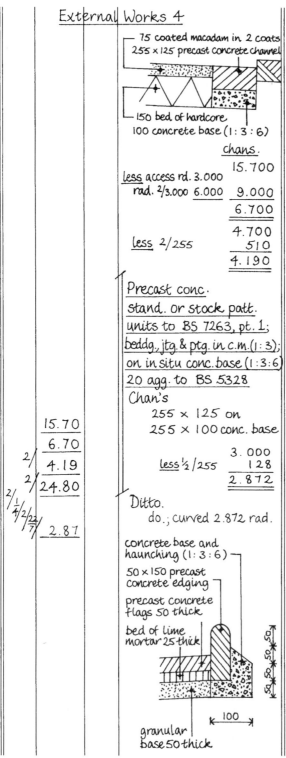

External Works 4

75 coated macadam in 2 coats
255 × 125 precast concrete channel

150 bed of hardcore
100 concrete base (1 : 3 : 6)

chans.

        15.700
less access rd. 3.000
  rad. 2/3.000 6.000 9.000
                6.700
                4.700
less 2/255   510
                4.190

Precast conc.
stand. or stock patt.
units to BS 7263, pt. 1;
beddg., jtg. & ptg. in c.m.(1 : 3);
on in situ conc. base (1 : 3 : 6)
20 agg. to BS 5328
Chan's
   255 × 125 on
   255 × 100 conc. base

15.70	
6.70	
2/ 4.19	
2/ 2/ 24.80	
2/ ½/	
¼/2/²²⁄₇/	
7/ 2.87	

           3.000
less ½/255   128
           2.872

Ditto.
  do.; curved 2.872 rad.

concrete base and
haunching (1 : 3 : 6)
50 × 150 precast
concrete edging
precast concrete
flags 50 thick
bed of lime
mortar 25 thick

50·50·50

100

granular
base 50 thick

The diagram helps to explain the construction and shows how the precast concrete channel strengthens and supports the macadam surfacing in the absence of a roadside kerb. It needs a minimum of a concrete base as shown in the diagram and, if it had been a wider road taking more traffic, a larger base and haunching at the back of the channel would be advisable.

Calculations in waste (side casts) help to determine the lengths of channel needed.

A descriptive heading is so framed that it can be used almost in its entirety to cover the precast edging that will be measured with the paths.

Precast concrete channels are measured in metres with a dimensioned description and include concrete foundations and haunching where provided (Q10:4.1.0.2).

Curved channels form a separate item, as they are more costly, stating the radius (Q10:4.1.0.2–3). The lengths occupy the circumferences of two quarter parts of circles or quadrants, hence the timesing figures of 2/¼/, followed by 2/²²⁄₇/2.87 (being $2\pi r$).

The second diagram shows the path edging with its concrete base and haunching, and the adjoining precast concrete flags bedded on lime mortar and a granular base.

External Works 5

Paths
excvn. width

flags	1.050
add edg'g. & conc. haunchg.	100
	1.150

26.50	Excvtg.
1.15	Topsoil for presvn.
	125 av. depth
3.00	&
1.15	Surf. trtmts.

The width of excavation for the paths is wider than the precast concrete flags, as it is necessary to add in the edging and the concrete haunching to it. The lengths have been scaled from the drawing in the absence of figured dimensions. The excavation of the topsoil and compacting the bottoms of excavations are both measured in m^2 (D20:2.1.1.0 and Q20:13.2.3.0), whereas the disposal of the excavated topsoil on the site to make up levels is measured in m^3, even although it is only 125 thick (D20:10.1.1.3). The three items with a common area can be grouped together and the third item converted to a cubic dimension in the manner shown to reduce the number of sets of dimensions.

Excvtg.
Topsoil for presvn.
125 av. depth
&
Surf. trtmts.
Compactg. btms. of excvns.
&
Excvtd. matl.
Fillg. to make up levs.;
≤ 250 av. th.;
arisg. from excvns.;
topsoil
Cub × 0.13 = _____ m³

Granular matl. of nat. sds., gravels or crushed rock; obtd. off site
Fillg. to make up levs.
≤ 250 av. th.
Cub × 0.05 = _____ m³
&
Surf. trtmts.
Compactg. fillg.

Similar comments apply to the granular material filling which is merely 50 thick. It is necessary to give a description of the granular material, so that the estimator knows the product that he is pricing. It is likely that a more detailed description will be contained in the project specification.

26.50	
1.05	
3.00	
1.05	

&
Precast concrete flags to BS 7263, pt. 1; nat. fin.; ld. bkg. jt. w. 150 lap; jts. 6th.; sym. layout; grtd. in l.m.(1:3) on 25 wet screeded bed of l.m.(1:3)
Pavgs.
600 × 600 × 50 th. & 600 × 450 × 50 th. units; on gran. base; to falls & crossfalls & to slopes ≤ 15° from hor.

BS 7263 covers precast concrete flags, although they are commonly called paving slabs, and the subheading describes the flags, layout and grouting. While the second part of the description is mainly concerned with the requirements contained in Q25:2.1.2.1–2. Working around the stop cock box is deemed to be included under Q25:C1b.

External Works 6

Path edg'g.

Precast conc.
stand. or stock patt.
units on in situ
conc. base a.b. & haunchg.
Edgings
     50 × 150 w. rdd. top
    on 100 × 50 base &
    50 × av. 75 hnchg.

| 26.70 | |
| 3.60 | |

The precast concrete edging has a similar heading to that adopted for the road channel and includes the concrete base and haunching. The lengths are scaled from the drawing and are slightly longer than those obtained previously for the path measured on its centre line. Edgings are taken as linear items with a dimensioned description as Q25:7.1.0.2.

Many alternative forms of paving are available from clay bricks to concrete and natural stone pavers and setts. Clay cobbles have been selected as the alternative to be described.

Baggeridge I5 brindled red dragfcd. textd. solid fired clay cobbles, 75 × 75 × 65 dp. to BS 6677, pt. 1; on compactd. sd. lyg. cos. 30 th. w. sd. filled jts. 10 wide; to reg. patt.
Pavgs.
    flexible; to falls &
    crossfalls & to slopes
    ≤ 15° from hor. on
    gran. base

26.50	
1.05	
3.00	
1.05	

An alternative form of paving has been included to show a different form of finish. The clay cobbles are laid on and jointed in sand on a granular base, the latter being measured separately as for flag pavings. Alternatively, in the rigid system they are usually laid on a bed of cement and sand (1:3) 20 thick, with the joints filled with mortar. The measurement procedure is contained in Q25:2.1.2.1.

Grassed area
        len.
        24.800
add curvd. end   2.745
        27.545
less shrub border 1.900
        25.645

Seed'g/turfg.

Cultivtg. by hand or pedestrian operated mach.
Surfs. of grd.
    100 dp.; evenly
    graded to fin. levs. to
    a fine tilth; weeding
        &
Surf. applns.
Fertilizg. gen. surfs. of cultivated soil
    basic price £0.40/kg;
    rate of 0.07 kg/m²

| 25.65 | |
| 7.70 | |

The grassed area is calculated in waste from figured and scaled dimensions from the drawing. Cultivating, surface applications, seeding and turfing are each measured separately in m² (Q30:1–4). The cultivating description is to include the depth, method of cultivating, degree of tilth and weeding details (Q30:1.1.0.1 and Q30:S2). The removal of stones is deemed to be included (Q30:C1).

Fertilising is one of the surface applications listed in Q30:D1 and is to state the type of fertiliser and rate of application, and is deemed to include working in (Q30:2.1.0.0 and Q30:C2).

External Works 7

Seedg.

Gen. surfs. of cultivtd. soil; in 2 nr. operatns. w. apprvd. grass seed, basic price £2.70/kg.; rate of 0.07 kg./m². maintg. for 12 mths. after seedg.

25.65
7.70

The seeding description is to include the rate of application, kind and quality of material and is deemed to include raking or harrowing in and rolling (Q30:3.1.0.0 and Q30:C3). Note the inclusion of basic prices for these materials. Details regarding the timing of the operations are likely to be included in a preamble clause or a project specification. The grassed areas to the two irregular corners are suitably adjusted.

Ddt last three items

3/14/ 2.75
      2.75

(bellmth.

1/2/ 1.40
     1.40

(spld. cnr.

An alternative to seeding is turfing and the following item shows the different approach

Turfing w. gen. purp. utility turf w. perennial ryegrass to BS 3969, table 1, & trtd. w. sel. weedkiller not more than 3 mths. nor less than 4 wks. before liftg.

Gen. prepd. surfs. of grd.
rollg., waterg., weedg., cuttg., settlement returfg. until establishd.; ld. w. brkn. jts

25.65
7.70

Turfing enables a grassed surface fit for use to be provided much sooner than with seeding. The quality of turf does however vary considerably and meadow turf can contain a significant quantity of weeds. This area is being turfed with turf which contains perennial ryegrass, which is a course hard wearing grass, and hence the type of turf chosen will vary with the purpose it is to serve. Once again maintenance work is included in the description.

Shrubs to BS 3936, pt. 1;

Berberis Darwinii

ht. 450; balled fibrous rt. system; excvtg. pit 450 x 450 x 450 dp.; breakg. up btm.; refillg. w. excvtd. topsoil mxd. w. 20% apprvd. compost; removg. surplus excvtd. mat.; waterg. as specfd.

11

Shrubs are enumerated giving the botanical name of the shrub and describing the height and root system, size of pit to be excavated, material used for refilling, removal of surplus excavated material and any watering requirements as Q31:5.1.1.5–6.

External Works 8

		<u>Hedge plants to BS 3936, pt. 1</u>

Beech
 ht. 450, spaced 375 apt.
 in 2 rows stagd.; incl.
 excvn. 150 wider than rt.
 sprd.; remove surplus
 excvtd. mat.; refillg. w.
 topsoil & fertilsr.;
 waterg. as specfd.

2/ 3.00

Hedge plants can be enumerated or measured in metres. When measured in metres, the description is to include the height, spacing, number of rows and layout (Q31:6.1.2.5–6). Alternatively, reference could be made to the project specification clauses for detailed information.

<u>Trees</u>
<u>Nursery stock to BS 3936</u>

Malus Tschonoski
 lt. std.; gth. 40–50, ht.
 1.800–2.000, clr. stem
 1.200–1.500; adeq. compact
 rt. system; supptd. by
 impregntd. wrot. swd. 50
 ×50 stake, 2.000 lg. &
 PVC tree tie; excvtg. pit
 1.000 × 1.000 × 600 dp.;
 brkg. up btm.; refillg.
 w. excvtd. topsoil incorp.
 one third well rotted
 farmyd. manure as specfd.;
 removg. surplus excvtd. mat.;
 waterg. as specfd.

6

Trees are suitably classified in the manner prescribed in the *SMM7 Library of Standard Descriptions*. Information is required covering the botanical name, sizes and root system, including the girth, height and clear stem. Further information includes supports and ties, refilling with special materials and watering details (Q31:3.1.1–2.4–6). A large pit needs to be excavated to give sufficient space for root growth and a substantial quantity of farmyard manure has been added to the backfill.

Protectn.
Tree guards
 plastic tubes, perf.;
 1.200 lg. × 150 dia.;
 stapld. to swd. stakes

6

Tree guards are enumerated with a dimensioned description (Q31:10.1.1.0).

<u>Chain link fencg; to BS</u>
<u>1722, pt. 1; rolled stl. angle</u>
<u>posts; green plastic ctd.; grade</u>
<u>A mesh, line & tying wires;</u>
<u>galvd. stl. fittgs. & accessories</u>
<u>back fillg. ard. posts. in</u>
<u>conc. mix ST6</u>

Fencg.
 1.400 hi.; type PLS 140A;
 50 × 50 × 5 angle posts w.
 rdd. tops at 3.000 ccs.
 driven 600 into grd.

2/ 32.50

Fences are measured in metres, stating the type, height of fence and spacing, height and depth of supports (Q40:1.1.1.0). Supports are posts, struts and the like occurring at regular intervals (Q40:D1). The height of fencing is measured from the surface of the ground to the top of the infilling (Q40:D3). Details are required of the kind and quality of materials, construction, surface treatment before delivery and size and nature of backfilling (Q40:S1–4).

External Works 9

Chain link fencg. a.b.

E.o. for
  special suppts.;
  60 × 60 × 6 L.strng.
  posts, 2.000 lg., 600
  into grd. & 50 × 50 × 5
  L strut, 2.000 lg. boltd.
  to strng. post; surrdg.
  btms. of posts & struts
  w. conc. mix ST6

2/ 2

Special supports, such as end posts, angle posts, straining posts and their supporting struts, are enumerated as extra over the fencing in which they occur, stating the size, height and depth (Q40:2.1–5.1.2). Fencing work is deemed to include excavating holes for supports, backfilling and disposal of surplus materials and earthwork support (Q40:C1). This leaves only the concrete to be described.

Close boarded fencg.; to
BS 1722, pt. 5; sawn swd.
posts, rails, pales, gravel
bds. & centre stumps;
pressure impregntd. w.
preservative; backfillg.
ard. posts in conc. mix ST6

Fencg.
  1.050 hi.; type BW105
  posts at 3.000 ccs.;
  600 into grd.

0.40
7.70
9.10

Close boarded fences are measured in metres with a full description as Q40:1.1.1.0 and Q40:S1–4. The posts are included in the description of the fencing, stating the size, height above ground level and depth below surface. In this example the description is reduced in length by referring to the relevant British Standard. The integral gate posts are enumerated as extra over the fencing as Q40:2.3.1.0.

E.o. for
  integral gate posts;
  200 × 200 × 1.650 lg.;
  600 into grd.; conc. a.b.

2/ 2

Galv. steel
Pr. of ornamental gates
  to BS 4092, pt. 1,
  3.000 × 900 o/a; cat.
  ref. 'X' of manufac. 'A',
  incl. asscctd. fittgs. &
  1 ct. primer at wks.

1

Gates are enumerated stating the type, height and width as Q40:5.1.1.0. Gates are deemed to include gate stops, gate catches and independent gate stays and their associated works (Q40:C5), and hence these do not need describing. Reference has been made to a manufacturer's catalogue and an alternative approach would be to give basic prices.

Ornamental gates
  ditto, 1.100 × 900 o/a;
  cat. ref. 'Z' of manufac.
  'A' do.

1

Paintg. prev. primed
met.; prep. & ③

Gates
  ornamental type;
  EXT.

2/ 3.00
   0.90

2/ 1.10
   0.90

Painting of gates is measured overall, regardless of the voids on each side as M60:7.3.0.0 and M60:M12. Priming has not been included as the British Standard requires protective treatment at the manufacturer's works.

17 Bill Preparation and Production

This chapter starts with an examination of the traditional method of bill preparation and then leads on to an investigation of the more recent processing techniques with their merits and demerits.

WORKING UP

A description is given of the final stages leading up to the preparation of bills of quantities for building work by the traditional or group method after the dimensions have been taken off in the manner described and illustrated in earlier chapters. The term 'working up' is applied to all the subsequent operations collectively and consists of the following processes:

(1) Squaring the dimensions and entering the resultant lengths, areas and volumes in the third or squaring column on the dimensions paper.
(2) Transferring the squared dimensions to the abstract where they are written in a recognised order, ready for billing, under the appropriate work section headings, and are subsequently totalled and reduced to the recognised units of billing, preparatory to transfer to the bill.
(3) In the bill of quantities, the various items of work which together make up the complete building are listed under the appropriate work section headings, with descriptions printed in full and quantities given in the recognised units of measurement, as laid down in the *Standard Method of Measurement of Building Works* (SMM7). The bill also contains rate and price columns for pricing by contractors when tendering for the project, as illustrated later in the chapter.

BILLING DIRECT

The traditional working up process which was used for many decades in quantity surveyors' offices was very lengthy and tedious, and various ways of shortening the process have been developed. One of the first methods to be introduced was to 'bill direct', by transferring the items direct from the dimension sheet to the bill, thus eliminating the need for an abstract, and so saving both time and money, and referred to as 'direct billing' or 'billing direct'.

The direct billing system is most suitable where the number of like items is limited and the work is not too complex in character. This approach has its origins in the earlier trade by trade method of measuring completed work. The surveyor works through a single trade or work section and produces a draft bill with fully developed descriptions and headings in addition to quantities. Once the calculations have been checked, the bill can be produced directly from the draft. It offers a simpler system when measuring one trade at a time, which is not however the usual practice, and there is increased risk of duplication or omission when several quantity surveyors are measuring the work. The majority of building services items are either enumerated or linear items and hence lend themselves to direct billing, as illustrated in *Measurement of Building Services*, and similarly with alteration work.

With the object of speeding up the working up process and reducing expensive staff time, further methods using the 'cut and shuffle' system and computers were developed. These improved methods will be described later in this chapter.

SQUARING DIMENSIONS

The term 'squaring dimensions' refers to the calculation of the numbers, lengths, areas and volumes and their entry in the third or squaring column on the dimensions paper. The following examples illustrate the method of squaring typical dimensions on dimensions paper.

Dimensions			
12/		Sn. swd. a.b.	Notes
4.50	54.00	Flr. membrs.,	Linear item: Total
		50 × 175	length is 54 metres,
			and no further reduction
			is required in the abstract.
		Excvtg.	
5.78		Topsoil for presvn.	Square or superficial
5.48	31.67	150 av. depth	item: area is 31.67 m^2
			which will be reduced
			to 32 m^2 in the abstract,
			as the fraction
			exceeds one-half.
19.50		*In-situ conc*	
0.75		(21 N/mm^2–20 agg.)	Cubic item: Volume of
0.23	3.36	Fdns.	concrete is 11.46 m^3
		poured on or	which will be reduced
30.00		agst. earth	to 11 m^3 in the abstract,
0.90			as the fraction is
0.30	8.10		less than one-half.
	11.46		Note method of casting
			up a number of dimensions
			relating to the same item with
			the total quantity entered in
			the squaring column.
			Deductions following the
			main items can be dealt with
			in a similar manner, as shown
			in example 1.

Steelwork will often be transferred to the abstract in kilogrammes where the items will subsequently be reduced to tonnes, to the nearest 2 decimal places.

When there are timesing figures entered against the item to be squared, it is often simpler to multiply one of the figures in the dimension column by the timesing figure before proceeding with the remainder of the calculation. Alternatively, the total obtained by the multiplication of the figures in the dimension column is multiplied by the timesing figure. Where a number of measured items have the same area, but some are superficial and others cubic items, it is permissible to bracket them and insert the additional timesing dimension for the depth or thickness as appropriate in the description column, as illustrated in examples 1, 3 and 27.

The squaring must be checked by another person to eliminate any possibility of errors occurring. All squared dimensions and waste calculations should be ticked in coloured ink or pencil on checking and any alterations made in a similar manner. Amended figures need a further check. Where, as is frequently the case, calculators are used for squaring purposes a check should still be made. The traditional measurement processes of abstracting and billing are now considered in outline, so that the reader is familiar with their main characteristics, although it will be appreciated that they are now becoming mainly items of historical interest.

ABSTRACTING

Transfer of Dimensions

As each item is transferred to the abstract, the description of the appropriate dimension item is crossed through with a vertical line on the dimension sheet, with short horizontal lines each end, so that there shall be no doubt as to what has been transferred.

The following example illustrates this procedure.

	5.78		Excvtg.
	5.48	31.67	Topsoil for presvn.
			150 av. depth

Subdivisions of Abstract

The abstract sheets are ruled with a series of vertical lines spaced about 25 mm apart and are usually of A3 width.

Each abstract sheet is headed with the project reference, sheet number and work section, and possibly the subsection of the work to which the abstracted dimensions refer. The majority of bills and abstracts are subdivided into work sections in the manner adopted in the *Standard Method*.

Each work section is usually broken down into a number of subsections as appropriate. The usual approach is to adopt the subdivisions contained in the *Standard Method* and the *SMM7 Library of Standard Descriptions* (mainly levels 1 and 2). The following examples illustrate this approach with regard to work sections D20, E10 and F10

D20 *Excavating and filling*: subdivided into site preparation; breaking up; excavating; earthwork support; disposal; excavated material/topsoil/hardcore (filling); herbicides; surface treatments.

E10 Mixing/Casting/Curing *in-situ* concrete: subdivided into plain and re-inforced concrete with separate subheadings for each concrete mix.

F10 *Brick/block walling*: subdivided into brickwork and blockwork with separate subheadings for each type of brickwork and blockwork (e.g. common, engineering and facing bricks and blocks as specified).

General Rules of Abstracting

It is most important that the entries in the abstract should be well spaced and it is necessary for the surveying assistant doing the working up to look through the dimension sheets, before starting the abstracting, in order to determine, as closely as possible, how many abstract sheets will be required.

The items will be entered in the abstract in the same order as they will appear in the bill, as far as practicable, since the primary function of the abstract is to classify and group the various items preparatory to billing, and to reduce the dimensions to the recognised units of measurement. Descriptions are usually spread over two columns with the appropriate dimension(s) in the first column and any deductions in the second column. The total quantity of each item is reduced to the recognised unit of measurement such as the m, m^2, m^3 or tonne.

It is good practice to precede each description in the abstract with the prefix C, S, L or Nr., denoting that the item is cubic, square, linear or enumerated, and this procedure reduces the risk of errors arising with regard to units or quantities.

The order of items in each work section of the abstract normally follows the sequence indicated in the first column of the classification table in SMM7. In some instances the measurement rules require items which are generically similar to be given in different units of measurement, a typical example being 'formwork to sides of foundations' which may be measured in either m^2 or m according to height (E20:1.1.1–4.0). In such cases the order adopted should be in the sequence: cube, square, linear and enumerated item. Where otherwise similar items occur in different sizes, the cheapest and normally the smallest item is given first. Items of this type are often best grouped under a single heading with each size entered in a separate column, as shown in the following example:

L/Clay pipewk. of s&s pipes w. flex. jts. to BS EN 295–1

Pipes: nom.	sizes in trs.	
100	150	225
45.63 (4)	32.45 (3)	35.30 (3)
38.60 (5)	26.28 (3)	
15.40 (5)		
52.32 (6)		

Alternatively where similar items of varying size are encountered they can sometimes conveniently be entered on the abstract sheet in the following way:

Cop. pipewk.	to BS EN 1057,	w. comp. jts.
Nr/E.O. for made bend; size	15 nom.	
1 (28) 2 (29) 1 (30)	Nr/Ditto., 22 size	nom.
	2 (30) 1 (31) 1 (32)	

The number in brackets after each dimension represents the page or column number of the dimension sheet from which the dimension has been extracted, for ease of reference.

Deductions are entered in the second column under the main heading of the item under consideration, as illustrated in the following example.

S/Fcg. bwk, in Himley mixt. russet wirecut fcg. bks. in stret. bd. in g.m. (1:1:6) & ptg. w. nt. weather strk. jts. a.w.p. Walls h.b.th; facewk. O.S.	
95.64 (13) 152.36 (14) 76.19 (16) 8.32 (18) 2.40 (70) 334.91 42.31 292.60 $= 293\,\text{m}^2$	Ddt 6.95 (53) 8.91 (56) 26.45 (62) 42.31

In the last example the area is expressed in m^2, as the correct unit of measurement for this class of work. Furthermore, the dimensions have been crossed through to indicate that the deduction has been made and the adjusted area transferred to the bill.

When measuring some items such as glass in panes $\leq 0.15\,m^2$ and iron drain pipes in runs ≤ 3.000 in length, the number of items involved has to be stated in the billed description. A convenient method of dealing with this type of item in the abstract is indicated in the following example:

S/ Std. pl. glass; 4 clear float to BS 952 To wd. w.l.o. putty & sprigs in panes; areas $\leq 0.15\,m^2$	
18 = 1.47 (49) (18 panes with a total area of 1.47 m²) 54 = 4.02 (49) (54 panes with a total area of 4.02 m²)	

When enumerated items are to be written in the bill following an associated linear item, such as mitres and fitted ends with larger section hardwood skirtings and the like, the best method of dealing with them in the abstract is as follows:

L/ Wrot hwd. sktgs., etc. a.b. 38 × 225; mo.; plug'g to masonry		
85.20 (38)	Nr/E.O. for angles	
	16 (38)	Nr/E.O. for ends
		12 (38)

On completing the entry of all items on the abstract, all entries will be checked, columns of figures cast, deductions made, totals reduced, all the latter work checked and the totals finally transferred to the bill.

BILLING

Ruling of Bill of Quantities

It is desirable that one of the rulings detailed in the British Standard Specification *Stationery for Quantity Surveying* (BS 3327: 1970) should be used to ensure that a uniform method of setting out the information in a bill of quantities is obtained – although this Standard has now been withdrawn.

This British Standard specifies the rulings for both single and double bill papers (with one or two sets of pricing columns). The single bill with right hand billing is the most widely used for building as well as civil engineering work.

The single bill paper as prescribed by BS 3327, which is widely used in practice, is now illustrated. The widths of columns vary slightly on the face and reverse sides of each bill sheet. There can be a binding edge on the left hand side of column 1 measuring about 15 wide, reducing the widths of columns 1 and 7 to 10.

1	2	3	4	5	6	7

Nr. of column	Use of column	Width (mm) face side	reverse side
1	Item nr. or reference	19	14
2	Description	100	100
3	Quantity	24	24
4	Unit	14	14
5	Rate (£)	18	18
6	£	21	21
7	p	14	19

The double billing paper, with two sets of pricing columns, is mainly used in connection with final accounts for bills of variations, one set of columns being used for omissions and the other set for additions. It can also be used in the preliminaries bill for separate indications of fixed and time related charges.

Referencing of Items

It is essential that items in a bill of quantities which are to be priced by a contractor shall be suitably referenced. With bills of quantities for building works a common practice is to letter the items alphabetically on each page to avoid the use of the large numbers which arise if all the items are numbered consecutively throughout the bill. Thus the third item on page 20 of the bill of quantities could be referred to as item 20/C (page 20, item C). Another approach is to number the pages of each bill separately, hence item D on page 15 of bill 4 will be 4/15D.

A further alternative is to use the SMM work section references as illustrated in table 17.2.

Entering Quantities in the Bill

When transferring quantities to the bill in m^3, m^2 or m, they are to be billed to the nearest whole unit. Fractions of a unit which are less than one-half are disregarded and all other fractions (one-half or over) are taken as whole units. Where the unit of billing is the tonne, quantities shall be billed to the nearest two places of decimals.

Where the application of this principle would cause an entire item to be eliminated, the item is to be given as one unit (General Rules 3.3).

Units of Measurement

The words used in describing work measured in one, two or three dimensions are linear, square and cubic respectively. These words are now little used in practice and the following abbreviations are given in SMM7: metre(s): m; square metre(s): m^2; cubic metre(s): m^3; tonne(s): t; enumerated items: nr. (General Rules 12.1).

General Rules of Billing

The order of billed items will be the same as in the abstract, as far as practicable, and they will be grouped under suitable work section and subsection headings as described earlier in the chapter. The work section headings will generally follow the order and terminology adopted in the *Standard Method*, such as D20 Excavating and Filling and E10 Mixing/Casting/Curing *in-situ* concrete. It is recommended that the work section titles contained in SMM7 should be modified according to the actual content of the work and also to read grammatically. For example where M40 Stone/Concrete/Quarry/Ceramic Tiling/Mosaic only includes quarry and ceramic tiling, it should appear in the bill of quantities as M40 Quarry and Ceramic Tiling. Each work section should commence on a new page of the bill of quantities. This is beneficial to the contractor when forwarding sections of the bill to subcontractors for pricing. There will sometimes be a number of preamble clauses inserted at the head of each bill work section, relating to financial aspects of the work in the bill section sometimes and giving guidance to the contractor in his pricing of items. In addition preamble clauses may be used to give detailed material and workmanship requirements with a view to reducing the length of subsequent billed descriptions and eliminating the specification, as illustrated later in the chapter and dealt with in more detail in *Advanced Building Measurement*. However, a separate project specification is frequently included as part of the bill of quantities when the preamble clauses can be omitted and there will be extensive cross referencing to the specification in the descriptions of the billed items, as illustrated in chapter 9 and appendix 4.

Each item to be priced in the bill is referenced by letters and/or numbers in the first column. It will be noticed that all words in the billed descriptions are written in full without any abbreviations and this procedure should always be followed to avoid any possible confusion arising.

Provision is generally made for the total sum on each page of the bill to be transferred to a collection at the end of the work section or work group. The totals of each of the collections are transferred to a summary at the end of the bill, the total of which will constitute the contract sum. This procedure is preferable to carrying forward the total from one page to another in each work section, since the subsequent rectification of pricing errors may necessitate alterations to a considerable number of pages.

Billed descriptions should conform to the requirements of the *Standard Method*, follow in a logical sequence and be concise, yet must not, at the same time, omit any matters which will be needed by the contractor to assess the price for each item in a realistic manner.

The following example illustrates the normal method of entering items which are 'written short' with the descriptions set back. The contractor is able to price the enumerated items immediately after the linear item with which they are associated.

	Wrought mahogany Skirtings, picture rails, architraves and the like					
F	25 × 150 moulded; screwed with brass cups and screws; plugged to masonry	71	m			
G	Extra over for mitres	16	nr.			
H	Extra over for ends	12	nr.			
J	25 × 200 ditto.	22	m			
K	Extra over for mitres	8	nr.			
L	Extra over for ends	4	nr.			

Note the use of the word 'ditto.' when a similar item occurs to avoid unnecessary repetition of descriptions.

On completion the draft bill must be very carefully checked against the abstract and the abstract suitably marked in coloured ink or pencil as each item is dealt with. Particular care should be taken to ensure that all the quantities, units and descriptions are correct and that proper provision has been made for section and subsection headings, transfer of totals to collections and summary and a satisfactory sequence of items obtained.

Further checking arises in connection with the printer's proof which must be carried out extremely carefully. It is also good policy to calculate the approximate areas and volumes of major items of work such as excavation, disposal of excavated material, brick and block walling, roof coverings, painting and also the total number of fittings such as number of windows, doors, sanitary appliances and manholes and to compare them with the actual billed quantities, to ensure that no major errors have occurred.

Preliminaries Bill

The first work section in a bill of quantities is often termed the 'Preliminaries Bill' and covers many important financial matters which relate to the contract as a whole and are not confined to any particular work section, and the contractor is thereby given the opportunity to price them. Section A of the *Standard Method* details most of the items which would appear in such a bill.

SMM7 introduced significant changes in the rules for preliminaries. Employer's requirements (A30–37) are now clearly separated from the contractor's general cost items for management and staff, site accommodation, services, and facilities, mechanical plant and temporary works (A40–44).

Thus the preliminaries section of a bill of quantities contains two separate and distinct types of item, and these are now outlined:

(1) Items which are not specific to work sections but which have an identifiable cost which is best considered separately for tendering purposes, such as contractual requirements for insurances, site facilities for the employer's representative and payments to the local authority.

(2) Items for fixed and time related costs which derive from the contractor's expected method of carrying out the work, such as bringing plant to and from the site, providing temporary works and supervision.

The fixed and time related subdivision given for a number of preliminaries items enables tenderers to price the elements separately should they wish to do so. Tenderers also have the facility at their discretion to extend the list of fixed and time related cost items to suit their particular methods of construction (*SMM7 Measurement Code* 3.2).

The first four items in work section A of SMM7 (Preliminaries/General Conditions) relate to the project particulars, list of drawings from which the bills of quantities were prepared and those to be used for the contract, details of the site and existing buildings which could have an influence on cost, such as the existence of contaminated soil, and services and a description of the work, including any unusual features or conditions (A10–13). A useful description of a building is given in the *SMM7 Measurement Code* (A13). In this way the contractor is able to quickly obtain a general impression of the project and its main implications.

The next item contains a schedule of clause headings of the standard conditions of contract, any special conditions or amendments, appendix insertions, employer's insurance responsibility and performance guarantee bond/ collateral warranties where applicable (A20).

Items A30–37 prescribe in detail the employer's requirements or limitations including tendering and subletting; provision, content and use of documents; management of the works; quality standards and control; security, safety and protection; method, sequence and timing; facilities, temporary work and services; and operation and maintenance of the finished building. There is provision for the separation of all these items into fixed and time related charges.

The contractor is given the opportunity to price a wide range of general cost items in A40–44, to which the contractor can add such further items as are considered necessary. The cost of these items may be separated into fixed and time related charges. For example, the schedule of mechanical plant items in A43.1.1–8, ranging from cranes to concrete and piling plant, provides a useful checklist for pricing purposes. Some items can appear either as employer's requirements or contractor's general cost items, since work such as temporary hoardings may on occasions be fully defined by the tender documents and in other instances be left to the contractor's discretion.

Items A50–55 cover work undertaken or products supplied by or on behalf of the employer; nominated subcontracts containing descriptions of the work, provision for main contractor's profit and details of any special attendance required; details of materials to be supplied by nominated suppliers with provision for main contractor's profit; work by statutory authorities and undertakers; provisional work and dayworks. A single item is provided in the preliminaries bill for general attendance on all nominated subcontractors and suppliers. Prime cost and provisional sums are described in chapter 2 and dayworks are covered later in this chapter. The use of prime cost items is illustrated in examples 19.4, 20.1, 22.2 and 24.8 and provisional sums in examples 24.1 and 26.9.

General Summary

Where the bill of quantities is annotated in the SMM7 work sections, the General Summary can take one of two different forms. The simplest approach is to provide monetary collections at the end of each group of work sections, such as D: Groundwork and F: Masonry, the totals of which are then transferred to the General Summary. An example of this format of General Summary is illustrated and shows the discontinuity in the groups of work section references which some surveyors may find off-putting after being accustomed to using a numerical sequence of bill numbering.

The second approach is to provide collections at the end of each work section with the totals transferred to the General Summary. This makes for a very long and fragmented General Summary as, for example, Group E (*In-situ* concrete/Large precast concrete) could be subdivided into a number of separate work sections, such as E10 Mixing/Casting/Curing *in-situ* concrete;

GENERAL SUMMARY

		page	£	p
D	GROUNDWORK			
E	*IN-SITU* CONCRETE/LARGE PRECAST CONCRETE			
F	MASONRY			
G	STRUCTURAL/CARCASSING METAL/TIMBER			
H	CLADDING/COVERING			
J	WATERPROOFING			
K	LININGS/SHEATHING/DRY PARTITIONING			
L	WINDOWS/DOORS/STAIRS			
M	SURFACE FINISHES			
P	BUILDING FABRIC SUNDRIES			
Q	PAVING/PLANTING/FENCING/SITE FURNITURE			
R	DISPOSAL SYSTEMS			
S	PIPED SUPPLY SYSTEMS			
T	MECHANICAL HEATING SYSTEMS			
V	ELECTRICAL POWER/LIGHTING SYSTEMS			
	TOTAL CARRIED TO FINAL SUMMARY			

FINAL SUMMARY

		£	p
1.	PRELIMINARIES/GENERAL CONDITIONS		
2.	PRIME COST AND PROVISIONAL SUMS		
3.	MEASURED WORK*		
4.	CONTINGENCIES		
Add	Insurance against injury to persons and property Insurance of the works against clause 22 perils		
	TOTAL CARRIED TO FORM OF TENDER		

*Possibly subdivided into Substructure, Superstructure and External Works.

E20 Formwork for *in-situ* concrete; E30 Reinforcement for *in-situ* concrete; E40 Designed joints in *in-situ* concrete; and E1 Worked finishes to *in-situ* concrete. This arrangement has not proved popular in practice.

It is customary for the total of the General Summary to be transferred to a Final Summary, which may take a number of different forms of which the format illustrated is typical. The total of the Final Summary is carried to the Form of Tender.

Preambles

The purpose and nature of preamble clauses at the head of each work section in the bill were described in chapter 2. It would perhaps be helpful to the student if a few typical clauses were provided and the Excavating and Filling Bill has been selected for this purpose.

A. Trial holes have been excavated in the positions indicated on Location Drawing G.16, on which are also shown the levels and details of soil, which consist of an average of 225 of topsoil overlying well graded sand. Groundwater level was 295.40 AOD on 15 March 19–.
B. The quantities of excavation, including disposal of excavated material have been measured nett and the Contractor is to allow in his billed rates for any increase in bulk and for any additional excavation that may be required for working space, other than that provided in accordance with the *Standard Method of Measurement of Building Works*.
C. Excavation rates shall include for grubbing up and removing any tree roots and similar obstructions that may be encountered.
D. Excavations shall be inspected and approved by the Architect and the representative of the Local Authority before any concrete or filling is laid, and any variations measured by the Quantity Surveyor.
E. If any excavations are carried out to a greater width or depth than directed or required, then the Contractor is to make up or fill in to the required width or depth with concrete (1 : 12) at his own expense, where directed by the Architect, and no payment will be made for the additional excavation.
F. Hardcore shall be clean, hard brick or stone broken to pass a 75 diameter ring and be well graded and compacted in layers not exceeding 150 thick, with a 800 kg vibrating roller running at a slow speed with a minimum of three passes over each layer.

Note: This is a selection of clauses which are not necessarily comprehensive and they will vary from one contract to another. As mentioned previously, such preamble clauses are frequently incorporated into a project specification. This is often in a separately bound volume from the priced sections of the bill of quantities, but nonetheless remains as part of the bill of quantities, even though in some cases its contents may have been compiled by the architect or engineering consultant where Co-ordinated Project Information practices have been followed. Typical modern practice is to prepare specifications using standard reference systems such as the National Building Specification (NBS) and the National Engineering Specification (NES).

Daywork

For the valuation of variations which cannot properly be measured and valued at billed rates or rates derived from them, the Contractor will normally be allowed to charge daywork in accordance with Clause 13.5.4.1 of the JCT Conditions of Contract.

The basis of charging is normally the prime cost of such work as calculated in accordance with the *Definition of Prime Cost of Daywork carried out under a Building Contract* together with percentage additions to each section of the prime cost at the rates inserted by the Contractor in the Bill of Quantities. The rates for plant shall be those contained in the *Schedule of Basic Plant Charges*.

The Contractor will therefore add to the provisional sums contained in the Bill the percentage addition required for overheads and profit. The provisional sums and the percentage additions are then monied out and the total carried to the Summary of the Bill of Quantities, in the manner shown in the Schedule of table 17.1. In the bill of quantities a statement will precede the details contained in table 17.1, explaining when variations will be valued on daywork and the basis of charging, as already outlined.

Note: The tendered percentages contained in the latter part of the schedule refer to work which will be undertaken under different conditions from those prevailing during the contract period, since the contractor will have taken his workforce, materials and plant off the site following practical completion of the work. Hence it is likely that he will wish to insert higher percentage rates than those in the main part of the Schedule.

Specialist Bills

On occasions, specialist work such as an electrical installation, arising from a prime cost sum in the main bill, may be measured in detail and the main contractor may be given the opportunity to price this work in addition to specialist subcontractors. The arrangement of the specialist bill is similar to a normal bill with comprehensive preliminary and preamble clauses covering such matters as the relationship of the specialist to the main contractor, the provision of general services and facilities by the main contractor, details of special attendance, programme and method of payment.

Table 17.1   *Daywork schedule.*

				£	p
Labour Provide the Provisional Sum of £3000.00 for Labour.				3000	00
Add for overheads and profit.		%			
Materials Provide the Provisional Sum of £2000.00 for Materials				2000	00
Add for overheads and profit.		%			
Plant Provide the Provisional Sum of £1000.00 for Plant.				1000	00
Add for overheads and profit.		%			
The foregoing Provisional Sums and percentage additions apply to daywork ordered by the Architect prior to the commencement of the Defects Liability Period. Should any dayworks be ordered after the commencement of the Defects Liability Period it is proposed that the Definition shall also apply to such work and the Contractor is invited to insert the percentage additions that would be required to the Prime Cost for overheads and profit.					
Labour		%			
Materials		%			
Plant		%			
DAYWORK CARRIED TO SUMMARY		£			

MODIFIED TRADITIONAL PROCESSES

New measurement and processing techniques have been introduced progressively during the last three decades and are now being used to an increasing extent, since they have resulted in a speeding up of the measurement operations and a reduction in the overall cost of preparing bills of quantities.

Over the years many quantity surveyors have experimented with a number of systems designed to eliminate part of the working up process. One such system eliminated the abstract by direct billing/billing direct as described earlier in this chapter.

Another method was to abstract from standard dimensions paper on to specially ruled sheets or cards, designed to receive only one item per sheet or card. Where repeat items occurred the quantity was recorded with the original item, enabling the total quantity to be obtained eventually. As the abstract was prepared the sheets or cards were sorted into bill order. On completion of the abstract, any necessary editing was done and the bill typed direct from the abstract, resulting in the elimination of both billing and the checking of the bill.

Each of these systems suffered some limitations in use and could not therefore be applied universally.

Cut and Shuffle

General Arrangements
The system of 'cut and shuffle' was developed in the early 1960s and by the late 1970s was probably one of the most widely used methods of entering dimensions and descriptions. It has been aptly described as a 'rationalised traditional' procedure. Unlike abstracting and billing there is no universally accepted format and many different paper rulings and methods of implementation are used in different offices. Some offices even use more than one system to suit different types of work.

However, the following characteristics apply to most systems:

(1) Dimensions paper is subdivided into four or five separate sections which can subsequently be split into individual slips.
(2) Only one description with its associated dimensions is written on each section.
(3) Dimension sheets are subsequently split into separate slips and sorted into bill work sections or elements and eventually into bill order.
(4) Following the intermediate processes of calculation and editing, the slips form the draft for typing the final bill of quantities.

The cut and shuffle system is designed to eliminate the preparation and checking of the abstract and draft bill. Hence there is only one major written operation, namely taking off, compared with the three involved with abstracting and billing. By 1997 this system had been largely superseded by

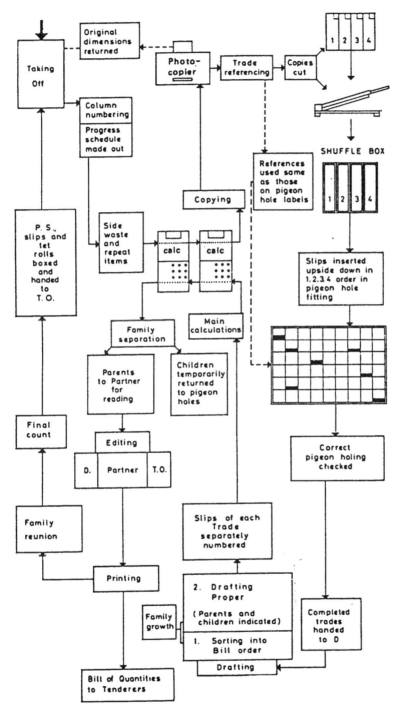

Figure 44 *Cut and shuffle procedure (Source: B. D. Henderson, Cut and Shuffle System paper, RICS Annual Conference, 1961).*

361

1000 gauge polythene
lapped 150 at joints
Tanking and damp-proofing
horizontal; laid on
concrete bed

4.50	
3.80	(Ward A
20.00	(do.
6.00	
5.50	(Pass.
2.00	
23.80	(Ward B
6.00	
5.50	(do.
3.80	
7.60	(Pass.
2.00	
8.90	(Office
7.70	
6.90	(Toilet
5.40	

1000 gauge polythene
Tanking and damp-proofing
a.b. (Col. 233)

20.00	(Ward C
6.00	
5.50	(do.
4.20	
7.60	(Pass.
2.20	

Cement and sand (1:3 - 4.5);
one coat; trowelled

Floors
25 thick
lev. or to falls only
≤15° from horizontal
on polythene

Sq. — J40.1 & 2

Cork floor tiles; Wicanders
ready sealed, medium density;
300 x 300 x 4.8 mm; fixed
with adhesive, in accordance
with manufacturer's
instructions; straight
butt jointed both ways

Floors
>300 wide;
lev. or to falls only
≤15° from horizontal;
to screed

Sq. — J40.1 & 2

EXAMPLE 28

computerised methods of bill production, although a BCIS survey of UK quantity surveying practices found that a number of offices were still using 'cut and shuffle' methods.

Detailed Procedure
One method of carrying out the system is described in detail below, with the procedure illustrated diagrammatically in figure 44.

(1) Taking off is usually carried out on A4 size sheets of dimensions paper, ruled vertically into four sections and thus accommodating a maximum of four items per sheet. Dimensions are entered on one side only of each sheet and each column is generally stamped with the project reference number and numbered consecutively. Ditto items must include a reference to the column number of the main item, where full particulars can be found. A typical cut and shuffle dimension sheet is illustrated in example 28.

(2) As sections of the taking off are completed, the side casts are checked and repeat dimensions calculated.

(3) When a taking off section is complete, each column is marked with the work section reference and column number. A copy of each dimension sheet is obtained generally either by using NCR (no carbon required) paper or by photocopying. However, some systems operate without the need to produce a copy.

(4) The taker off retains the copy and the original sheet is cut into four slips each containing one item. Some surveyors use sheets that are already perforated.

(5) The slips are shuffled or sorted into work sections such as Excavating and Filling and Mixing/Casting/Curing *in-situ* concrete. Similar items are collected together and the whole of the slips placed as near as possible in bill order.

(6) When all the slips for an individual work section have been sorted they are edited to form the draft bill, with further slips being inserted as necessary to provide headings, collections and other relevant items. The correct unit of billing is entered on the 'parent' or primary item slip and the 'children' or repeat item slips are marked 'a.b.' (as before). As each section is edited it is passed on for squaring.

(7) Quantities are then squared, totals cast and reduced to the nearest whole unit, with the reduced quantity entered on the parent slip. The calculation process is then checked.

(8) Parent and children slips are separated. The parent slips form the draft bill and are ready for typing.

(9) Any further checks on the draft bill are then carried out and final copies made and duplicated.

(10) The children slips are then replaced to provide an abstract in bill order for reference purposes during the post contract period.

The principal merits and demerits of the cut and shuffle system are now listed.

Merits
(1) It is claimed that the time taken from the start of taking off to the production of the finished bill is shorter than by traditional methods; working up time is definitely reduced.
(2) Dimensions and descriptions are only written once and not three times as with traditional methods, and it also eliminates the preparation and checking of the abstract and billing.
(3) The working up section completes almost the whole process after the taking off stage. As soon as the taking off for each section of work is completed, it can be shuffled, edited, typed, read back and duplicated, so that only a short time after all taking off is completed, the bill is ready for distribution. Previously it was often left to the taker off to bill the abstract after completing the taking off.

Demerits
(1) It is often found that although the time spent in working up is much reduced, the taking off is prolonged. The main reasons for this are the comprehensive system of referencing required on each slip and the need to write full descriptions without abbreviations for all parent items.
(2) After the bill has been produced, the cut and shuffle system is not so well suited to final account work, since without a comprehensive index the location of work can be time consuming.

Standard Descriptions

Since the mid-1960s many quantity surveyors have been using standard descriptions in bills of quantities by arranging the component parts of item descriptions into a graded structure. This followed a recommendation of a working party of the Royal Institution of Chartered Surveyors to develop a logical structure of bill item descriptions to assist the construction industry in obtaining a common approved standard.

The basic approach to the graded structure of a billed item description is to adopt a number of levels, each of which contains alternative words or phrases. The following example will serve to illustrate a practical application of this concept, applying SMM7 requirements as far as practicable, although the sequence of items tends to vary with different work sections. Nevertheless, SMM7 if applied systematically as described in chapter 2 and implemented in the worked examples, does produce a standardised approach to the preparation of billed item descriptions. This can be further assisted by the use of the *SMM7 Library of Standard Descriptions*, as is also illustrated in the worked examples in this book. The actual wording will vary extensively according to whether or not there is a detailed project specification, entailing the use of numerous cross references in the bill descriptions.

The following schedule shows the application of the *SMM7 Library of Standard Descriptions* in the formulation of measured item descriptions.

Level	Content	Example
1	Work section heading in bold large type	**F10 BRICK/BLOCK WALLING**
2	Specification information in bold type but smaller than level 1	**Facing brickwork in Ibstock Laybrook Parham Red stock bricks in stretcher bond in gauged mortar (1:1:6) and pointing with neat flush joint as work proceeds**
3	Basic item in normal type	Walls
4	Variable information such as size in lower case and indented	half brick thick; facework one side

Note: In practice the descriptions in levels 1 and 2 are usually underlined instead of being printed in bold type, and this approach has been adopted throughout the worked examples in this book.

The standard description format would then appear in the bill of quantities in the following way:

Item	Description	Qty.	Unit	Rate	£	p
	F MASONRY					
	F10 BRICK AND BLOCK WALLING					
	Facing brickwork in Ibstock Laybrook Parham Red stock bricks in stretcher bond in gauged mortar (1:1:6) and pointing with neat flush joint as work proceeds					
	Walls					
A	half brick thick; facework one side	46	m²			

The use of standard descriptions aids communication through the consistent approach, with resultant benefits to those concerned with the pricing and use of priced bills. The drafting and interpretation of descriptions are simplified and the editing of the draft bill is replaced by routine checks.

COMPUTERS AND BILL PRODUCTION

General Introduction

The RICS Quantity Surveyors Division Report *QS2000* in 1991 postulated that probably the biggest impact of computers on quantity surveying practice had been on improving the speed and efficiency of professional services. It also forecast that information flows in construction will increasingly be made electronically with, for example, the production of tender documents becoming increasingly automated, coupled with the use of computer aided design (CAD) systems. The application of computer technology will also assist in the development of cost modelling and cost and market forecasting.

A number of reputable systems of computerised measurement and bill production are available and their selection could be influenced by a number of factors, such as general appearance of output, logicality of processes, whether the system is subject to continual review and upgrading, extent of user base, whether user requirements are met, and competitiveness of the price of the system. Two systems were used extensively by the quantity surveying organisations approached by the author, namely CATO and Masterbill and, in 1997, both firms had introduced Windows '95 versions in CATOPro and Masterbill '97 respectively, offering increased versatility. In the descriptions that follow, I have adopted the Masterbill system to illustrate the operation of the process, with assistance from the Quantity Surveying Department of North Yorkshire County Council, a user of the system. Unfortunately, restrictions of space preclude consideration of more than one system.

Setting up a Project

The main steps in the process, wherein the suite of Masterbill programs are retained permanently on hard disk, are as follows:

(1) The project information screen is completed with the basic project information, such as the project title, project number and description of the works.
(2) The library to be used is loaded. This can either be the Masterbill Standard Library, the user's own standard library or a previous project of a similar nature. If the latter is used, the dimensions, bill titles and prices will require deleting immediately after the details have been loaded.

(3) Control lists are set up with particular reference to the following features:

   (a) *Parts*: this facility enables parts of the project to be identified separately, as for example, new extension, external works and drainage.
   (b) *Elements*: elemental listings for cost analysis purposes, probably using the BCIS element list.
   (c) *Phrases*: this facility stores standard phrases which can be added to the ends of descriptions, such as fixing with screws.

(4) New library descriptions can be inserted at this point by choosing a code which will insert the new description into the library at the appropriate location and either typing in the new description from scratch or copying and altering an existing description. It will often be simpler to insert new library descriptions during the take off when the nature of the required descriptions becomes apparent.

Take Off

There are five alternative approaches to taking off:

(1) Precoded input where written take off is done on paper and codes are written down with shortened descriptions, the input into the computer being done at a later stage.
(2) Direct entry of written uncoded dimensions.
(3) Direct input where taking off is done through the keyboard by the taker off, which North Yorkshire have found to be the most efficient of approaches (1)–(3), and as illustrated in figure 45 showing the dimensions for a concrete plain tiled roof covering, with the dimensions totalled.
(4) Digitiser linking where a separate digitiser package is used for measurement and quantities calculated from drawings, and the results are imported.
(5) AutoCAD linking where measurements are imported directly from CAD, and quantities are measured and calculated from drawings 'on screen', and results are imported.

The dimensions are stored in dimension files, and usually identified by the taker-off's initials and the part and element being taken off, to create a unique file. While descriptions are selected a level at a time on the screen from the library. As each successive level of the library is selected, the system moves on to the next level until four description levels have been selected. At this stage the inputting of measurements can begin. A fifth level is available for extra over items. It is possible to make the library smaller, more adaptable and more manageable by inserting variable markers throughout the library descriptions, to allow the input of sizes or words, such as timber sizes or trade references.

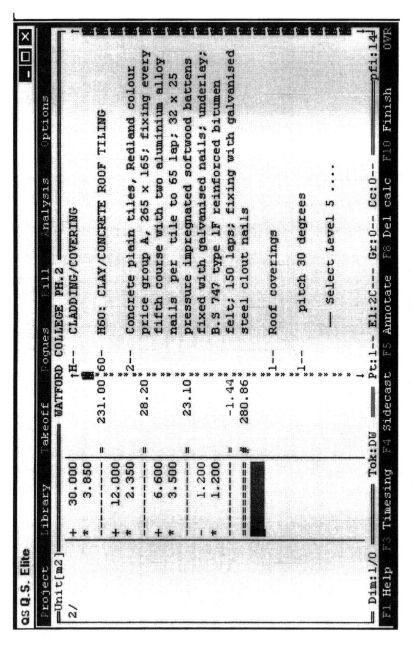

Figure 45   *Masterbill concrete roof tiling dimensions on screen.*

The Masterbill system allows traditional approaches to the task of taking off using timesing, sidecasts and annotations on a vertical dimension paper format on screen. This ensures that non-computer-literate surveyors recognise the method of taking off, thereby reducing reluctance to the use of modern technology. Dimensions and descriptions can be replicated from previous take off items and amended to suit. The system has two styles of 'anding on' dimensions, one which looks to the previous dimension item and another which looks at the first dimension item in a string of dimension sheets. The latter approach could be used for example where a m^2 item is converted into a m^3 quantity by a common factor and then the m^2 quantity is required again.

Dimension stores can be copied to create new stores which are useful when identical items appear in different parts of the work, and they can also be deleted, multiplied and undeleted when previously deleted, without any loss of information. Items measured in the wrong element can be amended to the correct element, even within an elemental take off file. Dimensions within dimension stores can be listed on the screen and access can be gained to any dimension, and can be printed out either complete, as a single dimension or as a range of dimensions. If alterations are made, all dimensions can be amended to suit and additional quantities added to them, with 'andings on' automatically corrected.

Abstract/Bill of Quantities

Individual bills are selected from a previously prepared list of combinations of parts, such as by merging several types of external works to give the external works bill. 'Default bill preferences' allow the user to configure normal requirements for all bills on the project, such as firm price or fluctuating price contract. Draft bills allow takers off to view and print just a particular take off section in bill format for checking, if required. Bills of quantities can be set up in different combinations, such as in workgroups, elements or a mixture of both, and different parts previously set up for take off purposes can be merged to form a single bill.

The bill can be viewed on the screen and a particular item identified for amendment. The bill can also be searched by requesting a particular bill page, as illustrated in figure 46, using a Windows '95 version embracing excavating and filling work, or a dimension or even a string of words. If requested by the contractor, the bill of quantities can be provided on disk for input directly into a computerised estimating system. Alternatively, the bill of quantities can be provided on disk, whereby the tenderer can enter the rates directly into the system thereby saving time after receipt of tenders.

The bill can be priced as a pretender cost check, as shown in figure 46, and on receipt of tenders the rates submitted by the various tenderers can be added. This enables a comparison to be made between the estimate and

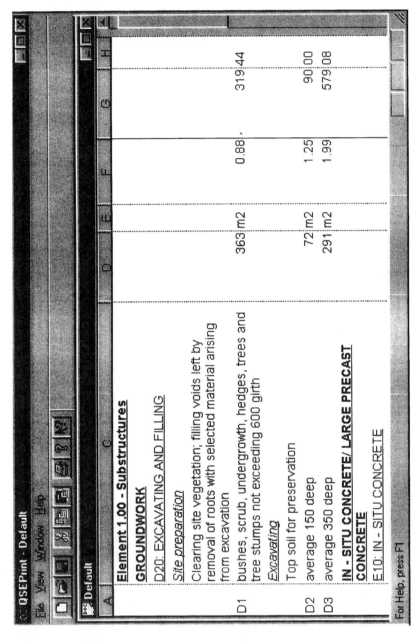

Figure 46 *Masterbill part of substructures bill on screen.*

the various tenders. A request for all prices outside chosen tolerances to the pretender estimate quickly highlights high or low rates. The tenders can be analysed to give costs in an elemental format which can be exported directly on to a spreadsheet for further analysis. In addition, a full post contract package is available on the system to accommodate valuations, recording of variations and costs and the production of final accounts and financial statements.

Table 17.2 illustrates a Masterbill '97 sample bill page relating to substructures and encompassing a number of different work sections. Level 2 specification headings are underlined adopting the same approach as has been used in the worked examples in this book. A number of different print faces are available. Estimated prices have been inserted as pretender cost checks.

Table 17.3 (pages 374 and 375) shows computerised bill items of *in-situ* concrete for workshops in Nottingham, where a comprehensive project specification has been prepared as a separate document. Hence the measured item descriptions are reduced significantly in length by the inclusion of the appropriate specification clause references or codes. The format of the bill follows the recommendations contained in the *SMM7 Library of Standard Descriptions*, as used in the worked examples throughout this book. The measured work is entered under the appropriate SMM7 work section headings (E10, E20, E30, E41 and E60), each being in block capitals underlined as level 1 headings and specification headings underlined as level 2 headings. Level 3 headings describe the basic type of work such as foundations or beds, followed by level 4 variable information, such as sizes, suitably indented. The priced totals at the bottom of each of the two pages are transferred to a collection on the final page, where the combined total for *in-situ* concrete will be transferred to the Bill Summary.

Principal Masterbill Features

Masterbill list the principal features of their system and these include the following impressive characteristics:

- Simplicity of operation: starting with a sheet of on-screen dimension paper
- Uses traditional quantity surveying techniques throughout
- Mouse or keyboard driven: direct or written take off or mixture of both
- No coding (screen selection of items)
- Automatic coding of rogues
- Take off schedule facility
- Change the dimensions on one screen and the bill is changed automatically
- Change the schedules and all items using the scheduled quantities will be automatically changed as well
- Direct and immediate interaction between dimensions, abstract and bills

Table 17.2   *Masterbill sample billing sheet for substructures.*
SAMPLE B.Q. PAGE (Windows '97 version)

					£	p
	Element 1.00 − Substructures					
	GROUNDWORK					
	D20: EXCAVATING AND FILLING					
	Site preparation					
	Clearing site vegetation; filling voids left by removal of roots with selected material arising from excavation					
D1	bushes, scrub, undergrowth, hedges, trees and tree stumps not exceeding 600 girth	363	m^2	0.88	319	44
	Excavating					
	Top soil for preservation					
D2	average 150 deep	72	m^2	1.25	90	00
D3	average 350 deep	291	m^2	1.99	579	08
	***IN-SITU* CONCRETE/LARGE PRECAST CONCRETE**					
	E10: MIXING/CASTING/CURING					
	IN-SITU CONCRETE					
	Reinforced *in-situ* concrete; BS 5328; designed mix C20, 20 aggregate, minimum cement content 240 kg/m^3; vibrated					
	Foundations poured on or against earth or unblinded hardcore					
E1	generally	244	m^3	63.00	15 372	00
E2	generally; (PROVISIONAL)	100	m^3	63.00	6 300	00
	Reinforced *in-situ* concrete; sulphate resisting; BS 5328, designed mix C25, 20 aggregate, minimum cement content 290 kg/m^3; vibrated					
	Foundations poured on or against earth or unblinded hardcore					
E3	generally	518	m^3	65.25	33 799	50
	Beds; poured on or against earth of unblinded hardcore					
E4	thickness 150–450	854	m^3	78.00	66 612	00
	To Collection			£	123 072	02

Table 17.3 *Typical billed sheets for* in-situ *concrete.*

Item				Bill 3			(E10)	
	BULWELL ENTERPRISE WORKSHOPS Job EED42						£	p
	E10: MIXING/CASTING/CURING _IN-SITU_ CONCRETE							
	Plain Spec. 130A Slabs							
A	150 to 450 thick	1	m^3					
	Plain Spec. 131A Beds							
B	not exceeding 150 thick; sloping over 15 degrees; poured on or against earth or unblinded hardcore	13	m^3					
C	not exceeding 150 thick; poured on or against earth or unblinded hardcore	37	m^3					
	Plain Spec. 134A Filling hollow walls							
D	not exceeding 150 thick	2	m^3					
	Reinforced Spec. 130A							
E	Foundations	126	m^3					
	Beds							
F	150 to 450 thick	156	m^3					
	E20 FORMWORK FOR _IN-SITU_ CONCRETE Formwork Spec. 610 Sides of foundations							
G	250 to 500 high	253	m					
H	250 to 500 high; sloping over 15 degrees; left in	528	m					
J	500 to 1 m high	15	m					
		To Collection £						
	Job Ref. BEC page 3/54							

(continued)

Table 17.3   (*Continued*)

Item		BULWELL ENTERPRISE WORKSHOPS Job EED42	Bill 3			(E20) £	p
		Sides of ground beams and edges of beds					
A		not exceeding 250 high	257	m			
B		250 to 500 high	250	m			
		E30 REINFORCEMENT FOR *IN-SITU* CONCRETE					
		Reinforcement Spec. 160A Bars					
C		12 nominal size; straight	2.82	t			
D		12 nominal size; bent	0.67	t			
		Reinforcement Spec. 160B Bars					
E		8 nominal size; links	2.17	t			
		Reinforcement Spec. 200 Fabric					
F		Ref A252	1540	m^2			
		E41 WORKED FINISHES/CUTTING TO *IN-SITU* CONCRETE					
		Spec. 350A					
G		Power floating	782	m^2			
		E60 PRECAST/COMPOSITE CONCRETE DECKING					
		Spec. 110 Slabs					
H		150 thick	53	m^2			
			To Collection £				
		Job Ref. BEC page 3/55					

- Reuse original dimensions for partial or complete remeasurements
- Multi-master libraries available
- Reuse of previous project data
- Full cost analysis feature
- Multi-sortation of documents in user defined format, such as elements, blocks, workgroups, or any combination
- Runs on industry standard computers and operating systems; Windows compatible
- Stand alone or networked
- Digitising facility with direct input to dimensions and bill
- Direct link to AutoCad
- Export data to spreadsheets and other programs.

Conclusions

The Masterbill '97 and CATOPro systems are intended to support direct take off, without any intermediate manual measurement processes. They have further been designed to work with digitisers and CAD, and users have the benefit of a range of standard phrase libraries to suit their chosen method of measurement. They also provide sortation facilities for subsequent post measurement analysis. In general, these computer based measurement and billing systems eliminate the working up process, with varying further savings in measurement depending on the type and complexity of project. The large user base of both systems shows very clearly the widespread use of computerised taking off and billing, with the added advantages generated by the Windows versions.

ALTERNATIVE BILL FORMATS

This final chapter concludes with a conspectus of the various alternative bill formats that have been introduced in the last three decades, mainly aimed at securing greater value to the contractor by simplifying the pricing of the work at the tender stage and by providing greater benefits at the construction stage.

Elemental Bills

As described in chapter 1, an elemental bill of quantities is divided into appropriate building elements instead of the normal work sections. Hence Excavating and Filling, Mixing/Casting/Curing *in-situ* concrete and Brick/Block Walling are replaced by such bill headings as Substructure, External Walls, Internal Walls and Floors. There are benefits to be gained by using the RICS Building Cost Information Service (BCIS) elements and subelements to secure standardisation and assist with cost planning and cost analysis work. Within each element work may be billed in order of work sections or grouped in building sequence. The principal objective is to secure more precise tendering by making the location of the work more readily identifiable and to provide a closer link with the cost plan.

In practice this bill format was not very well received, since it involved considerable repetition of billed items, and where work was let to a subcontractor, it was necessary to prepare an abstract of items prior to obtaining quotations. In like manner an estimator had to examine the elements in some detail to collect together all like items before he could assess the total quantities of each activity and/or material. For example, there can be brick walls in both Substructure and External Walls elements.

Operational Bills

Operational bills were developed by the Building Research Establishment in the 1960s and they subdivided the work into site operations as distinct from trades or elements. Labour and sometimes plant requirements were described in terms of the operations required, together with a schedule of the materials for each operation. Operations were defined as the work done by a man, or by a gang of men, at some definite stage in the building process. The sequence of operations was often illustrated in the bill by means of a precedence diagram, which showed their interrelationship.

It was considered by the main protagonists of the method that the contractor could more readily appreciate the implications of the design at the time estimates were prepared, and that the more detailed information would be of considerable value to the successful contractor on the site, without curtailing the contractor's freedom to select the best constructional techniques.

It involved a fundamental divergence from a traditional bill, with the separation of labour and material items and required significant changes in the rules of measurement prescribed in the *Standard Method*. The whole of one work section did not appear together and so it was customary to print each work section on differently coloured paper to facilitate the location of items.

Operational bills were both bulky and costly to produce. On the other hand, typing and printing of the bill could be started at an earlier stage and there was a reduction in the amount of editing necessary. It also simplified the work of making interim valuations. They generally increased the work of the contractor's estimating department, although the estimator obtained a clearer picture of the work involved and this should enable him to produce more realistic estimates, and also to ease considerably the task of obtaining quotations for materials. From the site organisational viewpoint, the contractor was helped since his programme was linked automatically with the bills and his manpower requirements were readily assessed from the operations.

In practice the response to this particular bill format was very disappointing. The need for a complete set of drawings prior to the preparation of the bills, coupled with a delayed start on the site, were probably felt to outweigh the possible advantages of a faster and cheaper completion. There was also the natural reluctance to make such a radical change from traditional methods, and it would not conform to normal estimating practice.

Activity Bills

The activity bill was a development of the operational form but without the separation of labour and materials. It was subdivided into sections based on activities or operations derived from a network analysis. The work was measured in accordance with the *Standard Method*, although on site and off site activities were usually separated and special equipment and components and the work of nominated specialists could be grouped in separate bills.

Some activities need to be completed before others could be started. For example, foundation trenches need to be excavated before the concrete foundations can be laid, which must in turn precede the building of brickwork. A graphical representation in the form of a network analysis showed the interrelationship and sequence of activities. Network analyses for the smaller and less complex projects could be produced by the design team, but those for larger and more complicated contracts generally required liaison with the contractor, which could militate against competitive tendering.

Annotated Bills

It is possible for the bill of quantities to give the contractor full details of the quantity, type and quality of materials and labour, and for an accurate and complete set of drawings to show him precisely where and how the work is to be executed. Nevertheless, there are always some billed items whose location in the works is not readily identifiable, and it is most useful to have a note against them in the bill giving their location. This approach has resulted in the production of annotated bills.

Annotations may be prepared in a separate document from the bill of quantities or they may be bound in at the back of the bill, although in either case they must be carefully cross referenced. Another alternative is to interleave annotations into the bill so that the notes appear opposite the relevant bill items. This provides the clearest and most helpful method.

General Conclusions

These more specialised types of bill format each have their own particular attributes and, in certain situations, a valuable role to perform in tendering procedures and contract administration. Nevertheless, the traditional work section format remains the one most widely used and is likely to remain so for the foreseeable future.

On the other hand, tendering arrangements have been subject to considerable change away from *open tendering* – previously mainly practised by public authorities who advertised publicly and invited tenders from all respondents – towards *selective tendering* – whereby a limited number of contractors of known ability and financial standing are invited to submit tenders – and *negotiated tendering* – where only one contractor is involved and the benefit of contractor participation at the design stage can be achieved. Within the negotiated tender approach there are a number of alternatives including two stage tendering and continuity contracts, which embrace serial, continuation and term contracts; all of which are described in *Quantity Surveying Practice* by the present author.

The National Joint Consultative Committee for Building (NJCC) has advised that contractual difficulties may be encountered where contractors tender for certain categories of works based on drawings and specifications alone, although it is accepted that bills of quantities may not always be warranted for small or relatively simple building works. It is further recommended that where tenders are invited without bills of quantities it may well be desirable instead to require the tender price to be broken down into the major constituent parts for ease of cross checking and comparison with other tenders and to require the pricing of a schedule of rates to assist in the valuation of any variations.

The main benefits of bills of quantities are that they provide a common basis for all tenderers and relative certainty of price for the client. For this to be achieved the bills must be based on a substantially complete design. If bills are prepared where this is not the case it is likely that the contractor will be misled as to the true nature of the project and its costs. It is also probable that the client will not achieve certainty of price because of the need to include too high a proportion of the work in the form of provisional and prime cost sums and approximate quantities. As a consequence of such factors the proportion of bills of quantities contracts is declining, but the need for measurement is not, this function being transferred instead to the contractor.

Appendix 1:
List of Abbreviations

a.b.	as before
a.b.d.	as before described
acc.	accordance
adeq.	adequate
adhes.	adhesive
adj.	adjoining/adjustment
agg.	aggregate
agst.	against
alum.	aluminium
appvd.	approved
apt.	apart
a/r	all round
arch.	architect
archve.	architrave
ard.	around
arisg.	arising
asp.	asphalt
assctd.	associated
attchd.	attached
av.	average
bal.	baluster
B.C.	bayonet connection
bd.	bond/board/bead
bdg.	boarding
bellmth.	bellmouth
b.i.	build in
bit.	bitumen
b & j	bed and joint
bk.	brick
bkg.	breaking
bkt.	bracket
bldg.	building
blindg.	blinding

blk.	block
blkwk.	blockwork
bott.	bottom
b & p	bed and point
br.	branch
brr.	bearer
b.s.	both sides
b & s	bed and surround
bt.	boot
btm.	bottom
btwn.	between
bwk.	brickwork
cap.	capacity
cast.	casement
cat.	catalogue
cav.	cavity
ccs.	centres
centrg.	centring
chan.	channel
chbr.	chamber
chfd.	chamfered
chrom.	chromium
chy.	chimney
c.i.	cast iron
circ.	circular
cistn.	cistern
₵	centre line
clg.	ceiling
c.l.m.	cement : lime : mortar
cln.	clean
clr.	clear
c.m.	cement mortar
cnr./crnr.	corner
col.	column
comb.	combined
comm.	common
commsg.	commissioning
comp.	composite
compactg.	compacting
compd.	compound
compg.	comprising
conc.	concrete
conn.	connection

constn.	construction
cont.	continuous
cop.	copper
copg.	coping
cos.	course
cov.	cover
covd.	covered
cplg.	coupling
cr.	contractor
c/s	cement/sand
csd.	coursed
c.s.g.	clear sheet glass
c.s.v.	circular soffit ventilator
ct.	cement/coat
cub.	cubic
cultvtg.	cultivating
cupd.	cupboard
c.w.	cold water
dble.	double
ddt.	deduct
descd.	described
dia.	diameter
diagm.	diagram
dim./dimsd.	dimensioned
disch.	discharge
disp.	disposal
dist.	distance
distbn.	distribution
ditto./do.	as before
dn.	down
dom.	domestic
dp.	deep
dpc	damp-proof course
dr.	door
drg./dwg.	drawing
drn.	drawn/drain
ea.	each
edg'g	edging
effect.	effective
e.m.l.	expanded metal lathing
emuls.	emulsion
enam.	enamelled

eng.	engineering
Engl.	English
e.o.	extra over
equip.	equipment
E/S	ensuite
e.w.s.	earthwork support
ex.	extra
exc.	excavate
excvn.	excavation
excvtd.	excavated
excvtg.	excavating
exp.	exposed
expd.	expanded
ext.	external
facewk.	facework
f.bd.	fibreboard
f/c	final coat
fcd.	faced
fcg.	facing
fdn.	foundation
f.f.	first floor/fair face
fillg.	filling
fin.	finish/finished
fittg.	fitting
Flem.	Flemish
flex.	flexible
flgd.	flanged
flr.	floor
fltd.	floated
follg.	following
formg.	forming
fr.	frame
furn.	furniture
fwd.	forward
fwk.	formwork
fxd./fxg.	fixed/fixing
g.	gauge
galv.	galvanised
gen.	general
g.f.	ground floor
g.l.	ground level
G.L.S.	general lighting service

glzg.	glazing
g.m.	gauged mortar
gran.	granular
grd.	ground
grp.	group
grtd.	grouted
grtg.	grouting
grvd.	grooved
grve.	groove
gtg.	grating
gth.	girth
gyp.	gypsum
H.A.	highway authority
h.b.	half brick
h.c.	hardcore
hd.	hard/head/hand
hdd.	headed
hdrl.	handrail
hg.	hung
hi.	high
hipd.	hipped
hkd.	hooked
hnchg.	haunching
holl.	hollow
hor.	horizontal
h.p.	high pressure
hps.	heaps
h.r.	half round
hsd.	housed
ht.	height
h.w.	hollow wall
hwd.	hardwood
imgry.	ironmongery
impreg.	impregnated
inc./incl.	including
incorp.	incorporating
individ.	individual
in-situ	in position
instln.	installation
instrs.	instructions
insul.	insulation
int.	internal

intersec.	intersection
invt.	invert
irreg.	irregular
isol.	isolated
jb.	jamb
jst.	joist
jt.	joint
jtd.	jointed
jtg.	jointing
junctn.	junction
k.o.	knock out
k.p.s.	knot, prime and stop
la.	large
L.A.	local authority
lapd.	lapped
ld.	laid
len.	length
lev.	level
lg.	long
ling.	lining
l.m.	lime mortar
l.o.	linseed oil
lt.	light
ltwt.	light weight
L.V.	low voltage
lyg.	laying
lyr.	layer
mach.	machine
manufac's	manufacturer's
mat./matl.	material
max.	maximum
m.c.b.	miniature circuit breaker
mech.	mechanical
med.	medium
met.	metal
m/gd.	make good
m.gth.	mean girth
min.	minimum/mineral
mldg.	moulding
mo.	moulded

mors.	mortices
m.s.	mild steel
m/s	measured separately
mtce.	mortice
mtr.	mortar
mull.	mullion
MUPVC	modified unplasticised polyvinyl chloride
mxd.	mixed
narr.	narrow
nat.	natural
nl.	nail
nom.	nominal
nr.	number
nt.	neat
o/a, ov'll	overall
o.b.	one brick
obtd.	obtained
o'hg.	overhang
O.P.ct.	ordinary Portland cement
opg.	opening
opp.	opposite/opposing
optd.	operated
ord.	ordinary
o.s.	one side
overfl.	overflow
pan.	panel
patt.	pattern
pavg.	paving
P.C.	prime cost
ped.	pedestal/pedestrian
perf.	perforated/perforation
perm.	permanent
pic.	picture
pl.	plate
pla.	plaster
plabd.	plasterboard
plgd.	plugged
plg'g	plugging
pln.	plain
pltd.	plated
p.m.	purpose made

pol.	polished
poly.	polymer
posn.	position
pr.	pair
prec.	precast
pre.conc.	precast concrete
prep.	prepare
presvn.	preservation
prev.	previously
prof.	profit
proj.	projection/projecting
prov.	provide
provsnl.	provisional
p & s	plugged and screwed
pt.	paint/part/point
ptg.	pointing
ptn.	partition
purp.	purpose
PVC-U	unplasticised polyvinyl chloride
quad.	quadrant
qual.	quality
rad.	radius
rd.	round/road
rdd.	rounded
reb.	rebated
rec.	receive
rect.	rectangular
ref.	reference
reg.	regular
reinfd.	reinforced
retn.	return
rf.	roof
rg.	ring
rl.	rail
ro.	rough
rt.	right/root
rubd.	rubbed
r.w.	rainwater
RWG's	rainwater goods
san.	sanitary
scr.	screw

scrd.	screed
scrn.	screen
sd.	sand
s.e.	stop end
sec.	section
sel.	selected
serv.	service
sh.	shape
sht.	sheet
sk.	sunk
sktg.	skirting
smth.	smooth
sn.	sawn
soc.	socket
soff.	soffit
sold.	soldered
solv.	solvent
sp.	spoil/space
spec.	special
specfd.	specified
specfn.	specification
spld.	splayed
sprd.	spread
sprgs.	sprigs
sq.	square
s.s.	stainless steel
s & s	spigot and socket
s.s.o.	switched socket outlet
st/stne	stone
stagd.	staggered
staty.	statutory
std.	standard
stl.	steel
stret.	stretcher
strg.	string
strng.	straining
strt.	straight
struct.	structural
sty.	storey
suppt.	support
surf.	surface
surrd.	surround
susp.	suspended
s.v.	sluice valve/stop valve

s.w.	surface water
swd.	softwood
sym.	symmetrical
syn.	synthetic
tankg.	tanking
tapg./tapd.	tapering/tapered
tbr.	timber
t.c.	terra cotta
textd.	textured
t & g	tongued and grooved
th.	thick
thro./thrtd.	throated
tk.	tank
tr.	trench
t & r	treads and risers
trans.	transome
trpd.	trapped
trtd.	treated
trtmt.	treatment
trwld.	trowelled
tub.	tubular
u/c	undercoat
u/ly.	underlay
uncsd.	uncoursed
uprt.	upright
upstd.	upstand
u/s	underside
vac.	vacuum
vert.	vertical
vit.	vitreous
w.	with
W	watts
wd.	wood
wdw.	window
wethd.	weathered
wh.	white
wk.	work
wkg.	working
w.p.	waterproof

wrot.	wrought
wt.	weight
x-sec.	cross section
xtg.	existing
yd.	yard

Appendix 2:
Mensuration Formulae

Figure	Area
Square	$(\text{side})^2$
Rectangle	length × breadth
Triangle	$\frac{1}{2} \times$ base × height or $\sqrt{[s(s-a)(s-b)(s-c)]}$ where $s = \frac{1}{2} \times$ sum of the three sides and a, b and c are the lengths of the three sides.
Hexagon	$2.6 \times (\text{side})^2$
Octagon	$4.83 \times (\text{side})^2$
Trapezoid	height $\times \frac{1}{2}$ (base + top)
Circle	$(22/7) \times \text{radius}^2$ or $(22/7) \times \frac{1}{4}$ diameter^2 (πr^2) $(\pi D^2 / 4)$ circumference = $2 \times (22/7) \times$ radius $(2\pi r)$ $(22/7) \times$ diameter (πD)
Sector of Circle	$\frac{1}{2}$ length of arc × radius
Segment of Circle	area of sector − area of triangle

Figure	Volume	Surface Area
Prism	area of base × height	circumference of base × height
Cube	$(side)^3$	$6 × (side)^2$
Cylinder	$(22/7) × radius^2 × length$ $(\pi r^2 h)$	$2 × (22/7) × radius × (length + radius)$ $[2\pi r(h + r)]$
Sphere	$(4/3) × (22/7) × radius^3$ $(4/3\pi r^3)$	$4 × (22/7) × radius^2$ $(4\pi r^2)$
Segment of Sphere	$(22/7) × (height/6) ×$ $(3\ radius^2 + height^2)$ $[(\pi h/6) × (3r^2 + h^2)]$	curved surface $= 2 ×$ $(22/7) × radius ×$ height (h) $(2\pi rh)$
Pyramid	$\frac{1}{3}$ area of base × height	$\frac{1}{2}$ circumference of base × slant height
Cone	$\frac{1}{3} × (22/7) × radius^2 × height$ $(\frac{1}{3}\pi r^2 h)$	$(22/7) × radius ×$ slant height (l) (πrl)
Frustum of Pyramid	$\frac{1}{3}$ height $[A + B + \sqrt{(AB)}]$ where A is area of large end and B is area of small end.	$\frac{1}{2}$ mean circumference × slant height
Frustum of Cone	$(22/7) × \frac{1}{3}$ height $(R^2 + r^2 + Rr)$ where R is radius of large end and r is radius of small end. $[\frac{1}{3}\pi h(R^2 + r^2 + Rr)]$	$(22/7)×$ slant height $×$ $(R + r)$ $[\pi l(R + r)]$ where l is slant height

For Simpson's rule and prismoidal formula see chapter 3.

Appendix 3:
Metric Conversion Table

Length

$1 \text{ in.} = 25.44 \text{ mm}$ (approximately 25 mm)

$$\left(\text{then } \frac{\text{mm}}{100} \times 4 = \text{inches} \right)$$

$1 \text{ ft} = 304.8 \text{ mm}$ (approximately 300 mm)
$1 \text{ yd} = 0.914 \text{ m}$ (approximately 910 mm)
$1 \text{ mile} = 1.609 \text{ km}$ (approximately $1\frac{3}{5} \text{ km}$)
$1 \text{ m} = 3.281 \text{ ft} = 1.094 \text{ yd}$ (approximately 1.1 yd)
$(10 \text{ m} = 11 \text{ yd approximately})$
$1 \text{ km} = 0.621 \text{ mile}$ ($\frac{5}{8}$ mile approximately)

Area

$1 \text{ ft}^2 = 0.093 \text{ m}^2$
$1 \text{ yd}^2 = 0.836 \text{ m}^2$
$1 \text{ acre} = 0.405 \text{ ha}$ (1 ha or hectare $= 10\,000 \text{ m}^2$)
$1 \text{ mile}^2 = 2.590 \text{ km}^2$
$1 \text{ m}^2 = 10.764 \text{ ft}^2 = 1.196 \text{ yd}^2$ (approximately 1.2 yd^2)
$1 \text{ ha} = 2.471 \text{ acres}$ (approximately $2\frac{1}{2}$ acres)
$1 \text{ km}^2 = 0.386 \text{ mile}^2$

Volume

$1 \text{ ft}^3 = 0.028 \text{ m}^3$
$1 \text{ yd}^3 = 0.765 \text{ m}^3$
$1 \text{ m}^3 = 35.315 \text{ ft}^3 = 1.308 \text{ yd}^3$ (approximately 1.3 yd^3)
$1 \text{ ft}^3 = 28.32 \text{ litres}$ (1000 litres $= 1 \text{ m}^3$)
$1 \text{ gal} = 4.546 \text{ litres}$
$1 \text{ litre} = 0.220 \text{ gal}$ (approximately $4\frac{1}{2}$ litres to the gallon)

Mass

$1 \text{ lb} = 0.454 \text{ kg}$ (kilogram)
$1 \text{ cwt} = 50.80 \text{ kg}$ (approximately 50 kg)
$1 \text{ ton} = 1.016 \text{ tonnes}$ (1 tonne $= 1000 \text{ kg} = 0.984 \text{ ton}$)
$1 \text{ kg} = 2.205 \text{ lb}$ (approximately $2\frac{1}{5}$ lb)

Density

$1 \text{ lb/ft}^3 = 16.019 \text{ kg/m}^3$
$1 \text{ kg/m}^3 = 0.062 \text{ lb/ft}^3$

Velocity

$1 \text{ ft/s} = 0.305 \text{ m/s}$
$1 \text{ mile/h} = 1.609 \text{ km/h}$

Energy   1 therm $= 105.506$ MJ (megajoules)
                     1 Btu $= 1.055$ kJ (kilojoules)

Thermal   1 Btu/ft^2 h°F $= 5.678$ W/m^2 °C   (where W $=$ watt)
conductivity

Temperature $x°F = \frac{5}{9}(x - 32)°C$
                          $x°C = \frac{9}{5}x + 32°F$
                          $0°C = 32°F$ (freezing)
                          $5°C = 41°F$ (cold)
                          $10°C = 50°F$ (rather cold)
                          $15°C = 59°F$ (fairly warm)
                          $20°C = 68°F$ (warm)
                          $25°C = 77°F$ (hot)
                          $30°C = 86°F$ (very hot)

Pressure $1\ lbf/in.^2 = 0.0069\ N/mm^2 = 6894.8\ N/m^2$
                       $(1\ MN/m^2 = 1\ N/mm^2)$
                       $1\ lbf/ft^2 = 47.88\ N/m^2$ (newtons/square metre)
                       $1\ tonf/in.^2 = 15.44\ MN/m^2$ (meganewtons/square metre)
                       $1\ tonf/ft^2 = 107.3\ kN/m^2$ (kilonewtons/square metre)

For speedy but approximate conversions:

$$1\ lbf/ft^2 = \frac{kN/m^2}{20}\text{, hence } 40\ lbf/ft^2 = 2\ kN/m^2$$

and

$$tonf/ft^2 = kN/m^2 \times 10\text{, hence } 2\ ton/ft^2 = 20\ kN/m^2$$

Floor loadings   office floors – general usage: $50\ lbf/ft^2 = 2.50\ kN/m^2$
                  office floors – data-processing equipment: $70\ lbf/ft^2 = 3.50\ kN/m^2$
                             factory floors: $100\ lbf/ft^2 = 5.00\ kN/m^2$

Safe bearing $1\ tonf/ft^2 = 107.25\ kN/m^2$
capacity of $2\ tonf/ft^2 = 214.50\ kN/m^2$
soil $4\ tonf/ft^2 = 429.00\ kN/m^2$

Stresses $100\ lbf/in.^2 = 0.70\ MN/m^2$
in $1000\ lbf/in.^2 = 7.00\ MN/m^2$
concrete $3000\ lbf/in.^2 = 21.00\ MN/m^2$
                      $6000\ lbf/in.^2 = 41.00\ MN/m^2$

Costs

$£1/m^2 = £0.092/ft^2$

$£1/ft^2 = £10.764/m^2$ (approximately $£11/m^2$)

$£5/ft^2 = £54/m^2$

$£10/ft^2 = £108/m^2$

$£20/ft^2 = £216/m^2$

$£30/ft^2 = £323/m^2$

$£50/ft^2 = £538/m^2$

Appendix 4: Specifications for Internal Finishes

M10   CEMENT:SAND SCREEDS
To be read with Preliminaries/General conditions.

TYPE(S) OF SCREED

110   CEMENT : SAND SCREED TO HALL, LOUNGE AND DINING
Base: *In-situ* concrete
Construction: Bonded as clause 260
Minimum thickness 38 mm
Maximum thickness 58 mm
Mix:
Cement: Ordinary Portland or Portland blastfurnace.
Fine aggregate: To BS 882, grading limit M but with not more than 10% passing sieve size 150 microns
Mix proportions: $1 : 3-4\frac{1}{2}$
Admixture: Water reducing to BS 5075: Part 1, dosage to manufacturer's recommendations.
Finish: Trowelled as M10/540 to receive vinyl tiles
Soundness: Test to BS 8204: Part 1, Appendix B.
Maximum depth of indentation: 5 mm

GENERALLY/PREPARATION

210   SUITABILITY OF BASES: Before starting work ensure that:
Bases are sufficiently flat to permit specified levels and flatness of finished surfaces, bearing in mind the permissible minimum and maximum thicknesses of the screed/topping.
Concrete slabs have been allowed to dry out by exposure to the air for not less than 6 weeks.
Bases are clean and free from plaster, dirt, dust and oil.
260   BONDED CONSTRUCTION:
Shortly before laying screed completely remove mortar matrix from surface to expose coarse aggregate over entire area of hardened base slab using abrasive blasting or, for *in-situ* slabs only, pneumatic scabbling. Remove all dust and debris. Keep base slab well wetted for several hours before laying screed/topping.

FINISHING/CURING

530   SMOOTH FLOATED FINISH: Use a hand float, skip float or power float to give an even surface with no ridges or steps.
540   TROWELLED FINISH to receive applied floor finishes:
Float to an even surface with no ridges or steps. Hand or power trowel to give a uniform smooth but not polished surface free from trowel marks and other blemishes, and suitable to receive the specified flooring material.

395

Adequately protect the surface from construction traffic. If, because of inadequate finishing or protection, the surface of the screed is not suitable to receive the specified flooring material, it must be made good by application of a smoothing compound by and to the satisfaction of the flooring subcontractor. Allow for the cost of any such making good.

M20   PLASTERED COATINGS
To be read with Preliminaries/General conditions.

TYPE(S) OF COATING

211   LIGHTWEIGHT GYPSUM PLASTER:
Location: New blockwork
Background: New blockwork
Undercoat(s):
Premixed lightweight browning plaster to BS 1191: Part 2.
Thickness 13 mm
Final coat:
Premixed lightweight finish plaster to BS 1191: Part 2.
Thickness: 2 mm
Finish: Smooth as clause 780.
Accessories: Angle bead, stop beads.

280   PLASTERBOARD AND SKIM:
Location: All ceilings and stud partitions on first floor
Background: Timber joists at 450 mm and 600 mm centres and studs at 600 mm centres
Plasterboard backing: 9.5 mm gypsum baseboard with grey paper face, nail fixed as clause 610.
Skim coat:
Board finish plaster: To BS 1191: Part 1, Class B.
Proprietary reference: Thistle
Thickness: 5 mm applied in 2 coats.
Finish: Smooth
Accessories: Gyproc cove in Lounge

PREPARING BACKGROUNDS

480   ACCEPTANCE OF BACKGROUNDS: Before starting preparation and applying coatings ensure that:
Backgrounds are secure, adequately true and level to achieve specified tolerances, free from contamination and loose areas, reasonably dry and in a suitable condition to receive specified coatings.
All cutting, chasing, fixing of concealed conduits, service outlets, fixing pads and the like and making good of the background, are completed.

BACKINGS/BEADS/JOINTS/FEATURES

610   PLASTERBOARD BACKINGS:
Plasterboard: To BS 1230: Part 1.
Ensure that perimeter and unbound or cut edges of boards are fully supported by additional noggings as necessary.
Ensure that noggings, bearers, etc. required to support fixtures, fittings and services are accurately positioned and securely fixed.
Arrange boards with bound edges at right angles to supports, end joints staggered between rows and a gap of 3 mm between boards.

Working from the centre of each board, fix to all supports at not more than 150 mm centres with 2.6 mm diameter galvanised clout nails of length not less than 3 times the thickness of the board being fixed.
Set heads flush; do not break paper or gypsum core.

640 BEADS/STOPS:
Provide beads/stops at all external angles and stop ends except where specified otherwise.
Cut neatly, form mitres at return angles and remove sharp edges, swarf and other potentially dangerous projections.
Fix securely, plumb, square and true to line and level, ensuring full contact of wings with background.
After coatings have been applied, remove coating material while still wet from surfaces of beads/stops which are to be exposed to view.

660 SERVICE CHASES: Cover with steel mesh strip fixed at not more than 600 mm centres along both edges.

665 CONDUITS bedded in undercoat to be covered with 90 mm wide jute scrim bedded in finishing coat mix, pressed flat and trowelled in. Do not lap ends of scrim.

685 JOINTS IN GYPSUM BASE BOARD: Fill and scrim all joints (except where coincident with a metal bead) between boards.

690 JOINTS BETWEEN BOARDS AND SOLID BACKGROUNDS which are both to be plastered: fill and scrim.

PLASTERING

710 APPLICATION GENERALLY:
Apply each coating firmly to achieve good adhesion and in one continuous operation between angles and joints.
All coatings to be not less than the thickness specified, firmly bonded, of even and consistent appearance, free from rippling, hollows, ridges, cracks and crazing.
Finish surfaces to a true plane, to correct line and level, with all angles and corners to a right angle unless specified otherwise, and with walls and reveals plumb and square.
Prevent excessively rapid or localised drying out.

720 DUBBING OUT: If necessary to correct inaccuracies, dub out in thicknesses of not more than 10 mm in same mix as first coat. Allow each coat to set before the next is applied. Cross scratch surface of each dubbing out coat immediately after set.

770 DISSIMILAR BACKGROUNDS: Where scrim or lathing or beads are not specified, cut through plaster with a fine blade in a neat, straight line at junctions of:
Plastered rigid sheet and plastered solid backgrounds, Dissimilar solid backgrounds.

780 SMOOTH FINISH: Trowel or float to produce a tight, matt, smooth surface with no hollows, abrupt changes of level or trowel marks. Do not use water brush and avoid excessive trowelling and over polishing.

785 WOOD FLOAT FINISH: Finish with a dry wood float as soon as wet sheen has disappeared from surface to give an even overall texture.

Bibliography

Aqua Group. *Precontract Practice for the Building Team*. Blackwell (1992)

Aqua Group. *Tenders and Contracts for Building*. Blackwell (1990)

Building Project Information Committee. *Common Arrangement of Works Sections for Building Works (CAWS)* (1987)

Building Project Information Committee. *Production Drawings: A Code of Procedure for Building Works* (1987)

Building Project Information Committee. *Projection Specification: A Code of Procedure for Building Works* (1987)

Business Round Table. *Thinking about Building* (1995)

Chartered Institute of Building. *Code of Estimating Practice* (1983)

Construct IT: Bridging the Gap. HMSO (1995)

Co-ordinating Committee for Project Information. *Co-ordinated Project Information for Building Works: A Guide with Examples* (1987)

Fellows R.F. *1980 JCT Standard Form of Building Contract: A Commentary for Students and Practitioners*, Third Edition. Macmillan (1995)

Flanagan, R. and Norman, G. *Life Cycle Costing for Construction*. RICS (1983)

Greater London Council. *Preambles to Bills of Quantities*. Architectural Press/Butterworth Heinemann (1980)

Griffith's Building Price Book (published annually)

Institution of Civil Engineers. *The Engineering and Construction Contract (New Engineering Contract)*, Second Edition (1995)

Institution of Civil Engineers and Federation of Civil Engineering Contractors. *Civil Engineering Standard Method of Measurement (CESMM3)*, Third Edition. Telford (1991)

Institution of Civil Engineers, Association of Consulting Engineers and Federation of Civil Engineering Contractors. *ICE Conditions of Contract*, Sixth Edition (1991)

Joint Contracts Tribunal. *Standard Form of Building Contract with Quantities* (1980)

Latham Report. *Constructing the Team*. HMSO (1994)

Laxton. *General Specification for Major Building and Civil Engineering Projects*. Two Volumes. Schal Property Services/Butterworth Heinemann (1996)

Laxton's Building Price Book (published annually)

Murray G.P. *Measurement of Building Services*. Macmillan (1997)

National Joint Consultative Council for Building (NJCC). *Code of Procedure for Single Stage Selective Tendering* (1994)

Nisbet J. *Called to Account: Quantity Surveying, 1936–1986*. Stoke Publications (1989)

Property Services Agency, Royal Institution of Chartered Surveyors and Building Employers Confederation. *SMM7 Library of Standard Descriptions* (1988)

RICS, Building Cost Information Service. *Guide to Daywork Rates* (1994)

RICS, Building Cost Information Service. *Schedule of Basic Plant Charges for use in connection with Dayworks under a Building Contract* (1990)

Royal Institute of British Architects. *Project Selector*. Two Volumes (1997)

Royal Institution of Chartered Surveyors. *Definition of Prime Cost of Daywork carried out under a Building Contract* (1981)

Royal Institution of Chartered Surveyors. *Principles of Measurement (International) for Works of Construction* (1979)

Royal Institution of Chartered Surveyors. *The Procurement Guide* (1996)

Royal Institution of Chartered Surveyors, Quantity Surveyors Division. *QS2000: The Future Role of the Chartered Quantity Surveyor* (1991)

Royal Institution of Chartered Surveyors and Building Employers Confederation. *Bills of Quantities: A Code of Procedure for Building Works (SMM7 Measurement Code)*. CPI (1988)

Royal Institution of Chartered Surveyors and Building Employers Confederation. *Standard Method of Measurement of Building Works (SMM7)*, Seventh Edition (Revised 1998)

Seeley I.H. *Advanced Building Measurement*, Second Edition. Macmillan (1989)

Seeley I.H. *Building Economics*, Fourth Edition. Macmillan (1996)

Seeley I.H. *Building Technology*, Fifth Edition. Macmillan (1995)

Seeley I.H. *Civil Engineering Quantities*, Fifth Edition. Macmillan (1993)

Seeley I.H. *Quantity Surveying Practice*, Second Edition. Macmillan (1997)

Smith A.J. *Computers and Quantity Surveyors*. Macmillan (1989)

Smith A.J. *Estimating, Tendering and Bidding for Construction: Theory and Practice*. Macmillan (1995)

Society of Chief Quantity Surveyors in Local Government. *The Presentation and Format of Standard Preliminaries for use with the JCT Form of Building Contract with Quantities* (1981)

TBV Consult. *PSA/1 General Conditions of Contract for Building and Civil Engineering Works* (1994)

Turner A. *Building Procurement*, Second Edition. Macmillan (1997)

Wessex Building Price Book: SMM7, Second Edition (1997)

Williams J. *Spence Geddes: Estimating for Building and Civil Engineering Works*, Ninth Edition. Butterworth Heinemann (1996)

Index